Veiled Reality

Veiled Reality

An Analysis of Present-Day Quantum Mechanical Concepts

Bernard d'Espagnat
Université Paris-Sud, Orsay

Advanced Book Program

CRC Press is an imprint of the
Taylor & Francis Group, an **informa** business

First published 2003 by Westview Press

Published 2018 by CRC Press
Taylor & Francis Group
6000 Broken Sound Parkway NW, Suite 300
Boca Raton, FL 33487-2742

CRC Press is an imprint of the Taylor & Francis Group, an informa business

Visit the Taylor & Francis Web site at
http://www.taylorandfrancis.com

and the CRC Press Web site at
http://www.crcpress.com

A Cataloging-in-Publication data record for this book is available from the Library of Congress

ISBN 13: 978-0-8133-4087-6 (pbk)

Frontiers in Physics

David Pines, Editor

Volumes of the Series published from 1961 to 1973 are not officially numbered. The parenthetical numbers shown are designed to aid librarians and bibliographers to check the completeness of their holdings.

Titles published in this series prior to 1987 appear under either the W. A. Benjamin or the Benjamin/Cummings imprint; titles published since 1986 appear under the Westview Press imprint.

Editor's Foreword

The problem of communicating in a coherent fashion recent developments in the most exciting and active fields of physics continues to be with us. The enormous growth in the number of physicists has tended to make the familiar channels of communication considerably less effective. It has become increasingly diffucult for experts in a given field to keep up with the current literature; the novice can only be confused. What is needed is both a consistent account of a field and the presentation of a definite "point of view" concerning it. Formal monographs cannot meet such a need in a rapidly developing field, while the review article seems to have fallen into disfavor. Indeed, it would seem that the people who are most actively engaged in developing a given field are the people least likely to write at length about it.

Frontiers in Physics was conceived in 1961 in an effort to improve the situation in several ways. Leading physicists frequently give a series of lectures, a graduate seminar, or a graduate course in their special fields of interests. Such lectures serve to summarize the present status of a rapidly developing field and may well constitute the only coherent account available at the time. One of the principal purposes of the *Frontiers in Physics* series is to make notes on such lectures available to the wider physics community.

As *Frontiers in Physics* has evolved, a second category of book, the informal text/monograph, an intermediate step between lecture notes and formal texts or monographs, has played an increasingly important role in the series. In an informal text or monograph an author has reworked his or her lecture notes to the point at which the manuscript represents a coherent summation of a newly developed field, complete with references and problems, suitable for either classroom teaching or individual study.

Veiled Reality is just such a volume. It represents something of a departure from most previous volumes in this series, in that it deals with the frontier topic of the foundations of quantum mechanics, rather than the application of quantum mechanics to a topic at the frontier of physics. It differs also in that it is addressed not only to physicists at an early stage in their careers (the first-year graduate student, for whom many of us try to write) but to philosophers as well. Beginning with a chapter which could be described as "philosophy for physicists," it presents an in-depth analysis of present-day quantum mechanical concepts, an analysis of interest to physicists and philosophers alike. As a significant contributor to research on quantum measurement theories and the foundations of quantum mechanics, Professor d'Espagnat is highly qualified to write on these matters, and it gives me pleasure to welcome him to "Frontier in Physics."

Contents

Preface

This whole book centers on the conviction that whoever tries to form an idea of the world—and of man's position within the world—has to take the findings of quantum physics most seriously into account. Never mind the difficulty and the necessary recourse to scientific technicalities!

The foregoing great truth is not immediately obvious. In fact, if quantum physics opens wide vistas it is because of the mismatch between it and commonsense views; and at first, of course, this mismatch creates dissatisfaction. It may take some time before we realize that it indicates the subject is deep and fascinating. As for me, anyhow, I have not, as yet, forgotten the strange perplexity I felt when, as a beginner, I studied texts on quantum physics and, later, read articles on its foundations. These documents appropriately described the relevant facts and the theoretical formalism. But they were quite unclear—I thought—as to the actual meaning their authors attributed to the formulas they wrote down. Were these just good recipes for predicting the outcomes of experiments? Were they supposed to describe the very *things* the experiments bore on? At some places the argument seemed to go through only by virtue of an axiom that correct prediction of what is *observed* is all we may expect of science (all the rest just being empty words). But at other places (in introductions, summaries, applications to solid state physics, thermodynamics, or cosmology) the same authors would unabashedly express themselves as if the probabilities and other items they juggled with *did* refer to events "as they really happen." *There* they expected me to take such a meaning for granted. This experience was frustrating.

My discomfort was not allayed by the then available quantum measurement theories And it became especially acute when I read about one, then often proposed, interpretation of the quantum correlations at a distance the idea that entanglement could be removed through the merely technical change of describing systems by density matrices instead of state vectors On pondering these questions I convinced myself that any tentative realistic description (in the sense of conventional realism) implies the necessity of distinguishing between *proper* and *improper* mixtures, as I called them With the—negative but epistemologically significant—consequence that *entanglement*, or *nonseparability* as it is also called, is not just a bizarre feature of the wave function or state-vector formalism In slightly modified garments it is in fact present within *any* conventional (i e , hidden-variables-free) quantum mechanical description viewed as revealing how "things really are," be it based on the density-matrix notion or on any other concept When, at CERN (where we both worked), John Bell told me about his disproof of "local causality," I thought at first that his was merely another proof of the just mentioned result But he readily convinced me that, although local causality violation and entanglement are twin concepts, still his result was more general since it also holds true concerning quantum theories admitting of hidden variables In my eyes this generalization immediately took on a crucial importance because it demonstrated something that philosophy alone had vainly endeavored to prove: namely that our so-called commonsense views of the world (many of them, in fact, inherited from the eighteenth century) are erroneous and can be salvaged by no speculations, however wild, carried on within their realm Considered this way, entanglement and local causality violation concur in raising fascinating conceptual problems But, I must say, Bell and I amicably disagreed about the directions in which the solutions of these problems were to be hunted with the greatest chances of success

This book is meant to analyze further just the galaxy of the conceptual problems of this type In this, it parts from the mainstream of scientific literature Available books on quantum mechanics are mainly of two kinds Most deal primarily with the mathematical techniques, with perhaps a few, scanty words about the interpretation of the formalism A few do center on the epistemological (or even just the historical) side of the question

But when these refer to the underlying mathematical formalism, they do not go into the details needed for analyzing the
implicit philosophical basis and epistemological implications of
the more elaborate theoretical developments. Here I try to put
the two things together. My final objective is *interpretation* of the
existing theory (or theories). But I consider that proper views
concerning interpretation cannot be reached without going into
quite a detailed description of the formulas and physical arguments of which the theoretical models—including the most
recent ones—actually consist. Such a description is given in the
central part of the book.

In contrast, the first and three last chapters focus on the philosophical problems at issue. Chapter 1 is an introduction to them,
intended for the benefit of physicists. It consists of a reminder
of the essential proposals of past and present philosophical research concerning reality, experience, and the link between the
two, with special emphasis on philosophical problems posed by
physics. The last three chapters (14–16) constitute the essential
part of the book. They describe an attempt at determining the
philosophical implications and suggestions to be derived from
today's knowledge in quantum physics. There are some differences in style between these four chapters and the rest of the
book. They are due to the fact that physics and epistemology
are two really quite different approaches, which I strive here to
bring closer.

As my interest in conceptual and interpretation problems
should make clear, I am not one of those who reject realism—
even in its so-called naive version—on any a priori ground. In
fact if physics unambiguously proposed a picture of the world
obeying the prerequisites of *physicalism* (alias "physical realism"
the doctrine that physics ultimately is a good candidate for describing "reality as it really is" exhaustively and in detail), I
would tend to accept physicalism. My point, however, is that it
does *not*, and one of the main purposes of the book is to make
this point apparent. This is not totally easy. Elementary claims to
the effect that mere inspection of the formal structure of quantum mechanics (Heisenberg relationships and so on) suffices for
ruling out physicalism are in error. In fact, every now and then a
theorist springs up with a brilliant scheme intended for restoring
physical realism and some (not all!) do come quite near this end.

Several of these theorists are friends of mine and I thoroughly admire their work. I must therefore apologize to them collectively that, in this book, I made a critical analysis, not, actually, of their theories but of the plausibility of considering any one of them in particular as *the* true description of *what is*. This is not to say I drew no benefit from these investigations. Most chapters in this book testify to the contrary.

On the other hand, I could not draw a benefit from everything that was published in this domain; and here I did not even *try* to give an exhaustive review of such an extensive literature. In particular I systematically ignored all the theoretical proposals not meeting a certain condition of epistemological strictness explained below (Section 1.6). I tried to do justice to all those that do, but still cannot rule out the possibility that some of them escaped me. So, maybe I missed just precisely *the* rare bird that makes physical realism truly scientifically convincing. If future developments in physics should show this to have been the case, then my only excuse will retroactively have to be that I scanned the domain with zeal and impartiality.

We live in a time in which hosts of students in all sorts of applied fields have somehow to learn the basic ideas and calculation techniques of quantum mechanics and decently succeed in such a task. Undoubtedly, students in the philosophy of science would benefit following the same path, and it would be a pity if they, alone, should find this path inordinately steep. The difficulty is that many of them received a good training more in humanities than in natural sciences. It is with this in mind that, in this book—although my ambition is to tackle quite delicate questions, and to do so at the level of the experts—I decided to start more or less from the beginning, explaining, even if concisely, the essential elements of the quantum mechanical techniques, and give calculation details that experienced theorists are, of course, welcome to skip. The book is intended to be easily readable by first-year graduate students in physics and for its main part understandable by students in epistemology having a smattering of physics. It can to some extent be considered a sequel to my *Conceptual Foundations of Quantum Mechanics*, getting more to the core of the philosophical problem as well as taking into account the important theoretical developments that occurred since this book appeared. But, of course, it was seen that it should be quite self-contained; and, to this end, the small introductory scientific chapters (1–7) of *Conceptual Foundations* were taken over with but relatively short adjunctions.

Still, my main purpose in writing this book has been to argue in favor of my own views. In this sense, the book is not a textbook. I am well aware, of course, that the arguments it sets out do not settle matters once and for all. In particular, I grant here anew that its critical analysis (to be found in the central chapters) of a set of current physicalist proposals does not have the force a no-go theorem would have. A steady flow of new and brilliant suggestions aimed at reconciling quantum physics with physical realism will undoubtedly go on appearing: How could it be *demonstrated* in advance that none of them will, ultimately, prove satisfactory and convincing? This subordination to the passing of time is, however, the common lot of all scientifically based inquiries. Such considerations do not therefore alter my reasoned expectation that the general idea underlying the conclusions reached in this book will remain credible. In other words it seems to me that, when all is said and done, the notion of a *veiled reality* emerges with increased plausibility from what is explained in this book.

Bernard d'Espagnat
Paris and Orsay, 1994

Philosophy and Physics

1 This introductory chapter could also be called "Philosophy for Physicists."[1] Physicists are often put off by the manner in which contemporary philosophers deal with such apparently obvious notions as, for example, that of *object*. It seems that, when most philosophers hear the word "object," the notion of a subject *for whom* the object is an object immediately and spontaneously comes to their mind. Not so for physicists, with but very rare exceptions. And there lies a source of misunderstanding, since to the average philosopher it is so clear that "object" refers to "subject" that he or she does not even bother to mention such an obvious fact, and even less, of course, to explain *why* it is obvious.

What is the source of such conceptions so ingrained in philosophers' mode of thought? The answer is: the great classics of philosophy. And this is why a physicist who is in search of possible interpretations of quantum physics but reluctant to shut out the light of philosophy should not exclusively read contemporary philosophy books. On most questions, to "get the idea" at a level sufficiently deep to become able to discuss it, one must turn to the basic sources. Because of this, several of the following sections are devoted to a description and critical analysis—from the point of view of a physicist—of the aspects of the works of Hume and Kant that could be relevant to the problems we have to cope with in this book. When, in the end chapters, we try to put together the various pieces of information quantum mechanics provides us, and appreciate the relative weights of

---

1. Consequently, students in philosophy may bypass large parts of it. On the other hand, students in physics are encouraged to, at one time or other, read it in its entirety.

the different interpretations of it that theorists have put forward, the thus gathered knowledge will prove useful.

Also described in this first chapter are a few broad questions raised by quantum physics and general notions useful when trying to interpret it.

1.1 A Reminder of Some Basic Tenets in General Philosophy

In physics circles the motto "Remember that physics is an experimental science" is often heard. Prima facie, its meaning seems obvious, but it is not. In fact, it can be understood in two different ways. It can mean we must remember that, when all is said and done, it is only through the testimony of our senses that we can learn some (partial) truths about what exists "out there" independently of ourselves. In this interpretation physics is a science about what exists—reality—and human experience is just a *criterion* (the final one, it is claimed) for distinguishing true statements on reality from false ones. Alternatively, the maxim in question can mean that—again taking for granted that no reliable knowledge can be found outside experience—we must go to the bitter end of such a logic and ultimately conceive of physics as a science *about* human experience, all the rest being "metaphysics."

Essentially, the first interpretation is the "commonsense" one. It seems it was also the one the early empiricists took for granted. For example, in Locke's famous distinction between *primary* (position, shape) and *secondary* (color, taste) qualities, the first qualities are considered to be possessed by the objects while the second ones were, he claimed, merely human responses to excitations by the objects. This view sounds highly reasonable and is still entertained by the great majority of the scientists, including many physicists. For brevity's sake we shall henceforth refer to it as "conventional realism" or, still more concisely, as "realism" (although, as we shall see at the end of the book, there is at least one consistent conception for which the name "realism" is appropriate and which parts from the just characterized view).

Regarding the second interpretation of the motto, quite a number of physicists (mainly theorists) would at first feel inclined to endorse it, on the ground that our science of physical objects essentially bears on what we observe about and do to them. But the number of enthusiasts decreases dramatically when it is pointed out that if this second interpretation is not to be just a disguised form of the first one, it must also concern the primary qualities. This means, for example, that the relative positions of the planets in the solar system at a given moment, or that of a pointer on its dial, or that of a table relative to the surrounding walls, are merely elements of our experience, devoid of a one-to-one relationship with anything not boiling down to the experience in question. At this stage, indeed, the idea becomes most unpalatable to the majority of physicists. Some will even join the man in the street in declaring it just absurd. And I shall soon mention some nonpreposterous arguments they may resort to for backing up their position. But still, it is a fact that this "second interpretation," notwithstanding its apparent incredibility, is the one best attuned to a philosophical doctrine vindicated by many great thinkers throughout the ages. It also is a fact that, as we shall see later, if unreservedly accepted this view makes some of the most aching problems of physics dissolve. For these two reasons it cannot just be brushed aside. It must be considered seriously, which forces us to also take seriously the philosophical doctrine to which it is connected.

Concerning the latter we, first of all, must agree about the name by means of which we shall henceforth designate it. Many are available but, strangely enough, it is not easy to find one that really fits. "Empiricism," for example, is not really good, although it is quite often used. Many contemporary philosophers consider themselves empiricists. But it is clear from the above that this name admits of an inherent ambiguity: using it does not make it immediately clear whether the doctrine thereby referred to involves the *first* or the *second* interpretation of the motto that "physics is an experimental science", and in fact it sometimes happens that even after having read thick books of empiricist philosophers we are at a loss to determine which one of these two senses their author would endorse. The same is true of other often-used names, such as "positivism" and "phenomenalism."

Admittedly, professional philosophers can point to some valid arguments justifying such an ambiguity. It is true that "pure reason" should not tackle too abruptly most basic questions such as the one implied here, for the risk is high of getting trapped in a framework of oversimple definitions. It certainly is safer to approach these questions more obliquely, through a preliminary study of more tractable, purely epistemological problems. And even if this study does not, in the end, open on decisive answers to the basic questions at issue, still not everything is lost since its outcome is usually interesting in its own right. However, even though such arguments do have weight within a *general* philosophical context, they leave us wanting here since, as we shall see in detail, quantum mechanics raises acute definite problems that are of such a kind as to force us to "take sides" and remove the ambiguity. For this reason we cannot be satisfied with names such as empiricism or phenomenalism. What we need is a generic name for the common core of all philosophical theories that part with conventional realism in opting for the *second* interpretation of the motto: that is, conceive physics as a science *about* human experience. Only then shall we be able to state clearly to which side we think such and such quantum mechanical development turns the scales.

In fact there is one name that adequately fulfills this condition. It is "idealism." Unfortunately this word has been used to designate several doctrines, other aspects of which are now more or less obsolete. For this reason, and also because some of these doctrines were rather averse to science, even the philosophers of science who definitely declare for the second interpretation would, most of them, nowadays refuse to be labeled idealists. In a way this is a pity for, in a difficult field in which subtle analyses are needed, the unwritten law that a basically suitable word such as this one should be banned introduces an additional, artificial and unwelcome element of indefiniteness, and thereby of ambiguity. Here, for clarity's sake I shall infringe this overly strict taboo law and use the word "idealism" in all the cases in which I feel no other word would be quite explicit enough. In less crucial instances I shall designate the *same* general option by the name "phenomenism," an uncompromising word contrived in the hope it should raise neither epidermic rejection nor emotional allegiance. Both words will be used in a broad sense, to mean any doctrine stressing that our human means of perception, action, and explanation contribute decisively to the very way in which we shape the concepts we use for describing our experience—with the corollaries that we have no access whatsoever to anything that could be called reality per se; that we only know the phenomena, in the philosophical sense of the word; and that therefore even the so-called primary qualities (relative positions included) are in fact nothing more than *human* modes of apprehension. These views serve as starting material to a considerable part of contemporary philosophical research and this is one of the reasons why we will consider them somewhat in detail.

Here let it just be pointed out that the foregoing definition of what I mean by "idealism" or "phenomenism" clearly leaves room for distinctions. I propose to call *"radical* idealists," or radical phenomenists, the philosophers who claim that the very notion of existence has no meaning except in cases in which it can be referred to the notion of experience; and to use the expression *"moderate* idealism" (or "moderate phenomenism") to mean the doctrine of the thinkers who do not go quite that far, or who do not take sides on this issue. Readers, particularly scientists, should note that, as here defined, idealism is quite the contrary of an esoteric doctrine. In fact a large number of great scientists of the past, Helmholtz, Poincaré, Schrödinger, and others, were quite definitely idealists in the sense defined here. This, of course, is another reason for us to analyze the doctrines in question in some detail.

1.2 "For" and "Against" Idealism

Of course the people who developed idealism were philosophers. "Man is the measure of all things," Protagoras said in the fourth century B.C. Later Descartes considered a somewhat similar idea, although merely as an intermediate step in his argument. It is to Berkeley that we owe the first really elaborate idealist doctrine. His doctrine was rather of the radical kind (although it kept God as an independently existing Being) and the motto "esse est percipi" (to be is to be perceived) is known to summarize it partially but, on the whole, adequately. Kant again took up the idea. He may be said to rank as a moderate idealist since he explicitly accepted the notion of things-in-themselves, existing independently of any knowledge. Still, his view was that we can have no real knowledge of such things. It is well known that, according to him, neither space nor time nor causality have any autonomous existence: they all are but modes of our human sensibility and understanding. Later, most of the philosophers who adopted Kant's general approach were more radical in that they rejected the "things-in-themselves" concept.

It is of course out of the question to give here a detailed pedagogical description of the views of Berkeley, Kant, and the neo- and post-Kantians. Such accounts can be found in philosophy textbooks, and it is assumed here that the main lines of the subject are known. More to the point is a survey of questions of such a nature as to arouse the interest of scientists. These questions fall into two complementary sets: (a) what were the reasons, arguments, and so forth that led these philosophers to believe in idealism and (b) how did they respond, or could have responded, to some apparently obvious and devastating objections to it?

In order not to overdevelop this analysis, let us merely consider Kant's standpoint on such matters. Following Berkeley, Kant took as his starting point a remark the validity of which is hard to question, namely that any specific knowledge we claim to have of such and such an external object is obtained through our senses, hence is at best only indirect and questionable (think of the broken stick experiment), and that what we know directly and with certainty is therefore only the set of our *ideas* (a word that, in the language of the time, also covered sensory data). Concerning space, for example, Kant forcefully claimed that what is most certain about it is that we absolutely need this notion for ordering our sense-data (phenomena being the outcomes of the ordering in question).

This general remark has far-reaching consequences. For instance it can be made the basis of a sensible argument (not quite Kant's, perhaps, but never mind!) in favor of ideality of space. First, observe that, independently of whether or not the things-in-themselves are in space, we are forced, as Kant stressed, to set our representation of them into space. Then conclude that under these circumstances the hypothesis that the things-in-themselves *are* actually in space becomes arbitrary and redundant.

Admittedly this elementary argument could, prima facie, be countered in a simple way. It would be pointed out that, nevertheless, the hypothesis according to which the things-in-themselves are known more or less as they really are (i.e., approximately coincide with what our senses or scientific instruments reveal)—and are therefore *in* space, as we see them to be—must be a good one, since it explains in a straightforward, simple manner an enormous number of facts. But Kant would presumably have responded to this that "to explain" means "to find out the causes," that, in the present context, it means "to find out the detailed causes of the phenomena" and that the very notion of causality, as soon as it is meant to extend beyond that of mere usual succession in time, is one that, as Hume conclusively showed, can be derived neither from experience nor from the exercise of pure reason. For Kant, its only possible status is therefore that of an a priori mode of human understanding, which implies that it is not "in" the things-in-themselves but just "in" us. Consequently (a) giving up the view that the things-in-themselves are known and *in* space in no way implies giving up causality (concerning the phenomena), and (b) conversely, the many causal links that are found in the realm of the phenomena cannot be interpreted as an indication that the things-in-themselves are known and are in space.

Noteworthy in this argument is the quite central role played in it (as Kant himself acknowledged) by Hume's critical analysis of the causality concept. In view of this it is clear that, to form some sort of a value judgment concerning idealism in general—and Kant's version of it in particular—a detailed critical analysis of Hume's own argumentation is necessary. This will constitute the subject of the next section. For the time being, let me just enumerate a few comments on the preceding discussion.

1. It is well known that when Kant mentions space he means Euclidian space and when he mentions time he means Newtonian universal time. His claim that these concepts are a priori modes of our sensibility seems to imply (and in his mind certainly *did* imply) that physics must be entirely built up within the framework of the said concepts. Since the advent of relativity theory this, of course, is known to be false, and some serious reservations are thereby justified concerning the details of Kant's philosophical system.

2. A similar remark is in order concerning Kant's conception of causality. Quite in line with Leibniz's "principle of sufficient reason," and in keeping with the views of practically all the philosophers and physicists of his time (and even later), Kant conceived of causality as identical to determinism. Modern physical theories have made such a view obsolete so that, again, the fact Kant made causality as he understood it an a priori element of our understanding—and therefore of the mold into which physics has to be cast—must make us skeptical concerning the details of Kantianism.

3. More generally, Kant lived in a time when it was considered that science not only *could* attain certainty but in fact *had* already attained it in at least two domains, geometry and Newtonian mechanics. Now, as Popper [1] appropriately pointed out, there was a rather blatant contradiction between this and the skepticism Hume had derived from his analysis of causality, a skepticism that, as we know, stemmed from the need we have of this concept and the fact we have no way of rationally justifying our use of it. According to Popper, Kant seems to have considered that, in a way, no proposal could be too daring if it removed this difficulty. The one he put forward, that of considering causality as an a priori mold into which we *must* cast our experience, obviously did have this effect since it disposed of the very need of justifying our belief in causality.

Popper's own view on this is that the subsequent development of physics showed Hume's skepticism to be well founded, that the kind of certainty Kant attributed to science was delusory and that therefore the whole problem Kant strived to solve was a false one. Unquestionably there is some fair amount of truth in this judgment. Let us note, however, that for anybody who would at the same time endorse Hume's critical analysis and believe science can ultimately reveal "how reality really is" (reveal "God's blueprints of the Universe," in Hawking's language), the problem Kant pitted himself against would be quite a real one, even today. And let us also stress that Popper's remark does not really invalidate the above-reported Kant-like argument on the ideality of space since the notion of certainty plays no explicit role in it.

To sum up: some aspects of idealism, and especially of Kantianism, are definitely obsolete, including in particular the main motivations Kant had for developing his ideas. But on the other hand, and contrary to appearances, the realist's firm belief that the primary qualities are possessed by the objects ("really" are, on the whole, as we see them to be) and that space (or, nowadays, space-time) is an independently existing arena where the things (nowadays, events) *are*, does not seem to be well grounded either in logic (pure reason) or in factual knowledge. Or, to put it less bluntly, if, after all, it *is* somehow grounded in the latter, the way it is is by far not as obvious as it would seem, and the extent to which it is must be analyzed and discussed.

1.3 Discussing the Scope of Hume's Analysis

Everybody nowadays agrees about the relativity of space and time, the word "relativity" being taken in the sense imparted to it by Einstein's relativity theory. But in relativity theory space-time and events are not "relative," or, at least, they are not relative to man's knowledge. They can be, and usually are, thought of as existing quite independently of man's existence, just as space and objects were thought of as existing—in some absolute sense of the verb "to exist"—by most nineteenth-century physicists. The point is worth noting for, in the hope of making the ideal-ist's statement about the ideality of space and time somehow less unpalatable, it is sometimes suggested to replace in it the term "ideality"—too reminiscent of some antiquated doctrines—by the more modern "relativity." The idea is good if—and to the extent that—it prevents mixing up the general guiding ideas of current doctrines partaking in some way of idealism (moderate or radical) with some conceptions we said are obsolete in, for example, Kantianism. But it is not a far-reaching proposal. The reason is that the current guiding ideas just mentioned do include, quite in the lines of Kantianism, the view that the existence of space-time, events, and so on—indeed existence in general—is logically secondary with respect to that of knowledge. And no euphemism can remove the considerable rupture with our most deeply ingrained ways of thinking—one could even say the scandal—this standpoint implies.

It thus remains necessary to inquire, quite apart from semantics, whether and to what extent this rupture is well founded. And as we saw that the originally developed (and still much referred to) arguments in its favor rely in a crucial way on Hume's analysis of causality, we cannot spare having a somewhat detailed, critical look at the latter. It proceeds through several steps that will now be described and discussed.

Description

The first step is just the observation that, in most cases, what persuades us of the reality of some fact, existence, or presence is a reasoning based on a cause-and-effect relationship. For example, when we hear a human voice in a dark room we infer somebody is there; if we find a watch on a desert island we infer that somebody came there before, and so on.

Hume's second step is to observe that the cause-effect relationships we thus make such large use of originate exclusively from experience. It is experience, not reasoning, that taught us human voices are heard only when human beings are present (gramophones and transistors were not invented in Hume's times), and that watches do not just grow naturally, as trees and grass do. And, he stresses, this is also true in cases in which we are intuitively inclined to think we could discover the effects of some given cause by the operation of pure reason. We tend to consider that, were we suddenly introduced into the world, we could, by pure reasoning, predict a moving billiard ball will impart its momentum to a standing one it collides with. But we are wrong. A priori, it is just as conceivable that after the collision both balls will be found at rest, or that the moving one will bounce back or.   Only past experience tells us what process will really take place.

Hume's third (and main) step consists in searching for the basis of our confidence that such a collection of data from past experience does enable us to formulate a cause-and-effect relationship and apply it, for predicting future events. In modern language we would say he is on the lookout for a justification of induction. And, as is well known, he discovers that such a justification cannot be found. It is clear from the preceding statements that the cause-effect relationships cannot be inferred by pure reasoning or, in other words, that the validity of induction cannot be established this way. And if we try to justify it by referring to experience, that is, by asserting that since induction worked in many instances it should also work in other instances, we are immediately caught in a vicious circle; for this, obviously, already is an inductive argument so that we are postulating the validity of induction as an element in our very attempt at proving this validity. To put it in a nutshell, there is *no* justification for induction.

The preceding paragraphs sketched the way in which Kant and his successors turned this negative result of Hume into a powerful argument in favor of the ideality of causality and, therefrom, in favor of (moderate) idealism in general. Kant's proposal is, as we know, that the notion of cause-effect relationship be considered as being an a priori element of understanding, conditioning the very possibility of building up human experience (from raw sense-data). If this is the case, then of course there can be no question of requesting that it should be *derived from* experience, and Hume's difficulty is thus removed. But then, clearly, the overall picture is consistent only if the phenomena themselves—the cause and the effect—are considered as being "mental" rather than "physical" for otherwise we would be at a loss how to justify our applying the cause-effect relationship, now an essentially mental structure, to the phenomena in question.

Discussion

As we just saw, Hume's argumentation is quite basic with regard to the turn to be given to our thinking about the world. The ways in which we shall interpret scientific results in general, and particularly those of physics, depend very much on whether and to what extent we consider it as binding.

In this respect, one, perhaps significant, observation is that Hume essentially argues as a philosopher—a very good one, by the way!—not as a scientist. This is especially noticeable on his examples of cause-effect relationships. Most of them are of the type "Bread is known to nourish," "Spring induces fruit trees to blossom," and the like. And of course this makes it especially easy for him to stress the purely empirical nature of the relationships in question. Indeed it is, as he points out, a priori just as conceivable that bread should *not* nourish (it hardly nourishes tigers!) and that fruit trees should blossom only in winter. Even at places where he refers to billiard balls, about which Newtonian mechanics tells us a few things, he keeps silent on the fact that the *same* general laws (those of the mechanics in question) account for a large number of distinct cause-effect relationships. Can this be made use of for questioning the general validity of his argument?

This is in fact quite a delicate point. There is no question of asserting that a mere reference to the existence of physical laws can remove the motivations of Hume's misgivings concerning the validity of induction. To such a suggestion Hume would, of course, respond that these laws were themselves inferred from experimental evidence, that the evidence in question only bears on a finite number of past observations and that it is by postulating the validity of induction that we attribute universality to these laws; thus we would be caught in the same vicious circle as before if we attempted to justify our belief in the validity of induction by referring to this very universality of laws.

This response is purely epistemological, in the sense that it makes no reference whatsoever to ontology (questions on "what really exists"). At first sight it seems therefore that its strength cannot be altered by taking up a "realist" standpoint. However, we should not be too rash concerning this, for it is clear the vicious circle would be avoided if only our belief in the universality of laws could be grounded on some reason not referring to induction. Can realism provide one? The answer is no, not really. But still, defeat is not complete for we can at least *postulate* the universality of the structures of reality that give rise to the observed physical laws. And it seems that this postulate can be made a plausible one on the basis of a maximal simplicity argument. Most scientists, at any rate, would presumably agree that the one assumption that reality has general structures somehow reflecting in our universal laws (Maxwell's equations and the rest) is simpler (because we make it once and for all) than the set of the particular assumptions that, every time we predict an effect y from a cause x, we must make if we take up Hume's strict empiricism: and which are, concerning every one of these cases separately, that induction will work in *this* case.

The strength of the foregoing argument should not be overestimated. It depends basically on the notion of simplicity and it must be granted that this notion is too human-centered to be quite convincingly used in such very general matters. It seems therefore that for the argument in question to succeed, a kind of preestablished harmony should be postulated between reality and the human mind, implying that the *real* structures of reality somehow look simple to us. The trouble, of course, is that this postulate is a daring one, in that, again, it cannot be justified by referring to the success of science without falling back into the vicious circle we are trying to escape. Still, it is not unpalatable to scientists, even though most philosophers will claim it is both arbitrary and "repellantly metaphysical."

On the whole, a not unreasonable standpoint on these matters might be as follows. Grant that the impossibility of justifying, by means of a totally rational and assumption-free argument, our firm belief that the physical laws will remain valid tomorrow is a real failure of rationality. Grant also that this failure must make us quite unpretentious. But observe that it certainly does not force us to give up the said firm belief since it does not falsify it either, the aforementioned postulate being unquestionably self-consistent. If this view is taken it must be considered that, when all is said and done, radical idealism, moderate idealism, and conventional realism, all are a priori admissible doctrines and that, concerning their relative weights, philosophy alone does not provide all the information we would expect. In later chapters we shall investigate whether by any chance physics has, in this respect, something to say.

1.4 Existence and Intersubjectivity

One of the reasons why we physicists must bother about philosophical problems of the kind sketched in this chapter is this: a great number—in fact the majority—of the physicists whose activity compels them, at one time or other, to go into the very foundations of contemporary physics (for example, because they start writing a book) implicitly embrace the standpoint of phenomenism, that is, are led by the very nature of the problem they tackle to take up positions ultimately very much akin to those of either moderate or radical idealism. Some of them may be surprised by this statement and may not like it that this parallelism be stressed. But it is unquestionable. This is particularly clear at the stage when these authors start explaining that contrary to classical concepts the quantum ones cannot be viewed as *directly* describing independent reality. This, of course, is true but does it mean they describe it *indirectly*? And if so in what way? Some such authors (see, e.g., [2]) begin by stressing that although no quantum concept completely covers in all their aspects the elements of reality it serves to investigate, still each one provides us with reliable pieces of information and can be used, in association with other similar concepts, for building up increasingly faithful pictures of reality. Up to this point, of course, they do not compromise with idealism. But those of them who (as the quoted authors) cling to the Copenhagen "orthodoxy" cannot stop there. They must—and do—add that "moreover," according to the basic *complementarity* idea, quantum reality cannot be totally described by just one such picture, and that a description of it requires in fact a *duality* of mutually contradictory (but "complementary") pictures. Now we must ask: Is the thus completed conception still compatible with realism? Clearly the answer is no. Indeed, necessary as it is for exactness, this last additional specification quite definitely overthrows the "realistic message" the preceding sentences were supposed to convey. If the best we can do to describe reality is to resort at the same time to two mutually contradictory pictures, then quite obviously we cannot claim that we have described "reality as it really is." In fact, as will be shown below in more detail, the type of "reality" that contemporary physics describes when interpreted along the orthodox Copenhagen lines is merely an "effective" or "empirical" reality; that is, it is the set of all the *phenomena*, in the Kantian sense of the word.

Such circumstances make it necessary that we, as physicists, ponder some objections that realists tend to produce against any form, radical or moderate, of idealism and that they consider conclusive. These objections are, as a rule, brushed away as naive by nonrealist philosophers. Indeed, most of the latter seem to think them so naive as not even to deserve mention. This, however, is not the kind of approach that scientists consider fair. Let us therefore unabashedly have a look at the two main objections. They concern the problems of (past and present) existence and of intersubjective agreement.

Existence

Stated bluntly the objection is: "Phenomenism and all the contemporary philosophical doctrines that essentially take over the basic argumentation of idealism (without always acknowledging this debt!) define the phenomena as basically *mental events*, these events being composed of both sensations and a priori forms ready, in our mind, to be applied to sensations ('unde-termined objects of empirical intuition,' in Kant's complicated wording). This means that the very concept of phenomenon is essentially anthropocentric. But such a view is absurd since, as we all know, planet Earth, the stars, the universe, with all the phenomena taking place on them and in them, already existed at times when neither human beings nor any other living organ-isms existed."

The response idealists and their fellow travelers would give to this objection is that it is flawed since it implicitly raises the concept of time to the level of something externally given and absolute, in contradiction with the basic leading ideas of the philosophical doctrine under study. This response is a valid one. The objection was nevertheless worth considering. It puts into full light the fact that for phenomenism and related doctrines to be consistent, they must be accepted in their entirety, up to and including their most disconcerting features. In fact, for a follower of Kant, to assert that there was a time when the Universe existed and mankind was not yet in existence merely means we have a possibility of organizing our collective experience by describing it in this manner. The assertion cannot, and therefore should not, be given any meaning beyond this and, when understood this way, is in no way inconsistent with the anthropocentric nature of all phenomena.

Of course, the realist's objection can also be expressed in a somewhat more general form, by saying that idealism makes the human being the cause of everything that we observe—events, objects, tables, quasars, and so on—and that to imagine the hu-man generates stars is absurd. The idealist's response is that, again, the realist who states the objection takes up his own abso-lute conception of what a cause is and unjustifiably applies it to criticize the consistency of a theory one basic element of which is just that this conception must be changed. A tenet of (declared or crypto-) idealism is, as we saw, that the cause-effect relationship is one *between* phenomena, which themselves are compounds of sensations and mental (a priori) structural molds, in which the sensations are cast. To insist on applying the cause-effect con-cept to a relationship between two elements, one of which (the human being, conceived of as an unpersonal epistemological subject) is *not* a phenomenon, means making use of it outside its domain of validity, and is therefore faulty reasoning.

Still more generally, the realist may claim that we must not confuse things with our perceptions of things, or, in other words, that things "really exist" quite independently of us. But of course the idealist has a ready answer to this, based this time on a critical analysis of the domain of validity of the notion of meaning. As Heisenberg [3] puts it at the place where he describes Berkeley's standpoint: "If really our whole knowledge is derived from perception, to say that the things 'really exist' has no meaning; for if we have 'perceptions of things' it makes no difference whether the thus perceived things exist or not. Hence 'to be perceived' is equivalent to 'to exist.' "

Whatever our final judgment on the existence problem will be, it must be acknowledged that these arguments do have weight.

Intersubjective Agreement

The expression "intersubjective agreement" is a very general one. The agreement may be on moral values, on logical rules, on mathematical theorems, and so on. Here let it be specified that we shall only be interested in intersubjective agreement concerning observation of contingent facts: if there is a teapot on the table around which we sit, we all agree that there is a teapot on the table (not zero teapots, nor two, nor three, but precisely one teapot). It is in this sense only that we shall henceforth understand the expression under study.

Intersubjective agreement is a fact (at least in normal circumstances: let us disregard the cases of drunkards and so forth). Our question is this: Is its existence an argument in favor of realism or, in other words, does it constitute a valid objection against idealism and its variants?

The answer seems to be an unqualified yes, for if Alice and Bob agree that at noon they both saw no teapot on the table, that at five o'clock they saw one and that at six, again they did not see it, the assumption that at noon there really was no teapot there, that at five o'clock there really was one and that it was *really* removed at some time between five and six has all the appearances of being the best explanation of such a remarkable correlation sequence. If on the contrary, as the idealist claims, the statement that the teapot "really exists" has no meaning beyond that, for Alice, of describing the way she mentally organizes her sensations (and same, of course for Bob), then this must also be the case concerning the assertion that the teapot "is really there" (or not) at such and such times, and the fact that Alice and Bob agree that they always had the *same* sensations in this respect becomes puzzling: a kind of constantly renewed miracle, in fact. The realist is therefore very much entitled to press the idealist on this point, and ask what explanation the idealist has to offer that would be as simple as the one just stated.

Surprisingly enough, few, if any, idealist philosophers seem to have worried about this problem and the corresponding possible attack on their views. In the relevant literature, practically only intersubjective agreement concerning *noncontingent* facts (mathematics) is discussed. If we may try to reconstruct the system of defense these philosophers conceivably could have drawn up concerning the point under study here, it seems it might go somehow like this. They would try to respond that (a) generally speaking it is a weakness of the human mind to always be on the look for explanations, and (b) in particular the realist's alleged explanation is not valid. Since to a physicist both responses must look extremely disconcerting it is worthwhile considering each one in some detail.

Response (a) is, in a way, well illustrated by what Cassirer wrote in 1910 concerning the concept of atoms [4]. In 1910 quite a number of basic experiments had already been performed that conclusively pointed to the validity of this notion; and the quoted author himself mentioned the facts that: "we speak of the number of atoms contained in a certain volume of gas" and that: "we attribute to the atoms of all simple substances the same heat capacity." On this basis he readily acknowledged that the concept of atom has the "logical function" of relating several series of measured data to one another. He called it a "mediating concept" for this reason. But he nevertheless did not accept considering the assumption that atoms really exist as meaningful, and called it "metaphysical." That he took up this standpoint is all the more remarkable as in 1910 classical physics was at its apex, and it could not be surmised that quantum mechanics would later raise all the questions that now prevent us from considering the elementary particles—the atoms in the philosophical sense—as just classical corpuscles, existing "out there." Since, in Cassirer's time, such warning signs did not exist, how can we understand his position? Rather than building up his quite elaborate and subtle philosophical concept of "logical functions" would it not, for him, have been infinitely simpler to say "well, in view of all this, it is extremely likely that atoms exist, since their existence *explains* all these data" (whereas the "mediating concept" *atom* mediates but does not actually explain). The philosopher's implicit negative answer to this can only be based on the view that we physicists unduly crave explanations that ultimately can only be metaphysical and therefore unsound. The famous remark of Quine's that the theoretical entities involved in current physics (like electrons, say) are epistemologically in par with Homer's gods (both being convenient fictions, introduced for trying to explain empirical findings) points in the same direction.

A look at the possible ways in which philosophers might substantiate their response (b) (that the realist's alleged explanation of intersubjective agreement is not valid) will at the same time throw some additional light on the structure of their response (a). A first indication as to the way they conceivably could construct it may be found in a passage from Reichenbach [5]. This author tells us that "to explain" a fact can only mean to link it with some general law. The barking of a dog can be explained by assuming that a stranger is approaching the house, for it is a general law that dogs bark, as a rule, on such occasions. Similarly, when we explain the presence of fossil seashells in mountain soils by assuming that the sea level was once higher than it now is, we refer to the general law that sea animals do not live on land, and so on. If this definition is adhered to we must grant that the realist's alleged explanation that Alice and Bob both see the teapot on the table "because the teapot is really there," simple as it looks, is no explanation at all! It is grounded, not on a reference to a general law but on a "metaphysical" concept of reality[1]

Surprisingly, this criticism of the realist's explanation was (to some extent) explicitly stated, not by Reichenbach himself (nor, to my knowledge, by other philosophers) but by the physicist Erwin Schrödinger [6]. Not many users of the Schrödinger equation know that Schrödinger had a strong interest in philosophy. He is one of the very few thinkers who did pay attention to the question here considered (he called it the "arithmetical problem" to mean that it only arises because of the multiplicity of human minds). He assures us that the realist's explanation is not truly one. His argument is constructed as follows. First he notes that the alleged explanation rests on the postulated similarities between the mental pictures of Alice and the real world R on the one hand, and those between the mental pictures of Bob and that same real world R on the other hand. But then he claims: "Whoever thinks this way forgets that R is not observed." This rebuttal is expressed somewhat concisely but we should understand it as meaning that for the explanation to succeed, a cause-effect relationship must be postulated between R and the mental pictures of Alice (and similarly concerning Bob) and that this is getting far beyond the well-established domain of validity of the causality concept since R is not a "phenomenon."

Once again we observe how decisive in all such arguments is the restriction of causality to phenomena. And some may therefore think, perhaps legitimately, that the idealists (and philosophers of language and so on) who essentially argue along the lines depicted here are getting trapped—by their own fault—within an arbitrary, or at least too strict and somewhat Byzantine "rule of the game," which in the end prevents them from grasping the force of even most obvious and matter-of-fact arguments. Although this is not yet the proper place for definite proposals to be expressed on such matters (a detailed examination of contemporary physical theories is to be made first), still let me incidentally remark that the just stated depreciating appreciation would be even more firmly grounded were it not for the fact that *not* all the instances of intersubjective agreement are explainable by means of the realist's commonsense argument. Indeed, according at least to conventional textbook quantum mechanics, there are cases in which intersubjective agreement can definitely not be explained that way. Imagine for example that Alice and Bob both perform, one immediately after the other, a measurement of the position of one and the same electron, each using his own instrument. And assume further that before the first measurement the electron wave function is not a "delta function," which is by far the most general case. Then the rules of quantum mechanics unambiguously predict intersubjective agreement: when Alice and Bob later compare their notebooks they will discover that they both saw the electron at the same place. Before the first measurement, however, as the quantum mechanical formalism tells us, the electron was at no definite place whatsoever, and therefore, in particular, it was not *really* at *that* place. The, allegedly obvious, realistic "teapot-like" explanation of intersubjective agreement is, in this case, simply false.

We have here, I think, a nice, straightforward illustration of the fact that the sweeping, commonsense objections we intuitively formulate against the views of the idealists and their followers should not be taken at face value. However, this is not to say the idealists may completely disregard them. It certainly is true that the idealist, phenomenist, and other approaches developed by pure philosophers meet with a real problem concerning the question of intersubjective agreement and that on the whole the upholders of these doctrines took the matter too lightly. In the last chapters we shall consider it more in detail, in the light of quantum physics. Meanwhile, the existence of this problem should reinforce our intuitive reluctance to unhesitatingly take up the views of radical, or even moderate, idealism. And we should not forget that this reluctance may be backed up by some arguments that are not just intuitive and irrational. For example, when Kant points out that since we are anyhow forced to conceive of the things as being in space, the fact we conceive them that way is not a proof they "really" are in it, we could respond. Specifically we could retort that if the structure of the human mind forces us to conceive the objects in space this might well be just because they actually *are* in it and, in the course of its adaptative evolution, the mind eventually hit on the proper concept: the one—space—that corresponds to truth and works well just for this reason. Hence the matter is worth much more reflection and study, and this is why it will be taken up again only after we have had a systematic look at the presently existing physical theories.

1.5 A Note About Scientific Revolutions

In this chapter we have noted that the distrust most philosophers show concerning anything resembling an ontology—a tentative description of what "really is"—has its deep roots in great classical philosophical works, and we saw in detail what these roots are. But it would be misleading not to mention that a number of contemporary philosophers derive the distrust in question from another source as well, namely the existence of the so-called scientific revolutions. Gaston Bachelard [7] was presumably the first to draw attention to the latter (he called them: "epistemolog-ical breakings"). Later, the concept was rediscovered and elabo-rated on by Kuhn [8] and others. It gave rise to some unjustified extrapolations (that science is to a large extent arbitrary and so on) which it is proper to ignore, but its intrinsic validity is un-questionable. It is a fact that from time to time the advancement of science—and of physics in particular—went together with a radical change of basic concepts. The example of the replace-ment of Newton's gravitational force by space curvature suffices to illustrate this point. In view of it, it is not necessary to be a diehard phenomenist to feel suspicious concerning the possibili-ties of ever arriving, by means of scientific research, at some true knowledge of how "things really are."

This, at least, constitutes the claim of the contemporary philosophers just alluded to, and it must be said that this skepti-cal conclusion anticipates to some extent the results of the queries reported in the present book. However, it does not, I feel, make them superfluous, the reason being that in the opinion of a ma-jority of physicists (myself included), inference from past sci-entific revolutions is not sufficiently conclusive concerning the question at issue. The point is that, judging on quantitative as-pects (such as number of presently active physicists, public fi-nancial efforts, and so on) physics has developed tremendously in recent decades and can be considered as only now reach-ing its maturity. In contrast, the findings concerning scientific revolutions perforce rest on investigations bearing on a more distant past: the infancy of physics, in a sense. Hence, extrap-olating these findings to the future is risky. In the past—these physicists would say—when physics still consisted of several independent or quite loosely connected parts, a theory could be both successful and based on notions (such as the gravitational force) that are now recognized as not fitting reality. But, they would continue, we have no strong indications that such circum-stances will occur again, now that physics is tightly structured as we know it to be. Their point can be understood by com-paring the circumstances in question to some historical ones, such as what actually took place when America was discov-ered. As is well known, the success of Columbus' enterprise was due in part to the fact that he planned it on the basis of too low an estimate of Earth's radius, which made him believe his ships could reach China. Columbus's fortunate misappreciation could happen only because geography was still in its infancy in his time. In this branch, a similar process of error leading to discovery (I mean: to a dramatic upsetting of the whole picture) could not take place today. The same—these physicists would say—is nowadays presumably true of physics, which is now a mature science whose various parts tightly back each other up.

The fact that many contemporary physicists and astrophysicists have such a conception in mind, even if they do not state it explicitly, is presumably the reason why most of them do not let themselves be impressed by the scientific revolutions argument to which the philosophers confer such a great importance. For example, there can hardly be any doubt that when a scientist such as Hawking mentions, as we noted, the possibility of eventually discovering God's blueprints of the universe he implicitly argues in the manner just described.

1.6 Some Questions Raised by Physics

So far in this chapter some basic aspects of the problems this book deals with have been considered from the point of view of general philosophy. In this section some more specific questions and concepts are put forward. Their nature is still philosophical, and therefore fairly general, but they are especially relevant for the physicist. In particular they must necessarily be considered in connection with any physical analysis of the kind we propose to undertake.

Weak and Strong Objectivity

That scientific statements are objective is a truism. But this truism conceals from many eyes the fact that the very notion of objectivity is not an immediately obvious one and that, even in science, this word is used in at least two different senses. To prevent confusion it is important to clear up this elementary point before we embark in physics.

To this effect, it is sufficient to observe that the objective statements of physics are not all expressed in the same way. Some of them have a form that makes it possible to interpret them as informing us directly of attributes of the things under study. Statements such as: "Two oppositely charged bodies attract each other in such and such a way" or "The magnetic field of the Earth is so-and-so," or, more generally, *all* those of classical nonrelativistic physics (with the noteworthy exception of those of statistical thermodynamics) are obviously of this type. It is not necessary to understand them that way, and some people may have good extrinsic (say, philosophical) reasons not to, but at least they are of such a form as to *allow* for such an interpretation. I have proposed [9] to call them *strongly objective* statements.

But although all scientific statements are objective, as just noted, still not all of them are strongly objective. In most formulations of statistical thermodynamics for example, some reference is explicitly made (e.g., by means of the notion of *coarse graining*) to what it is possible to observe, or to what kinds of systems can be prepared. Similarly in conventional textbook quantum mechanics some basic laws are expressed, as we shall see, by referring to what actually will be observed in such and such circumstances. Such statements are quite appropriately said to be objective since, by definition, they are true for everybody. But their very form makes it impossible to understand them just as descriptions of how the things actually are. Of such statements I proposed [9] to say that they are *weakly objective* only.

Concerning these definitions a few remarks are in order.

1. Any strongly objective statement can quite trivially be replaced by a weakly objective one. Instead of asserting that two bodies carrying opposite electric charges attract each other, we can always say that they will (upon measurement) *be found* to attract each other. However, the converse is not true. Of course, some weakly objective statements can trivially be converted into strongly objective ones (take just the foregoing example and consider it in reverse order). But concerning the basic statements of statistical thermodynamics, such a replacement is quite certainly a highly nontrivial operation, and indeed seems to be viewed as an impossible one by most experts. In the quantum mechanical case we shall have opportunity to investigate the question at length below, with a comparable conclusion.

2. Clearly the distinction between strong and weak objectivity has nothing to do with that between truth and error. A statement such as: "Bodies carrying the same electric charge attract each other" is a false, strongly objective one. Similarly, it goes without saying (or at least it should) that the distinction has nothing to do with the one between materialism on the one hand and, on the other hand, the theory with a spiritualist flavor that makes individual consciousnesses act directly on matter (unfortunately, however, some commentators still occasionally mix up the two things).

3. Weak objectivity of concepts: statements that are not fully objective in the strong sense are sometimes made use of in order to define concepts, or meanings bestowed to familiar words. An example is Bohr's definition of a measuring instrument. As will be recalled in detail in Chapter 11 and later, the way Bohr uses the concept of an instrument of observation in the arguments he puts forward, implies that what he takes in under this name is a set of objects defined by reference not to their material structure but to the use made of them (an instrument pointer, for example, or a movable diaphragm, is, or is not, an instrument part, depending on whether or not it is made use of to gain information). The statement that such and such an object *is used* in such and such a way is one that refers to a collectivity of human beings, so that the objectivity of the definitions in which it plays a basic role cannot be strong. On the other hand, within the realm of a well-defined experiment, there is no ambiguity as to what we call 'the instrument' (or 'instruments'). The specification is the same for everybody and is, in this sense, objective. In general, let us call *weakly objective* the concepts or meanings defined by such references to operations carried out by human beings and *strongly objective* those whose definition does not involve such references (the distinction is clear enough for present purposes; later in the book we shall go more deeply into the problems it may raise).

4. A strictness condition for realism: to come back to the distinction between strong and weak objectivity, clearly it has bearings on that between conventional realism and phenomenism (alias idealism, moderate or radical). Here, as above, what, lacking a better expression, we call "conventional realism"[2] is a conception distinctly broader than just naive realism. It is the view that the ultimate attributes of the "things as they really are,"

2. Not to be confused with Henri Poincaré's "conventionalism." In this book, Poincaré's elaborate conception of reality will be referred to as "structural realism."

God's blueprints of the universe to use once again Hawking's metaphor, can and will (perhaps asymptotically in time only) be known by man, even though the concepts used for this will mostly (contrary to what naive realists believed) be extremely different from the commonplace, familiar one (we call this view physical realism when the vector of this knowledge is assumed to be physics). Clearly, whoever believes in conventional realism and insists on consistency must require that all the basic laws of physics be ultimately expressed by means of strongly objective statements *and concepts*. John Bell rightly emphasized this point [10, 11], speaking of "beables" and so on. Unfortunately, it is one that some of the physicists interested in the problem of "realistically" interpreting quantum mechanics are not clearly aware of, the reason being that the distinction between strong and weak objectivity is not a traditional one. Typically, such physicists would claim that they are in search of "objective" interpretations of quantum mechanics, and produce ones based on such concepts as those of "measurement" or "preparation procedure." Such schemes are flawed as the distinction between strong and weak objectivity immediately shows. If by "objective" what is meant is "weakly objective," then conventional quantum mechanics being weakly objective already, there is actually not much to search for. If, on the other hand, what is meant is "strongly objective," then the proposed schemes fail since they are based on weak-objectively defined concepts. More generally, for a scheme to constitute a truly "realistic" interpretation of quantum physics it is not sufficient that the statements it makes use of have the grammatical form of strongly objective statements. It must be grounded on notions and involve arguments none of which basically refer to human-made operations, to the limitations of human possibilities, and so on. In particular it must meet the conditions of the "truth criterion" to be introduced below. Schemes that do not must be banned from the theory by whoever believes in conventional realism and is strict enough on matters of consistency (we come back to this point in the next subsection).

5. The upholders of phenomenism (in the sense defined in Section 1.1) should not be identified as diehard operationalists making a point of only using weakly objective statements. Even Kant willingly accepted the conventional formulation—which is fully in terms of strongly objective statements—of Newton's theory. Only he interpreted these statements as bearing, not on the things-in-themselves but on the phenomena, that is, on what I called empirical or effective reality in Section 1.4. It is indeed a remarkable fact that so much of physics (and the quasi-totality of commonsense factual knowledge) can be expressed in terms of statements having the form of strongly objective ones (and which therefore we should consider as *being* strongly objective if, in accordance with a literal interpretation of the stated definition, we referred to their *form* only), but bearing on concepts the definition of which refers to human abilities. Kant would perhaps have explained this fact away by asserting that the possibility of expressing every basic law of physics by means of strongly objective statements is a necessary condition for the very possibility of building up this science. But if so, then the counterexample of conventional textbook quantum mechanics shows he would have been mistaken, so that a less sweeping explanation, more refined and owing something to physics, must be looked for. This a problem contemporary physicists have investigated, and their work will be reported on and analyzed in further chapters.

Due account being taken of all these remarks, it is clear that if the set of the basic laws of a theory contains weakly objective statements or concepts that cannot be replaced in a straightforward way by strongly objective ones, this theory must be viewed with different eyes by phenomenists and conventional realists. The former can quite consistently be satisfied with the theory as it is, and will consider that any attempt at amending it by making it "more" strongly objective is not only useless but even meaningless. On the contrary, the latter, as already noted, cannot rest content with the theory. They must change it—or change their philosophical standpoint.

A Truth Criterion

We all have in mind general ideas that we consider as being immediately and obviously valid, because we see them as either tautologies or mere definitions. One of these ideas is that a statement having one (at least) observable consequence that is found false cannot be right.

Admittedly, epistemologists have elaborated on this idea and made the test more sophisticated. Pierre Duhem convincingly showed that within science as it actually is, truly "crucial" experiments do not exist; and quite a number of other epistemologists further discussed the conditions under which a theory eventually gets considered as being falsified. Duhem's and his followers' argumentation consisted in pointing out that the consequences of a theory T can, as a rule, be derived only with the help of some, explicit or implicit, simplifying assumptions A, so that when one such consequence is falsified, rather than giving up the theory it often seems more advisable to modify some of the A's. Of course their remark also applies here since the observable consequences of a statement are necessarily derived with the help of a theory—call it T again—and of such simplifying assumptions A. And, in case of troubles, the latter can sometimes be changed in such a way as to salvage the validity of both T and the statement.

Such considerations are entirely well founded and backed by many indisputable facts in the history of science. Here, however, I have something else in mind. I propose to discuss the correctness and the applicability of the criterion just sketched (more precisely stated below) within the working assumption that both theory T *and all assumptions A that conceivably enter the picture* are strictly true. Theory T is of course assumed to be a physical, not a purely mathematical, theory (we shall particularize to quantum mechanics). More precisely, we assume T to be such that it grants the meaningfulness of some contingent factual statements such as "At such and such a space-time point such and such a physical quantity has such and such a value," and allows for observable consequences to be derived from them. For short, let such a theory be called a "type Ty" theory. The precise expression of the truth criterion of which we observed that it is quite deeply ingrained in our minds is then as follows.

Truth criterion. A factual statement can only be true if all the observable consequences derived from it by means of a "type Ty" theory considered as valid, are true.

Notwithstanding its apparently quasi-tautological character, the truth criterion is worth some comment. Its validity is subject to some tacit assumptions that, reasonable as they look, still are better being made explicit. One of them has to do with the fact that the expression "all observable consequences" implicitly carries with it a notion of counterfactuality that normally refers to that of free choice. To take a trivial example, if the considered theory is electromagnetism and if the statement whose truth is tested is that such and such a piece of iron is magnetic, then, according to the theory, the statement has several observable consequences, whose testing calls for different apparatuses. For testing the attraction the body is predicted to exert on other iron bodies, the experimentalist will decide to put some small iron filings near it. For testing the induction the body is predicted to create, he will decide to pass it through a coil, and so on. The truth criterion means that, if any of these tests fails, the body is not magnetic, and so its use implies the experimentalist has the liberty of actually performing or not performing a measurement and of deciding which one to do, by picking up the appropriate instrument (in the quantum case the necessity of free choice is even clearer since in some cases two different measurements cannot be combined into one, even in principle). Now the notion of free choice is, at least in our elementary conceptions, foreign to the realm of physics proper, so that we should see it as nonabsurd that in fact some elaborate theoretical developments in contemporary physics *do* simply ignore the truth criterion as formulated here.

Another, related reason on the basis of which the truth criterion could be questioned is that it implicitly refers to an assumed qualitative difference between past and future, namely that past is fixed once and for all, while future, as the experimentalist's freedom shows, is left open. Again, such a qualitative distinction between past and future is alien to the basic laws of physics. So that, again, we may see it as nonabsurd if some basic theories reject it, and thus escape the conditions set forth by the truth criterion.

On the other hand, a violation of the rule that we cannot in any way alter the past would mean that whatever "really exists" is incredibly different from the network of the physical phenomena since, clearly, there is a sense in which we cannot change the "phenomena" of the past (think of the well-known paradoxes that go with the idea we can). Similarly, denying that the choice of what measurement we shall perform in an hour's time is now still open seems to run counter to something quite basic in our

conception of empirical science. For these reasons the validity of
the truth criterion is, as a rule, implicitly postulated by all scientists, be they realists or phenomenists. However—and this brings
us back to a point mentioned earlier—realists and phenomenists
are not bound by it quite in the same way. The distinction has
to do with the question whether or not, in the criterion, some
measurements should be considered that could be performed *in
principle* (there is no law of physics that forbids them and the sequence of operations by means of which they would be made can
be quite precisely stated) but cannot be done *in practice* because
they are tremendously too complicated (they would induce a
sudden, appreciable increase of the entropy of the whole Earth,
for example). For future reference, let the observables that such
measurements would be measurements of, be called "sensitive
observables." What, within the truth criterion, is the "logical
status," so to speak, of such sensitive observables?

For a "conventional" realist there cannot be any doubt on
this: the truth criterion must take the sensitive observables into
account on a par with practically measurable ones since the opposite view would imply introducing a human-referred element
into the very formulation of the laws governing the world. And,
as we shall see, this obligation, in its turn, forces the conventional
realist to dispute the validity of some, otherwise quite attractive,
quantum measurement theories. On the contrary, an upholder
of phenomenism is not necessarily bound by such a tight consistency requirement. Since, for her, all the contingent, factual
statements, including those bearing on outcomes of measurements, are but elements in a system of ordering of human experience, she may, not unreasonably, consider that some consistent
way should somehow exist of setting the sensitive observables
outside the realm of those we must consider when applying
the truth criterion. In the later chapters of this book we shall
investigate whether and how such an idea can be employed.

What Is "Measurement"?

In ordinary language the word "measurement" has quite an
obvious meaning. We consider that a table, say, has some definite
length and width, and we perform measurements of these, just
to know them. Clearly this conception is based on realism. In
it, the view that the table *has*, by itself, a definite length even if
nobody knows it, is taken for granted: that is, it is considered

as (a) meaningful and (b) true. This idea is of course backed up by the empirical fact that if we, or somebody else, measure this length again the same value is found (to within controllable experimental errors).

In the case of tiny objects this repeatability condition quite often fails to be fulfilled. It may very well happen that if, say, the velocity of a freely moving little ball is measured twice in succession the two outcomes do not coincide. Of course, realism has a ready explanation for this. To perform any measurement, some measuring apparatus must be made to temporarily interact with the measured system, and it may therefore well be that the first measurement act disturbed the motion of the little ball and changed its velocity. Even though *this* case is still purely classical (nonquantum), a question arises as to what the outcome of the measurement corresponds to: the "initial" velocity of the ball, just before it interacted with the apparatus, or the "final" one, just after?

In quantum mechanics, as we shall see, the situation is even more subtle. In it there are serious reasons for questioning the validity of the realist's assumption that the measured quantity *has*, at any time, a definite (even if unknown) value. But it is felt nevertheless that the measurement outcome must somehow have *some* relationship with reality, and therefore, in most descriptions of quantum mechanics—either in textbooks or in articles—one of the two possibilities considered here is assumed true. Surprisingly enough, however, even at this elementary stage different views are held by different authors. Most of them realize that, as we shall see in Chapters 3 and 4, conventional textbook quantum mechanics naturally leads to the view that the measurement outcome refers to the *final* value. But not all of them are aware of this and some who are reject conventional textbook quantum mechanics just for this reason. They consider it obvious, or at least most natural, that the measurement outcome should refer to the *initial* value of the measured quantity (before it was "disturbed" by the apparatus) and they correspondingly build up their own quantum theories in such a way that it fulfills this requirement. Such a conception may be said to have its roots in Bohr's interpretation of quantum mechanics (in contrast with the conventional textbook one, which was inspired by John von Neumann as we shall see). Recent examples of this type of approach are Griffiths's and Omnès's theories (see Chapter 11).

There is an escape from the horns of this dilemma. It consists in going over to a radically operationalist standpoint and asserting that the measurement outcome represents nothing else than just what is read on the scale of the instrument. This standpoint is often implicitly taken up by theorists because of its convenience. And it seems a number of them feel that they thereby neatly separate physics from all uncertain, vague, and ill-defined problems of philosophy. Indeed they do, but at the price—and this they do not always realize—of paying allegiance to a specific philosophical doctrine, namely operationalism, and of thereby leaving undiscussed the acute problems specifically raised by quantum physics in the realm of the relationships between knowledge and the world.

One of the main reasons why the questions concerning measurement are so crucial in the field of quantum physics is that, as will become clear in what follows, both Niels Bohr's approach and the conventional textbook one use this notion for stating the "basic laws" of this science (even if they do not proceed in quite the same way). In view of this, it is not surprising that the physicists who hold to realism and are strict on questions of consistency should have expressed their dissatisfaction concerning the present state of quantum mechanics. One of John Bell's last articles [11] was entitled "Against Measurement."

It is now time that we turn to physics proper. The following chapters develop this topic in a most didactic and pedestrian way for, in order to grasp the possible bearings of this science on the selection of admissible views on the world, no intermediate step can safely be skipped.

References

1. K. Popper, *Conjectures and Refutations, the Growth of Scientific Knowledge*, Routledge and Kegan Paul, London, 1969.
2. G. Cohen-Tannoudji and M. Spiro, *La matière-espace-temps*, Fayard, Paris, 1986.
3. W. Heisenberg, *Physics and Philosophy*, Harper and Brothers, New York, 1958.
4. E. Cassirer, *Substanzbegriff und functionsbegriff*, Berlin 1910; Wissenschaftlische Büchergesellschaft, Darmstadt, 1969.
5. H. Reichenbach, *The Rise of Scientific Philosophy*, U. of California Press, 1951.

6. E. Schrödinger, *Mind and Matter*, Cambridge U. Press, 1958.

7   G. Bachelard, *Le rationalisme appliqué*, Presses Universitaires de France, Paris, 1949; *La formation de l'esprit scientifique*, Vrin, Paris, 1970; *Le matérialisme rationnel*, Presses Universitaires de France, Paris, 1972.

8. T. Kuhn, *The Structure of Scientific Revolutions*, U. of Chicago Press, 1962.

9. B. d'Espagnat, *Conceptual Foundations of Quantum Mechanics*, Addison-Wesley, Reading, Mass., 1971, 2nd ed. 1976.

10. J. S. Bell, *Speakable and Unspeakable in Quantum Mechanics*, Cambridge U. Press, 1987

11. J. S. Bell, "Against 'Measurement,'" in *Sixty-two Years of Uncertainty*, A.I. Miller (Ed.), Plenum, New York, 1990.

Matter Waves, Superposition, Linearity

2 Classical physics identified matter with particles and radiation with waves. Although these descriptions accounted for a large number of very important facts, it became apparent at the beginning of the twentieth century that other phenomena, such as blackbody radiation and photoelectric effect, are satisfactorily explained only if radiation is described in terms of particles. Some time later, in 1923–24, Louis de Broglie developed a theory according to which, conversely, matter has wave properties. As an outcome of this theory, he was able to write down a formula relating the corresponding wavelength with the momentum of the particle [formula (2.1)]. Three years later these predictions were quantitatively confirmed by Davisson and Germer's experiment of electron diffraction, an experiment in which an electron beam was passed through a regular lattice (crystal); the experiment showed the well-known diffraction pattern that is characteristic of waves. Meanwhile Schrödinger had written down a differential equation, since known as the Schrödinger equation, Eq. (2.11), of which the de Broglie wave is, in the free particle case, a solution. This equation yields quantization of the energy levels of bound systems, whence the name *quantum mechanics*.

2.1 The Schrödinger Equation

The outcome of the Davisson and Germer experiment and similar ones confirms the theoretical predictions that a monoenergetic electron beam should be described by a plane wave and that the wave vector **k** of the wave should be related to the momentum **p** of an individual electron by the formula

$$\mathbf{p} = \hbar \mathbf{k} \tag{2.1}$$

where, by definition of the notion of wave vector, the direction of **k** is that of the wave propagation and its magnitude is $|\mathbf{k}| = (2\pi)\lambda^{-1}$ (λ being the wavelength), and where $\hbar$ is Planck's constant divided by 2π.

The simplest possible form for the x dependence of a plane wave with wave vector **k** is

$$\psi(\mathbf{x}) = e^{i\mathbf{k}\mathbf{x}} \tag{2.2}$$

However, as the particle propagates, it is natural to associate it with a traveling wave, whose direction of propagation is that of the particle. This is achieved by writing for the wave function of the free particle:

$$\psi(\mathbf{x}) = e^{i(\mathbf{k}\mathbf{x} - \omega t)} \tag{2.3}$$

where, by definition, ω is the circular frequency of the wave. Now, if we assume that the Planck–Einstein relation

$$E = \hbar\omega \tag{2.4}$$

between the energy of a photon and the circular frequency of the corresponding wave is also valid for "material" particles such as those described by ψ, we can infer from Eq. (2.3) an equation for ψx, t) simply by noting the identities

$$i\hbar \partial\psi / \partial t = \hbar\omega\psi = E\psi \tag{2.5}$$
$$\hbar^2 \Delta\psi = -\hbar^2 \mathbf{k}^2 \psi = -\mathbf{p}^2 \psi \tag{2.6}$$

and making use of the relation

$$E = \mathbf{p}^2 / 2m \tag{2.7}$$

between the (nonrelativistic) energy and the momentum of a free particle. The result,

$$i\hbar \partial\psi / \partial t = -(\hbar^2 / 2m)\Delta\psi \tag{2.8}$$

is the Schrödinger equation for a free particle. It can be deduced formally from the corresponding classical equation (2.7) by the replacement

$$\mathbf{p} \rightarrow -i\hbar\nabla, \qquad E \rightarrow i\hbar\partial/\partial t \qquad (2.9)$$

and the prescription that both members of the operator relation thus obtained should be applied to ψ ψ is a function of the coordinate vector $\mathbf{x}$ of the particle and the time. It is called the *wave function* of the particle.

• *Remark 1. Phase and group velocities.* The velocity of the traveling plane wave (2.3), that is, the velocity of motion of the plane of constant phase

$$\mathbf{k}\mathbf{x} - \omega t = C$$

is (setting $|\mathbf{k}| = k$)

$$v_{ph} = \omega/k$$

which is different from the classical velocity v of the particle. The "group velocity" of the wave, however, defined in the usual classical way as the velocity of the center of a *wave packet* made of waves with wave vectors centered around $\mathbf{k}$ is

$$v_y \equiv d\omega/dk = d\hbar\omega/d\hbar k = dE/dp = p/m$$

It is obviously equal to v
  In the general case when the particle is not free, the Schrödinger equation, that is, the equation for the wave function $\psi(\mathbf{x}, t)$, is obtained formally by applying rule (2.9) to the identity

$$E = H(\mathbf{x}, \mathbf{p}) \qquad (2.10)$$

where H is the *Hamiltonian function*, that is, the total energy of the system, written in terms of $\mathbf{x}$ and the conjugate momentum $\mathbf{p}$ (the rule being to symmetrize between $\mathbf{x}$ and $\mathbf{p}$ whenever this proves necessary). In other words, the Schrödinger equation is

$$H\psi(\mathbf{x}, t) = i\hbar\frac{\partial}{\partial t}\psi(\mathbf{x}, t) \qquad (2.11)$$

where H, known as the *Hamiltonian* operator, is obtained from the classical expression $H(\mathbf{x}, \mathbf{p})$ of the total energy by making the substitutions

$$\mathbf{x} \rightarrow \mathbf{x}, \qquad \mathbf{p} \rightarrow -i\hbar\nabla \qquad\qquad (2.12)$$

after the appropriate symmetrization of $H(\mathbf{x}, \mathbf{p})$

That the Schrödinger equation (2.11) is correct is, of course, not a deduction. It is really a physical assumption. The validity of this assumption, however, is demonstrated beyond any reasonable doubt by the truth of its consequences.

When several particles with coordinates $\mathbf{x}_1, \mathbf{x}_2, \quad, \mathbf{x}_N$ are present instead of just one, the wave function ψ of the system they constitute is a function

$$\psi(\mathbf{x}_1, \mathbf{x}_2, \quad, \mathbf{x}_N, t) \qquad\qquad (2.13)$$

of these coordinates and the time. The Schrödinger equation for ψ is obtained from the expression for the total energy of the system, $H(\mathbf{x}_1, \mathbf{x}_2, \quad, \mathbf{x}_N, \mathbf{p}_1, \mathbf{p}_2, \quad \mathbf{p}_N)$, in exactly the same way as above.

If a system has a definite energy E, the substitution

$$\psi(\mathbf{x}_1, \mathbf{x}_2, \quad, \mathbf{x}_N, t) = \phi(\mathbf{x}_1, \mathbf{x}_2, \quad, \mathbf{x}_N)e^{-i(Et/\hbar)} \qquad (2.13a)$$

in Eq. (2.5) leads to the "time-independent Schrödinger equation"

$$H\phi = E\phi \qquad\qquad (2.14)$$

ϕ is then called the time-independent wave function of the system.

2.2 Superposition and the Born Rule

As already mentioned, diffraction patterns characterize waves, and the basic fact that corroborated the validity of the notion of matter waves was that the Davisson and Germer experiment proved electrons can be diffracted. In classical optics the simplest experiment displaying this phenomenon is the well-known Young two-slit experiment, in which a beam of light is passed through a diaphragm with two slits. On a screen set beyond the diaphragm a diffraction pattern (fringes formed by alternating zones of brightness and darkness) is then observed. For technical reasons it would be difficult to exactly reproduce this experiment with beams of electrons (or other particles), but other experiments equivalent to this one have been performed, notably with neutron beams passed through crystal interferometers [1]. To avoid going into the experimental technicalities that would be required for describing these actually performed experiments, let us just consider the simple *gedanken* experiment that would consist in passing a beam of particles (electrons, say) through a Young diaphragm with two slits. Let us choose a coordinate system in which the z axis is perpendicular to the slits, and let $p_1(z)$, $p_2(z)$ and $p_3(z)$ be the density distributions of the electron points of impact on the screen when slit 2 is shut, when slit 1 is shut, and when both slits are open respectively. The experimentally observed densities are such that

$$p_3(z) \neq p_1(z) + p_2(z) \tag{2.15}$$

and that two (complex) functions $\psi_1(z)$ and $\psi_2(z)$ can be found, such that

$$p_1(z) = |\psi_1(z)|^2 \tag{2.16}$$
$$p_2(z) = |\psi_2(z)|^2 \tag{2.17}$$
$$p_3(z) = |\psi_1(z) + \psi_2(z)|^2 \tag{2.18}$$

which, of course, is just what is to be expected if the electron beam had wave structure (in the case of a classical wave, ψ would correspond to the amplitude and p to the intensity).

The similarity of our electron beam with a wave is only partial, however, since, for instance, if the duration of the experiment with both holes open (the time of exposure of the screen) is kept constant and the intensity of the electron source is sufficiently lowered, what eventually happens is very different from what a classical wave theory would predict. The law $p(z)$ of distribution of the observed intensity, that is, of blackening of the emulsion on the screen (if the impacts are registered by such means) is altered appreciably, in a way that (a) is not reproducible (it varies from one experiment to the next) and (b) is such that it corresponds to the impact of zero or an integer number of particles in any given small area. This shows that the assumption that the source emits just classical waves instead of particles, and that what is observed on the screen is simply the local effect of these waves, is certainly not a correct hypothesis, although, if the time of exposure is now increased so as to allow for a large number of impacts in most areas of the screen, the cumulative density of impacts $p(z)$ again takes, within the usual approximation of statistics, the form $p_3(z)$ predicted by the wave picture.

The preceding discussion is very sketchy (more elaborate ones can be found in many textbooks). It is, however, sufficient to show that these diffraction experiments on particles raise problems that will not easily yield to an interpretation in terms of familiar concepts, such as those of waves or particles. Situations of this kind are not really exceptional in physics; and when they occur, it is usually not rewarding to try to force some model description upon them. A much better procedure, at least in the first stages of the inquiry, is to make no assumption whatsoever about the mechanisms at work and to simply describe the observed facts, it being understood that this description should be as general and as accurate as possible. The simple features of the description thus obtained are then assumed to be valid in similar cases involving different experimental settings and so forth.

If this is done in regard to the experimental facts recalled above, three general ideas, or principles, emerge. They are as follows.

(i) Elementarity of microstructures: Objects are normally endowed with properties that cannot be split indefinitely; consequently entities—called *particles*—apparently exist, which, in any experiment, manifest themselves as integer wholes.

(ii) Probability principle (Born's rule): In some situations at least, functions $\psi(\mathbf{x}, t)$ can be associated with particles in such a way that the probability of finding at time t a particle in a small region of volume dv around point $\mathbf{x}$ is

$$P(\mathbf{x}, t)\, dv = \frac{|\psi(\mathbf{x}, t)|^2\, dv}{\iiint |\psi(\mathbf{x}, t)|^2\, dv} \tag{2.19}$$

where the summation extends over all space. It is to be noted that, in view of this formula, the value of $P(\mathbf{x}, t)$ remains the same if $\psi(\mathbf{x}, t)$ is multiplied by any constant factor. Because of this, no generality is lost by requiring ψ to be "normalized to unity," that is, by setting the denominator in the right-hand side of Eq. (2.19) equal to 1. The probability $P(\mathbf{x}, t)$ then reads:

$$P(\mathbf{x}, t)\, dv = |\psi(\mathbf{x}, t)|^2\, dv \qquad (2.20)$$

These general ideas, suggested in a straightforward way by the experiments analyzed must now be supplemented by a theoretical assumption. Justified, as already noted, by the truth of its consequences, the latter will be that the function $\psi(\mathbf{x}, t)$ is just the one we called the wave function of the particle; that is, it obeys the Schrödinger equation (2.11). More generally, it will be assumed that systems of particles are described by wave functions such as (2.13). Formulas (2.19) and (2.20) are then easily generalized. Let it be noted that, as a consequence of Eq. (2.11), if a wave function is normalized to unity at some time, it is also normalized to unity at any other time.

Superposition principle: If $\psi_1(\mathbf{x}, t)$ and $\psi_2(\mathbf{x}, t)$ are two possible wave functions, that is, two solutions of the Schrödinger equation (2.11), any linear combination

$$a\psi_1(\mathbf{x}, t) + b\psi_2(\mathbf{x}, t) \qquad (2.21)$$

is also a possible wave function. This is an immediate mathematical consequence of the fact that the Schrödinger equation is homogenous and linear. It is called a principle only because of its great importance, particularly in matters concerning interpretation of the theory. The wave function (2.21) is said to be a *superposition* of the wave functions ψ_1 and ψ_2. Note, moreover, that since the differential equation (2.11) only involves the first time derivative of ψ, it determines ψ at any time t if ψ is known at one particular time t_0. An obvious, but quite far-reaching, consequence of this and of the superposition principle is that if, at time t_0, the wave function is

$$a\psi_1(\mathbf{x}, t_0) + b\psi_2(\mathbf{x}, t_0) \qquad (2.22)$$

at time t it must be equal to

$$a\psi_1(\mathbf{x}, t) + b\psi_2(\mathbf{x}, t) \qquad (2.23)$$

The fact that the form (2.22) of the wave function at t_0 entails this wave function has the form (2.23) at t is, as later chapters will show, of utmost importance for all the discussions bearing on the possible interpretations of quantum physics.

2.3 The Necessity for a More General Formalism

Wave mechanics, the formalism sketched in the foregoing sections, accounts with remarkable precision for practically all atomic and molecular phenomena, that is, essentially for all phenomena that take place in nature around us, where, because of prevailing conditions of temperature and so forth, pair creations and annihilations of nucleons and electrons seldom occur. This is because the conservation of the number of particles in a system makes it possible to describe the system by a wave function $\psi(\mathbf{x}_1, \ldots, \mathbf{x}_N, t)$ with a fixed number of variables. It is clear, however, that in problems in which pair creations and annihilations play a significant role, as is, for example, the case in high-energy physics, a formalism that attributes such an importance to the coordinates of the constituent particles must lead to some difficulties, since a change in time of the number of variables of a function is definitely not a standard mathematical operation. Moreover, even in low-energy physics there are physical observables, such as spin, the description of which by means of wave functions turns out to be, if not strictly impossible, at least very clumsy. These two reasons would suffice to justify taking interest in a more general formalism. But there is still another reason, which is that this new formalism (described in the next chapter) happens to be, in many instances, easier to handle mathematically than the one based on wave functions, even in the cases of stable spinless particles.

The formalism in question, due to Dirac, is known as the *formalism of bras and kets*. It describes the possible states of a system, not by wave functions but by abstract entities called state vectors, whose defining property is just that they obey the superposition principle. The mathematical notions that are prerequisites for understanding the said formalism are described in detail in practically all textbooks on quantum mechanics (see, e.g., Refs. [2], [3], or [4]). For the benefit of readers who do not have quite the patience to learn them as thoroughly as they would in these books, some of the most basic among the notions in question are briefly summarized in Appendix 1.

References

1. H. Rauch, W. Treimer, and U. Bonse, *Phys. Lett.* **A47**, 369 (1974).
2. A. Messiah, *Mécanique Quantique*. Dunod, Paris, 1958 (English translation: *Quantum Mechanics, Volumes 1 and 2*, Elsevier-North-Holland. Amsterdam, 1961).
3. K. Gottfied, *Quantum Mechanics*. W.A. Benjamin, Reading, Mass., 1966 (4th printing, with corrections, 1974).
4. C. Cohen-Tannoudji, B. Diu and F. Laloë, *Mécanique quantique*, Hermann, Paris, 1973 (English translation: *Quantum Mechanics, Volumes 1 and 2*), John Wiley and Sons, Ltd. Chichester Sussex, U.K., 1977).

Additional Introductory Reading

E. Squires, The Mystery of the Quantum World. Adam Hilger Ltd. Bristol, U.K., 1986.

The Rules of
Quantum Mechanics

3 Standard quantum mechanics—the theory that is made use of in practice and is, because of this, taught in university courses—is a theory that is weakly objective only. For this reason it is most conveniently and neatly expressed by means of a set of basic rules of calculation that enable physicists to make predictions concerning what will be observed.

3.1 Statement of the Rules

Before the rules are stated, two definitions should be given.

Definition 1. Physical System. Classical and quantum physics both make considerable use of the expressions "physical system" and "state of a physical system." Indeed, any sufficiently isolated thing—a voltmeter, an electron, or a molecule—is a physical system (the latter two, and more generally, systems that are of molecular size or smaller and for which a description in terms of classical physics is obviously insufficient, are often called microscopic systems). As a matter of fact, the concepts implied by the two expressions appear so obvious as to seem a priori or inherently meaningful notions. A word of caution is thus appropriate at the start. As results from the discussion in the following chapters, the concepts of systems and states, which are extremely useful in quantum mechanics, should nevertheless be handled with some care in that formalism. In particular, it should *not* be assumed that in all cases in which we intuitively speak of a physical system (and even of a "completely isolated physical system") this system (supposing the word is used) is necessarily in some definite, known or unknown, "state." More precisely, we may not always assume that a system possesses at all times a constant number of, known or unknown, properties (position, momentum, spin component along some direction, etc.) that can properly be considered as pertaining to it alone. Counterexamples to such an assumption are given below.

In what follows, the expression *physical system* is taken in its familiar, intuitive sense, as discussed above. For instance, a partially but sufficiently isolated atom (or electron, molecule, etc.) is a physical system or, for short, a system. As for the expressions *state* and *quantum state*, they are not always given the same definitions in the existing literature. At this stage, therefore, we avoid introducing them.

Definition 2. Ensembles. Quantum mechanics is essentially a probabilistic theory. Except in special cases it makes no predictions that bear on individual systems. Rather, it predicts statistical frequencies. In other words, it predicts, as a rule, the number n of times that a given event will be observed when a large number N of physical systems of the same type and satisfying specified conditions are subjected to a measurement process. The statistical frequency or *probability* of the event is then the number n/N (more precisely the limit, assumed to exist, of this ratio when $N \to \infty$). The collection of N systems is called a *statistical ensemble*, or, for short, *ensemble*. Each individual system is an *element* of the ensemble. The elements of the ensemble are noninteracting. Ideally, it should be imagined that to each element corresponds one particular apparatus by means of which the measurement is made. Only in the cases $n/N = 1$ or 0 is it possible to say that the theory makes a prediction that bears on individual systems, since in these cases the prediction is that the given event either will or will not be observed on any individual element. Therefore quantum mechanics is sometimes called an ensemble theory. This appellation is literally correct, but when it is used, it is necessary to be on guard against unwarranted inferences. As will be shown in detail in Chapter 13, quantum ensembles are conceptually extremely different from the classical image the very term *ensemble* naturally conveys.

The rules of quantum mechanics can now be summarized as follows (comments and alternative choices of axioms are postponed to later chapters).

Rule 1: A vector space H that has the structure of a Hilbert space is introduced for describing ensembles of physical systems of the same type.[1] There are instances in which, for the purpose of predicting the results of definite experiments, such an ensemble can be associated with a definite element of H Such an element, or *ket*, is then called a *state vector* Ensembles that can be described this way are called *pure cases.*

Rule 1 calls for some remarks.

• *Remark 1.* As Rule 6 in this section makes apparent, two kets differing only by a phase factor give rise to the same predictions. Such a phase factor has therefore no physical meaning although, in a superposition such as those considered in Rule 3, the relative phases are of course meaningful. The expression "definite element of H" should therefore be understood with this restriction.

• *Remark 2:* A frequent practice is to associate individual systems with state vectors. Often, this may be done without inconvenience. From a theoretical standpoint, however, the use of ensembles is preferable unless special precautions are taken. Rule 1 does not imply that every ensemble (or, in the language just alluded to, every system) can be described by a state vector. Indeed the majority of ensembles are such that no state vector can be attributed to them, not even in a loose language (see Section 7.2).

Rule 1a (Superposition principle): Any ket of H can usually describe an ensemble of physical systems.

Exceptions exist. They are known as *superselection rules.* When a ket $|\psi>$ can be written as a linear combination of several kets $|\phi_1>$,   , $|\phi_N>$, one says that $|\psi>$ is a *superposition* of $|\phi_1>$,   , $|\phi_N>$

Rule 2: If two nonidentical systems S and S' are elements of ensembles that can be described by means of the kets $|\psi>$ and $|\psi'>$ respectively, the ensemble of the compound systems composed of S and S' can also be described by a ket, denoted $|\psi> \otimes |\psi'>$

---

1. Readers not familiar with Hilbert spaces and related notions also used in this chapter are referred to Appendix 1.

or more briefly $|\psi > \quad |\psi' >$ (or $|\psi > |\psi' >$). This ket is defined as an element of the tensor product $H \otimes H'$ of the Hilbert spaces H and H' corresponding to S and S' respectively.[2]

Rule 3: The evolution in time of a state vector $|\psi >$ that describes an ensemble of systems subjected only to external forces is causal and linear. This means that if

$$|\psi(t) >= a_1|\psi_1(t) > + a_2|\psi_2(t) > \qquad (3.1)$$

then

$$|\psi(t') >= a_1|\psi_1(t') > + a_2|\psi_2(t') > \qquad (3.1a)$$

In the statement of this rule the term "external forces" means, as usual, that the influence (often called "reaction") of the system on the physical entities that create the forces can be neglected. If this is not the case, a larger system, including these entities, should be considered, and Rule 3 should be applied to the larger system.

It should also be noted that the usual Schrödinger equation

$$H|\psi(t) >= i\hbar \frac{\partial}{\partial t}|\psi(t) > \qquad (3.2)$$

where H is the Hamiltonian, is just a particular realization of Rule 3. The point is that Rule 3 embodies all the general properties of the Schrödinger equation that are of interest in the subsequent discussion.

Rule 4: To every measurable dynamical quantity[3] A there corresponds a Hermitean operator operating on the elements of the vector space defined in Rule 1.

For convenience we call simply *Hermitean* a Hermitean operator for which the eigenvalue problem can be solved (see J. von Neumann [1, Sec. 2.9]). Following Dirac we henceforth call

---

2. Note the difference with classical field theory, where instead we have vector *sums* of functions all defined on the same space.

3. The term *dynamical quantity* means, as usual, a physical quantity such as position, velocity, or momentum that is liable to change in time (with the exception of time itself).

a dynamical physical quantity an *observable*. Two observables are said to be *compatible*, or *commuting*, if the corresponding operators commute.

Rule 5: The measurement of an observable *A* necessarily yields one of the eigenvalues of the operator associated with *A*.

Henceforth the expression *"eigenvalue [eigenket] of observable A"* is often used as abbreviation for *"eigenvalue [eigenket] of the operator associated with A."*

Rule 6: Let the symbols a_k and $|k,r>$ denote, respectively, the eigenvalues and the normalized eigenvectors of observable *A* (*r* is a degeneracy index), and let $|\psi>$ be a normalized ket that describes an ensemble of physical systems *S* immediately before a measurement of *A* on *S*. Then the probability that this measurement will yield value a_k is given by

$$w_k = \sum_r | < k,r|\psi > |^2 \qquad (3.3)$$

When the kets are not normalized to unity, expression (3.3) must be modified in such a way that it is not affected by the arbitrariness in the normalization; that is, it must be replaced by

$$w_k = \sum_r \frac{| < k,r|\psi > |^2}{< k,r|k,r > < \psi|\psi >} \qquad (3.3a)$$

Again, what, in (3.3) and (3.3a), is meant by "probability" is just the relative statistical frequency with which, according to our predictions, the result a_k will be obtained upon measurement of *A*.

• *Remark 3.* The possibility of generalizing the concept of measurement and, in particular, of dissociating it from "the observer," or of reducing the observer to an ordinary physical system, is discussed in later chapters. What is meant here by the word "measurement" is the interaction of a system *S* with a generally complex entity *S'* incorporating, along with measuring apparatus and so forth, the observer. And what is meant by "the value yielded by the measurement" is just what this observer reads on the apparatus. At this stage we make no effort to describe the observer in the same way as an ordinary physical system, and for the time being the entity S' need not be analyzed further as to its constituents. In other words we formulate prediction rules that are for the time being independent of the more subtle question whether or not the observer (for whom the predictions are made) is to be identified with a physical system in the ordinary sense. At this stage we simply leave this question open. On the other hand, independently of the answer to be made to the said question, it is important to be aware that within the formulation of the quantum mechanical formalism developed here, the *measurement results* are attributed a basic role, and that, correlatively, they must, when obtained, be considered strictly true, "by definition" of the very notion of truth. The question whether and to what extent this conception may or should be altered is partly a philosophical question but also partly a physical one. It will be tackled in due course.

• *Remark 4.* The observables that have a continuous spectrum have eigenvectors that are not renormalizable. Concerning the measurement of these observables, some technical changes should therefore be made in expression (3.3). These are described in the textbooks. Then, when applied to a position measurement, Eq. (3.3) reproduces Eq. (2.19). A nonrigorous but simple way to derive this result is to define a "position eigenket" $|x'>$ corresponding to position x by the formulas

$$x|x' >= x'|x' > \qquad\qquad (3.3b)$$

and

$$< x'|x'' >= \delta(x' - x'') \qquad\qquad (3.3c)$$

where x is the position operator, and to define the wave function $\psi(x)$ by the relation

$$\psi(x) =< x|\psi > \qquad\qquad (3.3d)$$

Suppressing index r (no degeneracy) and identifying w_k with the probability density $P(x)$, we thus get

$$P(x) = |\psi(x)|^2$$

which is just Eq. (2.19) (since in [2.19] the denominator, here, is 1 because $\iiint |<x|\psi>|^2 dv = 1$)

Rule 7: For any observable it is always possible, at least conceptually, to invent a measuring device that satisfies the following condition: If the observable is measured twice on a given system, and if the time interval between the two measurements is vanishingly small, then both measurements yield the same result. Such a measurement will be called an *ideal* one or a *measurement of first kind*. All other measurements are measurements of the second kind.[4]

Theorem. Let A and B be two compatible observables. It is a theorem of Hilbert space theory (see Appendix 1) that there then exists a set $\{|k,j,r>\}$ of kets of H that are eigenkets of both A and B, that is, that obey

$$A|k,j,r>=a_k|k,j,r>$$

and

$$B|k,j,r>=b_j|k,j,r>$$

This theorem makes it possible to formulate the following rule.

Rule 8: Let again A and B be two compatible observables. If ideal measurements infinitely close in time are made of both A and B, then with the preceding definition of $|k,j,r>$, the probability of obtaining the pair of results a_k, b_j on an ensemble described by the normalized state vector $|\psi>$ is

$$w_{k,j} = \sum_r |<k,j,r|\psi>|^2 \tag{3.4}$$

independently of the order in which the measurements are made.

Although Rule 8 does not deal with the state of the system after the measurements take place, and hence, strictly speaking,

---

4. Note that some authors use the word "ideal" to designate what is called a "complete" measurement below.

does not rule out the possibility that this final state might depend on the order of these measurements, still it implies that, at least as far as the *outcomes* of these measurements are concerned, the order of succession of two ideal measurements made on compatible observables is irrelevant whenever the time interval between them is vanishingly small. Thus Rule 8 gives meaning to the notion of "simultaneous measurements" of such observables. This notion can obviously be extended so as to cover the case of an arbitrary number of compatible observables. In particular, a *complete* measurement on a system S is a set of simultaneous (ideal) measurements of the observables corresponding to a complete set (see Appendix 1) of commuting Hermitean operators acting on the vector space associated to S.

If the kets involved are not normalized, expression (3.4) should be modified in the same way as expression (3.3) was changed to expression (3.3a).

• *Remark 5:* It is important to realize that in expressions such as (3.3) the degeneracy index r can (and often does) correspond to the existence of observables that are not measured and are compatible with those that are. For example, in the case considered in the formulation of Rule 8, if observable B is not measured, the probability w_k of obtaining result a_k upon measurement of A is the same as if B were measured and the outcome were not recorded; that is,

$$w_k = \sum_j w_{k,j} = \sum_{j,r} |k, j, r|\psi > |^2 \qquad (3.5)$$

which is identical with formula (3.3), the role of r being played by the pair of indexes j, r

Rule 9: Let $A, B,$ be a complete set of compatible observables pertaining to some type of systems and let E be any well-defined ensemble of such systems. If, for some reason, the outcomes of simultaneous ideal measurements of $A, B,$ that will or could be performed on the elements of E are known (or assumed known) with certainty and are the same for all the elements of E, then E can be described by a ket.

Let $|\psi>$ be this ket. Let a_k, b_j be the (known) results of the measurements to be made on these systems. The probability of getting the results a_k, b_j is unity, and so by Eq. (3.3a),

$$1 = \frac{|<k,j, \quad |\psi>|^2}{<k,j, \quad |k,j, \quad ><\psi|\psi>}$$

This is the Cauchy-Schwarz inequality in the case in which it reduces to an equality. It follows that $|\psi>$ is proportional to $|k,j, \quad >$ Since a numerical factor is just a normalization constant without physical significance, the following proposition can be stated (for future reference we call it Proposition A).

Proposition A. The ket considered in Rule 9 is

$$|\psi>= |k,j, \quad >,$$

A special case of Rule 9 is the following: The prior knowledge of the outcomes of simultaneous measurements of A, B, to be made at time t comes from a previous measurement of these observables performed at a time t_0 immediately before t. In this case, Proposition A is known as describing the *collapse* or *reduction* of the wave packet taking place at time t_0. It can be restated as follows:

Rule 10 (Collapse theorem): Let E_0 be an ensemble of systems S. Immediately after simultaneous ideal measurements have been performed on a complete set of compatible observables pertaining to the systems S by a community of observers, this community can describe by a ket any subensemble E of E_0 that is composed of systems S on which the measurements have yielded identical sets of outcomes. This ket is the eigenket common to the eigenvalues found as outcomes of the measurements.

• *Remark 6:* This theorem is often stated as an independent postulate. We see here that it can be proved on the basis of the other quantum rules. The foregoing proof refers to Rule 9, but Remark 2 in Section 6.2 points out that the theorem can also be derived from a weaker assumption.

Finally, we must also consider the case in which an incomplete measurement is performed on a system. We then make the following postulate.

Rule 10a: Let E_0 be an ensemble of systems S that is described by a ket $|\psi>$ Immediately after we have performed simultaneous ideal measurements on an incomplete set of compatible observables $A, B,$ pertaining to the systems S, we can describe by a ket every subensemble E of E_0 that is composed of systems S on which the measurements have yielded identical sets $a_k, b_j,$ of results. This ket (to within a renormalization factor, of course) is the projection of $|\psi>$ on that subspace of the Hilbert space of S that is spanned by the eigenvectors common to $A, B,$ that correspond to eigenvalues $a_k, b_j,$

• *Remark 7* A great number of the problems quantum mechanics deals with require only the use of Rules 1–5. Consequently the physicists who are only interested in such problems never have to cope—at least not professionally—with the ticklish epistemological questions concerning the role of the observers and so forth that were alluded to in Remark 3. In fact, if quantum mechanics just consisted of Rules 1–5 it would, concerning these questions and the possible ways to answer them, show no great difference from classical physics (and correspondingly the notion of ensembles would not be needed in order to formulate it). On the other hand, the latter observation should be qualified in at least three ways. First, it is meaningful only to the extent that the considered systems can be thought of as isolated from their environment and, as we shall see this often is, much more than in classical physics, a questionable approximation. Second, Rule 2 entails a phenomenon called *entanglement* (on which more later) which has no counterpart in the realm of classical physics. Third, it is of course the case that in a number of problems the other rules (6–10) come into play. These are the main reasons why quantum physics calls for the consideration of epistemological problems that the physicists of the classical age could legitimately ignore.

3.2 Complements

Gathered in this section are a few further remarks, useful for either understanding the preceding rules and remarks more thoroughly or applying them to specific problems to be encountered later in the book. The remarks are numbered sequentially to those of Section 3.1.

• *Remark 8:* [5] Remark 5 may suggest a time evolution of the considered ensemble differing from the one described by Rule 10a. Consider for example an observable A having degenerate eigenvalues

$$A|k,r> = a_k|k,r> \tag{3.6}$$

and another observable B having the same eigenvectors but non-degenerate eigenvalues

$$B|k,r> = b_{k,r}|k,r> \tag{3.7}$$

As is easily shown, A and B are compatible. Remark 5 then suggests a possible, though roundabout, way of measuring A. This is to measure B first without recording the outcome and *then* to measure A. With the help of Rule 8 and formulas (3.3) and (3.5) it is easily checked that the probability of outcome a_k is then the same as if the measurement of B had not been done. However, the collapse theorem (Rule 10) shows that immediately after the measurement of B having yielded outcome $b_{k,r}$ we have to do with an ensemble in which a proportion

$$w_{k,r} = |<k,r|\psi>|^2 \tag{3.8}$$

of systems is described by ket $|k,r>$ Since these kets are also eigenkets of A, the A measurement may naturally be assumed not to change them, so that keeping only the systems having yielded outcome a_k amounts to selecting the subensembles described by the $|k,r>$ Hence, after this sequence of measurements we are left with an ensemble that cannot be described by one ket only but is a (proper) mixture of the thus described ensembles (see Section 7.2 for a definition of *proper*).

It was suggested by J. von Neumann [1] that this result should be generalized to the case in which no observable B is actually measured at all.[6] Rule 10a should then be replaced by

---

5. Can be bypassed on first reading.

6. Von Neumann was at this point only interested in the description of the *total* ensemble of systems, composed of all the systems S after their interaction, irrespective of which outcome a_k the latter has produced. For this ensemble he proposed a description that amounts to putting together all the E_k considered in

Rule 10a'. Again let E_0 be an ensemble of systems S that is described by a ket $|\psi>$ Immediately after we have performed an ideal measurement of an observable A whose eigenkets and eigenvalues are defined by Eq. (3.6), the ensemble E_k of systems on which outcome a_k has been obtained is a *mixture* with weights

$$w_{k,r} \left(\sum_r w_{k,r} \right)^{-1} \tag{3.9}$$

of ensembles each which is described by one of the kets $|k,r>$

As the preceding example of a "roundabout" way of measuring A clearly shows, the possibility cannot be ruled out of measuring procedures in which Rule 10a' instead of Rule 10a would apply. On the other hand it was proved by Balian [2] (see also [3]) that only Rule 10a is fully compatible with Rule 8, at least if it is assumed that every Hermitean operator in H corresponds to an observable. The point is that if a_k is a degenerate eigenvalue of A, Hermitean operators D commuting with A can be shown to exist with the following property· If Rule 10a is replaced by any other prescription (for example, Rule 10a'), the mean value of D when A has been measured first (without recording the outcome) is *not* what comes from applying Rule 8 to the pair A, D· so that with any rule other than 10a and with commuting operators such as A and D, Rule 8 fails and the order of the measurements is, in fact, relevant.[7]

• *Remark 9·* Let the eigenvalue equation of observable A be

$$A|k,r> = a_k|k,r> \tag{3.10}$$

as above. A decomposition of the unit operator can then be written (see Appendix 1) as

$$1 = \sum_{k,r} |k,r><k,r| \tag{3.11}$$

Rule 10a' below; so that it is natural to consider Rule 10a' as strongly suggested by his proposal.

7 Such operators D are, for example, projector operators onto eigenvectors of A pertaining to eigenvalue a_k and distinct from any one of the $|k,r>$

Applying formula (3.11) to the identity

$$A = A.1$$

then gives, with the help of (3.10)

$$A = \sum_{k,r} |k,r> a_k <k,r| = \sum_k a_k P_k \qquad (3.12)$$

where

$$P_k \equiv \sum_r |k,r> <k,r| \qquad (3.13)$$

is the projector onto the "eigensubspace" corresponding to eigenvalue a_k. Expression (3.12) is known as the *spectral representation* of observable A. With its help, Eq. (3.3) can be rewritten as

$$w_k = <\psi|P_k|\psi> \qquad (3.14)$$

or equivalently (see Appendix 1) as

$$w_k = Tr[P_k|\psi> <\psi|] \qquad (3.15)$$

• *Remark 10:* For observables with a continuous spectrum the sum in Eq. (3.13) must of course be replaced by an integral. In the case of the observable *position*, for example, the projection operator onto the subspace corresponding to a particle S lying, on the $x'x$ axis, somewhere between $x = a$ and $x = b$ is (see Remark 4)

$$\int_a^b dx' |x'> <x'| \qquad (3.16)$$

with of course

$$\int_{-\infty}^{+\infty} dx' |x'> <x'| = 1 \qquad (3.17)$$

These formulas are easily generalized to several variables. They imply that if an ideal position measurement is made on a particle described by state vector $|\psi>$ for the purpose of ascertaining whether or not it lies inside the (a,b) interval and if the outcome is positive, then, according to Rule 10a, the (unrenormalized) ket describing the particle after this measurement must be

$$|\psi_{1,\text{unr}} >= \int_a^b dx' |x' >< x'|\psi > \qquad (3.18)$$

so that the corresponding (unrenormalized) wave function is

$$\psi_{1,\text{unr}}(x) = < x|\psi_{1,\text{unr}} >= \int_a^b dx' < x|x' >< x'|\psi >$$

$$= \int_a^b dx' \delta(x - x')\psi(x') \qquad (3.19)$$

where Eq. (3.3c) has been used. Hence

$$\psi_{1,\text{unr}}(x) = \begin{cases} \psi(x) & \text{if } a < x < b \\ 0 & \text{outside.} \end{cases}$$

Such a measurement has the effect of chopping off the part of the wave function whose argument lies outside the (a,b) interval. Note that since

$$K \equiv \int_{-\infty}^{+\infty} |\psi_{1,\text{unr}}(x)|^2 \, dx$$

is not unity, after collapse the normalized wave function $\psi_1(x)$ must be written

$$\psi_1(x) = K^{-1/2} \int_a^b dx' \delta(x - x')\psi(x') \qquad (3.20)$$

Note, however, that although this process admits, of a simple (even if, admittedly, nonrigorous) mathematical description, its *conceptual* interpretation is in fact much less elementary than one might think. Indeed, as given, the very description of the purpose motivating the measurement is misleading. This is because it suggests that the measurement reveals where the particle really *was*. As will be seen (Section 4.4), this is not the case. The quantity $\int_a^b |\psi(x)|^2 \, dx$ is the probability, not that the particle *is* in (a, b) but only that it should *be found* within (a, b).

References

1. J. von Neumann, *Mathematical Foundations of Quantum Mechanics*. Princeton U. Press, Princeton, N.J., 1955.
2. R. Balian, *Am. J. Phys.* **57**, 1019 (1989).
3. J.S. Bell and M. Nauenberg, in *Preludes in Theoretical Physics, in Honor of V.F. Weisskopf*, A. De Shalit, H. Feshbach and L. Van Hove (Eds.), North-Holland, Amsterdam, 1966.

Comments

4 We have stated the rules of quantum mechanics in a condensed way that leaves many questions open. Some important specifications not yet given will be developed in later sections. Even at this stage, however, a few comments are in order.

4.1 On the Description of Ensembles by Kets

1. In many instances Rule 10 (reduction of the wave packet) is the one that is used to determine the ket of an ensemble of systems; filtration, in particular, can be roughly assimilated to a measurement. For example, when we attribute a wave function (essentially a plane wave) to particles emerging from an accelerator, we, in fact, just apply Rule 10 to a kind of implicit measurement of the momentum. Under these circumstances the necessity of introducing the more general Rule 9 may be questioned.

Let it therefore be noted that Rule 9 is, in fact, necessary. Consider a scattering experiment of, for instance, muons by hydrogen atoms. In order for the calculation to be at all possible, it is necessary that some wave function (or ket) be attributed, not only to the incident particle (muon) but also to the target (hydrogen atom). However, the experimentalist has not really measured the energy of the particular atoms on which the scattering is going to take place. He considers he knows the energy of these atoms only because of his knowledge that the temperature is small and that therefore these atoms should be in their ground state (with a very high probability). Rule 9 has been expressed in such a way that this knowledge should be sufficient to justify the attribution of a definite ket (or wave function) to the target atom.

2. This brings up the question whether spontaneous collapses exist. By "spontaneous" I mean collapses that would take place quite independently of any measurement process.

This question is liable of two quite different answers according to whether or not the idea of a small change in the Schrödinger equation is considered as being reasonable. If it is, then indeed the changes at issue can be conceived of in such a way as to allow for the possibility of spontaneous collapses without giving up the agreement between the experimental data and the predictions of the Schrödinger equation (we come back to this in Chapter 13). Here, however, we are interested in the opposite view, namely that, for describing stable atomic and molecular systems, the Schrödinger equation should be kept as it is, without alteration of any kind. The question then is: Under such conditions do microscopic systems exist, within which some spontaneous collapses take place that the general rules of quantum mechanics do not account for? Prima facie, good canditates for such a role seem to exist. They are the unstable particles or systems. And in fact, the idea is sometimes found in the literature that the decays of such particles or systems *are* instances of spontaneous collapse.

Although at first sight this idea looks quite simple and attractive, as expressed it cannot be upheld. This is because the decays in question must themselves be subjected to a theoretical analysis; that this analysis has indeed been made—revealing interesting, far from trivial behaviors, well confirmed by experiment; and that it was made using the very rules of quantum mechanics, without any collapse being introduced "by hand" in the picture. This is not an adequate framework in which to describe the relevant calculations, which can be found in numerous textbooks (see, e.g., Ref. [1. ch. XIII, Appendix D]). Let it just be mentioned that they bring into play in quite a significant way a quantum superposition of the initial, undecayed state with the decayed state (or states), and that assuming in addition some collapse or other to take place (presumably at some definite time in each element of the ensemble) would make it extremely difficult to account for such observed phenomena as line width, level displacement (including Lambshift), and superposition of parts of traveling waves emitted at different times in a decay process [2, 3]. Hence, when all is said and done it seems we can hardly avoid considering that "collapse" and "measurement" are quite tightly linked notions and that the first named one is inseparable from the latter (except, as previously mentioned, if we accept modifying the Schrödinger equation in a well-defined, ad hoc way). Of course we could arbitrarily decide to extend the name "collapse" to one of the results that emerge from the very calculations mentioned, namely to the fact that because of the infinite number of degrees of freedom of the involved fields there is no possibility of distinguishing the superposition of the decayed and undecayed states from a mixture. But this is a purely semantic move. It does not really bear on the point at issue.

4.2 The Completeness Hypothesis ("Standard Textbook" Quantum Mechanics)

A view most physicists adhere to is that scientific research should keep aloof from implicit metaphysical hypotheses. This prompts most of them to make it a principle that the only "existing" physical entities are those that could somehow be observed, directly or indirectly. In quantum mechanics the assumption is made, as we saw, that, with the exception of nondynamical quantities such as masses and electric charges as well as of time, all these observable quantities are described by Hermitean operators, with the help of which the quantum rules tell us all that can, or could, be observed, in any circumstances whatsoever. Taken together with the just stated principle, this assumption leads to the view that there is no finer specification of physical systems than those provided by kets. Let this be called the *completeness hypothesis*.

The fact that an ensemble E is described by a ket has of course—because of the quantum rules—a large number of observable statistical consequences concerning many different observables. If the completeness hypothesis were invalid, some subensembles of E could exist, on which these consequences would not be true. Such subensembles are called *nonquantum*. For example, there could then exist dispersion-free subensembles, that is, ensembles E' such that the measurement of any observable whatsoever would yield the same outcome on every element of E'. A full specification of such dispersion-free subensembles, if they existed, would obviously involve not only the state vector of the ensemble but also additional variables, conventionally called "supplementary" or "hidden." The completeness hypothesis is thus equivalent to the assumption that such hidden variables simply do not exist.

• *Remark 1.* In many articles or textbooks the proposal that such variables should exist is described as an assumption truly modifying quantum mechanics. The point of view adopted here is at variance with this. This view is that the quantum formalism is essentially neutral on the hidden variables issue as long as the completeness hypothesis is not made. To this opinion it might be objected, at first sight, that if somebody could, on an ensemble, know the values of a complete set of compatible observables and, in addition, the values of a few hidden variables, and if these values were the same on all the elements of the ensemble, the ensemble in question would be nonquantum (for example it could be dispersion-free) and this would constitute a violation of the basic rules of quantum mechanics (Rule 9 in particular). The answer to this objection is that (in the spirit of a "weakly objective" approach) the quantum formalism should be considered as a set of rules statistically predicting outcomes of experiments *actually* performed on ensembles that can actually be made available, not as rules predicting what would be observed by some hypothetical being endowed with knowledge about the Hidden. Then, deciding to add the completeness hypothesis to the quantum rules amounts to removing such a restriction to operationality. That is, it amounts to considering that the quantum rules (including Rule 9) accurately reflect genuine laws that govern the physical world and whose range cannot therefore depend on factual limits concerning the details human beings can know. This, let us note, implies in particular that these rules yield information even concerning measurements that could be imagined but not actually made.

• *Remark 2:* The no-hidden-variables hypothesis *is* usually—explicitly or implicitly—made in most textbooks and articles. For this reason the term *conventional* or *standard quantum mechanics* or *standard textbook quantum mechanics* will be used to refer to the form of the quantum mechanical theory that has the completeness hypothesis built in. But we shall also discover that the said hypothesis is not to be taken as a dogma. In fact, some reflection on the measurement processes shows it cannot be kept in its strict purity if such processes are to be described quantum mechanically (see Section 8.1). Moreover, interesting theories have been put forward that explicitly negate it, as we shall see in Chapter 13.

All the same, hidden-variables theories have long been viewed with suspicion by the great majority of physicists, who considered that allowing even implicitly for the existence of hidden variables amounted to introducing quite unwelcome complications into the theory. At the present stage we shall comply with this general trend and unless explicitly stated, keep to "standard" quantum mechanics.

• *Remark 3.* In the literature, a weaker definition of completeness can also be found, which may be more in accordance with what the Copenhagen "founding fathers" had in mind when they used this word. It is stated by Stapp [4] as being the assumption that "no theoretical construction can yield experimentally verifiable predictions about atomic phenomena that cannot be extracted from a quantum theoretical description." Let this be called *weak completeness*. That the assumption of weak completeness is indeed weaker than the stated completeness hypothesis is clear: weak completeness would *not* rule out a hidden-variables theory that would merely state that hidden variables "exist," without claiming that their existence necessarily has observable effects. In other words, it is an assumption the bearing of which is limited to the realm of science proper, viewed as a very strict critical epistemology would define it. Contrary to the stated completeness hypothesis, it has no implication concerning our "views of the world."

4.3 On Ascription of Dynamical Properties to Quantum Systems

The statement that such and such a physical quantity has such and such value is one of those most frequently found in scientific texts. It is practically impossible to dispense with statements of such a kind, but in quantum physics—where it is especially clear that we can take no "direct view" of the systems—they raise difficulties that will have to be considered at length later in the book.

Here let it simply be pointed out that when we say, "On physical system S the dynamical quantity A has value a," we want this statement to have a meaning, and preferably a non-metaphysical meaning. This implies we would like it to be fully explicable in terms of what can or could—at least in principle—be done and observed on S. Admittedly, not all physicists are completely strict on this point. And indeed, it sounds reasonable to consider that the notion of reality—in the sense of mere existence—is prior to those of thought or perception, and therefore to that of experience. Some theorists extend such a metaphysical realism to details, and in particular to some properties of systems, such as position. But if we reject metaphysical realism, and even if, as may be more reasonable, we accept it but deny it should be extrapolated to such details, then we must be stricter. This means that for the assertion "At time t, A has value a on S" to have meaning, some necessary conditions should be fulfilled. A loose one is that there should at least exist a possibility of measuring A on systems sufficiently similar to S. A somewhat stricter condition (we are here anticipating to some extent the content of later sections) is that we can at least *imagine* somebody who, without directly and actually measuring A, still, for some reason, would know that a measurement of A, *if* actually performed immediately after time t, *would* yield outcome a (this wording being meant to replace the loose one "the outcome of a possible measurement is *predetermined* and equal to a").

Is this "necessary condition of meaningfulness" also a sufficient one? As we shall see later, the view that it *is* is very much related to the "EPR (Einstein, Podolsky, and Rosen) criterion of reality" and there are nowadays some grounds for questioning the validity of this criterion. On the other hand if, instead of just being able to *imagine* somebody who somehow would know that the outcome would be a, we assume we *actually know* that this outcome would be a, then we may safely consider that a sufficient condition of validity of the assertion "A has value a" is thereby fulfilled. More precisely, there is then a sense (called the epistemological sense, see Section 15.3) of the verb "to have," in which the assertion in question is valid (although, as we shall find out in Sections 15.3 and 15.4, even then the said assertion cannot, in general, be attributed an ontological significance).

One last, quite obvious but still quite important, point concerning the notion of an observable having a value on a system is that, precisely, it concerns a system, not an ensemble of systems; and that therefore, if it holds good for a whole ensemble E, it holds good for any element of it individually, hence for any subensemble of E as well.

Keeping such notions in mind let us investigate whether or not we can—in well-defined circumstances—consider that some specified dynamical quantities *have*, on given systems, definite but unknown values. In classical physics—for example, in the classical kinetic theory of gases—this is what is assumed concerning properties such as the positions of individual molecules. The probabilities considered in such theories merely reflect our ignorance of these values. Our question is: Within the realm of conventional textbook quantum mechanics (completeness assumed), can an interpretation of such a kind be given to the probabilities or statistical distributions of this theory?

To study this point let us, for simplicity, consider a Hilbert space H with two dimensions only. Let A and B be two noncompatible observables whose operators operate in H. Let A constitute all by itself a complete set of commuting operators and let the eigenvalue equations of A and B be

$$A|a_i> = a_i|a_i> \tag{4.1}$$

and

$$B|b_j> = b_j|b_j> \tag{4.2}$$

respectively ($i = 1, 2$). Moreover let

$$|\psi> = c_1|a_1> + c_2|a_2> \tag{4.3}$$

with

$$|c_1|^2 + |c_2|^2 = 1 \tag{4.4}$$

describe an ensemble E of N systems. According to Rule (3.3) the number of cases in which value a_i is found upon measurement of A on these systems is approximately

$$N| < a_i|\psi > |^2 = N|c_i|^2 \tag{4.5}$$

and the conceptually simplest way of interpreting this result would, of course, be to assume that just before the measurement $N|c_i|^2$ elements of E already have a value $A = a_i$, that the measurement merely reveals. But what does the statement "A has value a_i" actually imply? To answer we must refer to the already stated necessary condition for the statement "A has value a on a system" to be meaningful. The fact we utter this statement and consider it meaningful implies we can picture somebody who somehow would know, concerning each one of these $N|c_i|^2$ systems individually, that a measurement of A, if actually performed on it, would yield outcome a_i. This "somebody" then can, as a consequence of the assumption, select these $N|c_i|^2$ systems just on the basis of what he knows concerning them. For him therefore Rule 9 is applicable to the subensemble E_i of E composed of these $N|c_i|^2$ systems. Consequently he knows that E_i can, before any measurement is done, be described by a ket, which then, of course, must be $|a_i >$ according to Proposition A in Chapter 3 (just after Rule 9). The assumption that on these $N|c_i|^2$ systems A has value a_i, if at all meaningful, therefore implies that before measurement the ensemble they constitute is describable by $|a_i >$ Incidentally, note our "somebody" is not saying: "I *define* the elements of E_i as the systems that will (or would) yield value a_i," which, as we shall see in the next section, would forbid him to apply Rule 9 to E_i. He asserts something else for, in fact, he proceeds in two steps. First he says: "of such, such and such systems, considered individually, I happen to know (as the case may be: among other things) that each one would yield outcome a_i." And second he says: "Let E_i be the ensemble of these systems." Under these conditions Rule 9 unquestionably applies to E_i.

On the other hand, if it were really true that E_i is describable by ket $|a_i >$, then, if B were actually measured instead of A, the number of cases in which the outcome of this measurement, made on all these $N|c_i|^2$ systems, would be b_j would be

$$N_{i,j} = N|c_i|^2| < b_j|a_i > |^2 \tag{4.6}$$

so that the total number of cases in which the outcome of the said measurement would be b_j would be

$$N_j = \sum_{i=1,2} N_{i,j}$$

$$= N[|c_1|^2| < b_j|a_1 > |^2 + |c_2|^2| < b_j|a_2 > |^2] \qquad (4\ 7)$$

However, when directly applied to the case at hand, which after all is just a measurement of B performed on an ensemble described by $|\psi >$, these same quantum mechanical rules yield

$$N_j = N| < b_j|\psi > |^2 \equiv N|c_1 < b_j|a_1 > +c_2 < b_j|a_{2>}|^2 \qquad (4\ 8)$$

If the quantum mechanical rules are generally true, this is undoubtedly the correct prediction And therefore, since the one provided by Eq (4 7) is in general different (because of the cross-terms), the hypothesis on which Eq (4 7) rests cannot be correct In other words the two assumptions that (a) completeness, and hence Rule 9 hold (in complete generality) and (b) some elements of E have $A = a_1$ and the others have $A = a_2$, are not mutually consistent If, as in conventional quantum mechanics, (a) is (explicitly or implicitly) assumed to hold, then we cannot consistently interpret the quantum probabilities $|c_i|^2$ as referring to our ignorance of a property that each system actually *has* (that of having $A = a_1$ or $A = a_2$) We can only interpret them as the probabilities that if a measurement of A is—or were—performed, results a_1 or a_2 will—or would—*be found*

The case in which the relevant Hilbert space H has dimensions larger than 2 and we are interested in degenerate values of the considered observable, A, is only slightly more complex and its study leads to the same conclusion The special case of the *position* measurement is important in connection with the interpretation of formula (2 19), yielding the probability that a particle P described by a wave function ψ (x) should be found in a small volume dv around point x In many textbooks $|\psi(x)|^2$ is referred to simply as the probability for P to *be* in dv, and in the simple one-dimensional case $|\psi(x)|^2 dx$ is similarly just referred to as the probability for P to *be* in dx Taken at face value this statement is equivalent to the assumption that the position x of P has a definite value, that this value is unknown, that only the probability density for it to be, say, x' is known, and that this probability density is $|\psi(x')|^2$ But then, by partitioning the x axis in, say, two intervals only and making use of the results proved in Section 3 2 (Remark 10) we can repeat the preceding argument practically without any change With the result that taking these intervals to be, say, $(-\infty, 0)$ and $(0, +\infty)$, if we finally decide to measure the momentum p instead of x and if the assumption in question were true, the probability density for getting p' as an outcome would be

$$|\int_{-\infty}^{0} \phi^*(p',x)\psi(x)\,dx|^2$$
$$+|\int_{0}^{+\infty} \phi^*(p',x)\psi(x)\,dx|^2 \qquad (4.9)$$

where $\phi(p',x)$ is the eigenfunction of operator p corresponding to eigenvalue p' Obviously this probability density differs from the correct one, which is just

$$|\int_{-\infty}^{+\infty} \phi^*(p',x)\psi(x)\,dx|^2 \qquad (4.10)$$

[in (4.10) the sum involving ϕ is just the scalar product $< \phi|\psi >$, where $|\phi >$ is an eigenket of p, expressed by means of wave functions].

It is thus clear that if we want to keep to "standard textbook quantum mechanics" (completeness included) we cannot interpret the expression $|\psi(x)|^2$ as the probability density for particle P to *be* at x, and must therefore interpret it as being merely the probability density for particle P to *be found* at x.

4.4 Assumption Q

In Section 4.1 (point 1) we observed that to open the way for an extensive use of quantum mechanics, Rule 9 or some rule equivalent to it is necessary. But of course this rule can be of a more general character. The motivation for trying to find such a generalization becomes apparent when it is observed that the ensembles considered in Rule 9 are rather special ones indeed. As we shall see in Chapter 6, there exist more general formulations of the quantum rules that can be applied just as well to these special ensembles—those described by kets—as to ensembles of a more general type (mixtures). Now Rule 9 goes along with the general idea that the elementary rules of quantum mechanics are *universally* applicable to ensembles described by kets. Therefore the generalization we are seeking should somehow assert that quantum mechanics is universally valid for all the ensembles to which the more general formalism just mentioned is applicable.

But indeed, the only condition that this new formalism explicitly postulates as regards the ensembles with which it deals is that these should not be specified with reference to future observations bearing on their elements. Hence we are led in a very natural fashion to formulate in a precise way the hypothesis that quantum mechanics has a kind of universal validity by stating it in the form of the following assumption:

Assumption Q. In their most general formulation the quantum rules are obeyed by any ensemble E of systems of a given type provided only that the very composition of E is not defined by selecting its elements according to the outcomes of not-yet-performed measurements.

• *Remark 1.* When time is thought of as "flowing," the restrictive phrase beginning "provided only that" seems odd: how could we select anything according to the outcomes of measurements that have not taken place as yet? But when a standpoint is taken in which events are somehow considered as "laid out" in space-time (as is, for example, the case in many elementary descriptions of classical relativity theory), the phrase in question is meaningful. It is meant to rule out ensembles defined within larger ones as composed of systems on which future measurements on some observables $A, B, C,$ will yield outcomes $a_k, b_l, c_m,$ Since, as discussed in Section 7 4, any quantum subensemble E' of an ensemble E described by some ket $|\psi>$ is also described by $|\psi>$ (at least when all Hermitean operators correspond to observables), the just mentioned "ruling out" is indeed necessary for the internal consistency of the formalism. This is clear, for consider an ensemble E described by $|\psi>$ and suppose that, in violation of the said ruling out, E' were defined as the subensemble E_k of all the elements of E upon which a measurement of A immediately after some time t will yield outcome a_k. This definition of E' obviously implies $w_k = 1$ for the probability w_k of getting a_k on E' in the measurement in question. But, on the other hand, $|\psi>$ is in general such that $\sum_r |<a_{k,r}|\psi>|^2 \neq 1$, where $\{|a_{k,r}>\}$ is a set of eigenfunctions corresponding to eigenvalue a_k. so that since, at time t, E', just as E, is described by $|\psi>$, Rule 6 is violated. Note, however, a point that may seem subtle at first sight but which a moment's reflection makes clear and which was alluded to already. It is that the restrictive phrase in question does *not* rule out ensembles whose composition is defined in some other way, even if some information bearing on the outcomes of possible future measurements eventually to be performed on their elements is then available because of this definition. In particular it does not rule out the subensembles E_k of E, selected in some way or other, on which, because of the very characteristics of this selection, it happens to be known that a (second) measurement of A performed on them just after the considered one would yield with certainty outcome a_k. Rule 9 must be understood as generally applying to situations of this kind. The nature of the selection in question is irrelevant (it may consist in some immediately preceding measurement of A, or, as in Section 4.3 above, in consequence of some additional assumption we want to test, or   ).

• *Remark 2:* In Assumption Q, as well as in Remark 1, the word "measurement" should be understood in a very general sense. For instance, if an ensemble of particles in some definite spin state is deflected by a Stern-Gerlach device D in such a way that several beams are produced, and if all the beams except one are intercepted, the surviving particles must be considered to have undergone a measurement of their spin component along the direction defined by D

4.5 Determinism and Indeterminism

As already pointed out, the completeness assumption, in either version, should not be considered as a dogma. As Louis de Broglie [5] and David Bohm [6] have demonstrated, it can be relaxed without altering in practice the predictions of quantum mechanics. Indeed these authors have independently of each other constructed an explicit, fully deterministic version of quantum mechanics appropriately called *pilot-wave theory* by the first-named author. In this theory, whenever individual observations performed on the elements of an ensemble E lead to different results, these elements are indeed different, even before the time when the observations are actually made and even if E is a pure case. They differ by the values of some variables that, as already mentioned, are called "hidden," essentially because they should not be confused with the usual *observables* of the theory In such a theory, any individual electron goes through but one slit of the device used in the two-slit experiment However it experiences the influence of hitherto unknown forces, which are different according to whether the other slit is open or closed

One basic feature of the pilot-wave theory is that the *quantum mechanical predictions are exactly reproduced for any ensemble E in which the hidden variables are distributed at random* This suffices to show how difficult it is to build up tests to discriminate the two theories on the basis of their observational predictions But of course this changes nothing to the fact that these theories are *conceptually* quite different, therefore, if a statement of a conceptual or interpretative nature has been proved within the framework of conventional quantum mechanics, this does not guarantee that it remains valid within the realm of the pilot-wave theory This holds true, in particular, concerning the statement made in Section 4 3 to the effect that the there-appearing $|c_i|^2$ cannot be interpreted as just ignorance probabilities The statement was proved at that place with the help of the completeness hypothesis (necessary to allow applying Rule 9 to E_i), since, in the pilot-wave theory, this hypothesis is rejected, we have no ground for expecting that the statement should be true in that theory Indeed it is not, since, in the theory in question, a particle with wave function $\psi(x)$ is always at some definite point on the x axis, with a probability $|\psi(x)|^2$ that this point is x No inconsistency is implied since there is no reason to assume that the hidden variables are distributed at random also in the subensembles of quantum ensembles

For any discussion on the conceptual foundations of quantum mechanics the mere fact that a deterministic theory can be exhibited that, in practice, is experimentally indistinguishable from "standard" quantum mechanics is of course of great importance On the other hand, it is shown below (Chapter 9) that any hidden-variables theory that reproduces the quantum predictions (including of course the pilot-wave theory) necessarily has *nonlocal* features On further analysis this must make one quite suspicious that any of these theories truly represent the world as it really is (see Chapter 13 for details)

As already said, unless otherwise stated, the no-hidden-variables hypothesis is made in what follows The theory then obviously has some intrinsic indeterminism built in, since identical measurements performed on identical systems do not always yield identical outcomes. This intrinsic "elementary" indeterminism is, however, balanced by a kind of statistical determinism because we usually have to do with large assemblies of particles and the law of large numbers then operates.

4.6 OTHER QUESTIONS

1. *Time arrow and collapse.* It obviously is a fascinating question to know whether or not the collapse rule induces a time arrow. At this stage, however, analyzing the arguments for and against this view would be somewhat difficult, and so the study of this question is deferred to Appendix 3.

2. *Undiscernability, second quantization, etc.* These important subjects are not touched on in this book.

References

1. C. Cohen-Tannoudji, B. Diu, and F. Laloë. *Mécanique quantique*, Hermann, Paris, 1973.
2. Z. Ou, X. Zou, L. Wang, and L. Mandel. *Phys. Rev. Letters* **65**, 321 (1990).
3. P. Kwiat, W. Vareka, C. Hong, H. Nathel, and R. Chio. *Phys. Rev.* **A 41**, 2910 (1990).
4. H.P. Stapp, *Am. J. Phys.* **40**, 1098 (1972).
5. L. de Broglie, J. Phys. **5**, 225 (1927).
6. D. Bohm, *Phys. Rev.* **85**, 166 (1952).

Complements

5 Two elementary points concerning quantum mechanics are briefly explained in this chapter: the mean value rule and the Heisenberg picture.

5.1 The Mean Value Rule

Rule 6 in Chapter 3 yields the probability [Eq. (3.3)] that a measurement of observable A yields outcome a_k when performed on an ensemble E of systems described by the normalized ket $|\psi>$ It thereby makes it possible to calculate the mean value $< A >$, that is, by definition, the quantity

$$< A >= \lim_{N \to \infty} \frac{n_k a_k}{N} \tag{5.1}$$

where N is the number of systems in E, and n_k is the number of those systems for which the measurement yields outcome a_k. Since

$$\lim \frac{n_k}{N} = w_k \tag{5.2}$$

we have

$$< A > = \sum_k w_k a_k = \sum_{k,r} a_k < \psi|k,r >< k,r|\psi > |$$
$$= \sum_{k,r} < \psi|a_k|k,r >< k,r|\psi > \tag{5.3}$$

Then, using the same symbol A for the observable and the corresponding operator and applying successively the eigenvalue equation

$$|A|k,r >= a_k|k,r > \qquad (5.4)$$

and the decomposition (3.11) of the unit operator

$$\sum_{k,r} |k,r><k,r| = 1 \qquad (5.5)$$

we get

$$< A > = \sum_{k,r} < \psi|A|k,r><k,r|\psi >$$
$$= < \psi|A|\psi > \qquad (5.6)$$

Similarly

$$< A^n >= \psi|A^n|\psi > \qquad (5.7)$$

When ψ is not normalized to unity, these expressions should of course be changed, the general formula then being

$$< A^n >= \frac{< \psi|A^n|\psi >}{< \psi|\psi >} \qquad (5.7a)$$

Expression (5.7a) has thus been deduced from Rule 5 (possible outcomes of a measurement) and Rule 6 (statistical frequencies). Conversely, Eq. (5.7a) *can be taken as an axiom*. Rules 5 and 6 are then merely theorems. Indeed they follow from Eq. (5.7a) (a detailed proof of this can be found in Chapter 5 of Ref. [1]). In other words Rules 5 and 6 of Chapter 3 can be replaced by the unique rule that follows.

Rule 1 (Mean value rule): Let E be an ensemble of systems that can be described by a ket $|\psi >$ Then, if a measurement of an observable A is made on each system, the mean outcome is given by Eq. (5.7a) (with $n = 1$).

The formulation of Chapter 3 that treats Rules 5 and 6 as basic and subsequently derives the mean value rule is equivalent to the present formulation, which does the reverse.

• *Remark 1.* In the language we have used, both formulations are expressed in terms of operational rules, that is, rules of the type "If such and such measurements are made, such and such results are obtained." The question may be asked whether such a wary formulation can be replaced by a more ambitious one that would claim to give information on the physical properties "really possessed" by systems or ensembles of systems, whether or not we observe them. This question will be dealt with in detail in Chapter 13. The point of the present remark is that the problems it raises are clearly of the same nature in both formulations, since the two are completely equivalent. It would, in particular, be erroneous to believe that replacement of Rules 5 and 6 by the mean value rule is by itself sufficient to eliminate from the theory any reference to the community of observers. Indeed it can be shown [1] that in conjunction with the mean value rule such a standpoint would imply that in the ensemble considered above there *are* $n_k = N w_k$ systems for which the physical quantity A has the value a_k even when no measurement is performed. Such a statement, however, is wrong, as already pointed out.

5.2 The Heisenberg Picture for Time Evolution

The description of the time evolution of systems given in Rule 3 of Chapter 3 is technically known as the "Schrödinger picture." It is not the only possible one. An important alternative description is the *Heisenberg picture*, which indeed was the one Heisenberg used when he first formulated quantum mechanics. The Schrödinger and Heisenberg pictures are two different but equivalent sets of rules. In other words, when applied to a given physical situation they lead to the same observable predictions. In some respects the Heisenberg picture has definite technical advantages. It is particularly convenient, for instance, in quantum field theory. As we shall see (Section 10.8 and Chapter 14), it also opens some vistas concerning the connections of quantum theory with general epistemology. In this book, however, we find it more practical to work, as a rule, with the equivalent Schrödinger picture. For this reason, the Heisenberg picture is only briefly sketched.

Let us, for simplicity (this assumption is not essential) consider a time-independent Hamiltonian. Then the Schrödinger equation (3.2), together with the initial condition that $|\psi(t_0)>$ is known, can be rewritten as

$$|\psi(t)> = U(t, t_0)|\psi(t_0)> \tag{5.8}$$

where U is unitary; that is,

$$U(t_0, t_0) = 1 \tag{5.9}$$

and

$$U^{\dagger}(t, t_0)U(t, t_0) = U(t, t_0)U^{\dagger}(t, t_0) = 1 \tag{5.10}$$

and where

$$i\hbar \frac{dU(t, t_0)}{dt} = HU(t, t_0) \tag{5.11}$$

The solution to this operator equation that satisfies the initial condition (5.9) is

$$U(t, t_0) = \exp[-i\hbar^{-1}H(t - t_0)]$$

where the operator e^A is defined by the formal expansion of the exponential.

   Let us now investigate whether we can describe the ensemble E under consideration, not by $|\psi(t)>$, but by a ket

$$|\phi(t)> = V(t)|\psi(t)> \tag{5.12}$$

where $V(t)$ is some unitary operator (the norms of ψ and ϕ are thus the same). Since, as shown, the formalism can be stated with the choice of the mean value rule as its basic rule, nothing is changed in its physical content as long as the predicted mean values of the outcomes of measurements to be performed at a time t are not changed. Since in the "old" description (using $|\psi>$) these mean values are given by

$$<A> = \frac{<\psi(t)|A|\psi(t)>}{<\psi(t)|\psi(t)>} \tag{5.13}$$

the expression for them in the "new" description (in terms of $|\phi>$) is obtained by expressing $|\psi>$ in terms of $|\phi>$ by means of Eq. (5.12), which leads to formula

$$< A > = \frac{< \phi(t)|A'(t)|\phi(t) >}{< \phi(t)|\phi(t) >} \qquad (5.14)$$

with

$$A'(t) = V(t)A[V(t)]^{-1} \qquad (5.15)$$

The Heisenberg picture corresponds to the special choice

$$V(t) = [U(t,t_0)]^{-1} \qquad (5.16)$$

whence, from Eqs. (5.8), (5.10), and (5.12),

$$|\phi(t) >= |\psi(t_0) > \qquad (5.17)$$

This is to say that in the Heisenberg picture (and as long as no measurement is made) the ket $|\phi>$ that describes the ensemble is constant in time. Correlatively, in this picture the observables are described by the operators (5.15), that is, by

$$A'(t) = U^{-1}(t,t_0)AU(t,t_0) \qquad (5.18)$$

which shows that these operators are time-dependent. The differential equation that is satisfied by these operators is easily derived from Eq. (5.18) (for simplicity, the assumption is made that A is independent of time in the Schrödinger picture). It is

$$i\hbar\frac{dA'(t)}{dt} = [A',H] \qquad (5.19)$$

This equation, in particular, shows in a direct way that an observable is a constant if it commutes with the Hamiltonian of the system.

As for Rule 6 [the probability that a measurement of A will yield result a_k. formula (3.3)], with $|\psi(t) >= U(t,t_0)|\phi>$ [Eqs. (5.8) and (5.17)] it reads as

$$w_k = \sum_r \frac{|< k,r|U(t,t_0)|\phi > |^2}{< \phi|\phi >} \qquad (5.20)$$

that is, as

$$w_k = \sum_r \frac{| < k, r, t | \phi > |^2}{< \phi | \phi >} \qquad (5.21)$$

with

$$|k, r, t > = U^{-1}(t, t_0) | k, r > \qquad (5.22)$$

Equations (5.14) and (5.21) show that the expressions used in the Schrödinger picture can be carried over to the Heisenberg picture. This implies, however, that now the kets associated with the possible values that can be found upon measurement of an observable are themselves time-dependent kets $|k, r, t >$. Moreover, the evolution in time of these kets depends on the Hamiltonian H representative of the forces to which one particular system (or ensemble of systems) is subjected.

Finally, while, as already noted, the state vector of a system that does not undergo a measurement does not change with time, obviously this state vector is suddenly modified—also in the Heisenberg picture—when a measurement takes place. Indeed, immediately after a measurement of A performed at time t_m and having yielded outcome a_k, the representative Heisenberg-picture state vector must (as a consequence of Rule 10a) be a linear combination of the $\{|k, r, t_m >\}$, the running index being r

5.3 Wave Functions

For describing pure cases the use of wave functions in place of kets is sometimes convenient. The mean value of an observable A and the probability that a measurement of A yields eigenvalue a_k can be expressed by means of wave functions. Formally these expressions may be obtained from formulas (5.6) and (3.3), respectively, by defining a symbol $|x' >$ playing the role of an eigenvector of the position operator x. As specified in Remark 4 of Chapter 3, such $|x' >$'s are assumed to obey the relation

$$< x' | x'' > = \delta(x' - x'') \qquad (5.23)$$

("orthogonality in a broad sense"). Similarly, they are assumed to yield a generalized "decomposition of the unit operator" defined by

$$\int |x' > dx' < x'| = 1 \qquad (5.24)$$

The wave function of a particle having state vector $|\psi >$ is then defined, as we know [Eq. 3.3d)] as the scalar product

$$\psi(x') = < x'|\psi > \qquad (5.25)$$

and an operator A acting on functions such as $\psi(x')$ is associated to the "same" operator A acting on the kets through the formula

$$A\psi(x') \equiv < x'|A|\psi > \qquad (5.26)$$

With the help of Eq. (5.24), Eq. (5.6) can then be written

$$< A >= \int dx' < \psi|x' >< x'|A|\psi > \qquad (5.27)$$

a formula that, using Eqs. (5.25) and (5.26), leads to

$$< A >= \int dx' \psi^*(x')A\psi(x') \equiv (\psi, A\psi) \qquad (5.28)$$

where the notation

$$(f,g) \equiv \int f^*(x)g(x)\,dx$$

has been used. A similar argument shows that Eq. (3.3) can be rewritten as

$$w_k = \sum_s |(\phi_{k,s}, \psi)|^2 \qquad (5.29)$$

where

$$\phi_{k,s}(x') = < x'|k,s > \qquad (5.30)$$

The generalization of these formulas to systems of n particles $n > 1$ is straightforward. The wave function is then a function of the coordinates $x_1, x_2, \quad , x_n$.

References

1. B. d'Espagnat, *Conceptual Foundations of Quantum Mechanics,* Addison-Wesley, Reading, Mass. 1971, 2nd ed., 1976, 3rd ed., 1989.

The Density Matrix Formalism

6 In quantum physics, we often need to consider ensembles that are not *pure cases*, that is, that cannot be described by state vectors. This is so, for instance, when an ensemble is composed of several subensembles that are described by nonequivalent state vectors. It also happens in most cases in which we consider ensembles of definite subsystems of some larger systems. The density matrix—or "statistical operator"—formalism is one that makes it possible to deal with such situations swiftly and efficiently.

The purpose of this chapter is not to give a systematic review of all the properties of statistical operators. Rather, it describes those properties that are the most useful for the particular applications we have in mind, that is, the description of mixtures, the various kinds of mixtures, and the problems associated with nonseparability and measurement theory.

6.1 The Statistical Operator (Density Matrix)

Definition 1. Let $E_1, E_2, \quad, E_\alpha, \quad, E_\mu$ be μ ensembles of physical systems, all of the same type, let N_α be the number of elements of E_α, and let $\hat{E}$ be the ensemble of all the $N = \sum_\alpha N_\alpha$ elements of the various E_α. Let us assume, moreover, that each E_α can be described by a normalized ket $|\phi_\alpha>$ (the kets $|\phi_\alpha>$ are not necessarily mutually orthogonal). Then the operator

$$\rho = \sum_{\alpha=1}^{\mu} |\phi_\alpha> \frac{N_\alpha}{N} < \phi_\alpha| \qquad (6.1)$$

is called the "statistical operator of ensemble $\hat{E}$ " Although, properly speaking, ρ is an operator, not a matrix, it is quite often also called the *density matrix* associated with $\hat{E}$

Some Mathematical Properties

The basic mathematical properties of ρ are as follows

(i) ρ is *Hermitean*
(ii) ρ is *positive definite*, that is,

$$< u|\rho|u >\geq 0 \qquad (6\ 2)$$

for any $|u >$
(iii)

$$\text{Tr}(\rho) = 1 \qquad (6\ 3)$$

where $\text{Tr}(A)$ means the trace of operator A (see Appendix 1 for a definition of this notion)

This follows from Eqs (6 1) and (5 5) since, $\{|n >\}$ being an orthonormal basis of the Hilbert space H of the system,

$$\begin{aligned}
\text{Tr}(\rho) &= \sum_{n,\alpha} \frac{N_\alpha}{N} < n|\phi_\alpha >< \phi_\alpha|n > \\
&= \sum_{n,\alpha} \frac{N_\alpha}{N} < \phi_\alpha|n >< n|\phi_{\alpha>} \\
&= \sum_\alpha \frac{N_\alpha}{N} < \phi_\alpha|\phi_\alpha >= 1 \qquad (6\ 4)
\end{aligned}$$

[Note also that, by the same token, for any $|\psi >$ and $|\phi >$

$$\text{Tr}(|\psi >< \phi|) =< \phi|\psi > \qquad (6\ 4a])$$

Properties (i–iii) have the following consequences

(iv) Every diagonal element of ρ in any matrix representation is *nonnegative* In particular, its eigenvalues are nonnegative [this is an immediate consequence of **(ii)**]
(v) The eigenvalues p_n of ρ satisfy

$$0 \leq p_n \leq 1 \qquad (6\ 5)$$

This follows from properties (iii) and (iv).

(vi) If ρ is a projection operator, it projects onto a one-dimensional subspace.

If ρ is a projection operator, then

$$\rho^2 = \rho \qquad (6.6)$$

The same algebraic equality holds in regard to any given eigenvalue

$$p_n^2 = p_n \qquad (6.7)$$

so that $p_n = 0$ or $p_n = 1$. Property (iii), however, entails that

$$\sum_n p_n = 1 \qquad (6.8)$$

so that only one eigenvalue is equal to one, all the others being zero. The general expression $\rho = \sum_n |n > p_n < n|$ thus reduces in this case to one term only. *Q.E.D*

(vii)

$$\mathrm{Tr}(\rho^2) \le 1 \qquad (6.9)$$

and the equality sign can hold only if ρ is a projection operator.

This inequality is a consequence of the facts that (a)—as easily verified—$\mathrm{Tr}(\rho^2) = \sum_n p_n^2$ and (b) no point of the hyperplane (6.8) with nonnegative coordinates p_n lies outside the hypersphere

$$\sum_n p_n^2 = 1 \qquad (6.10)$$

Concerning the equality case, note that when the numbers $p_1 \quad , p_n$ satisfy both Eqs. (6.8) and (6.10) they also satisfy

$$\sum_n p_n(1 - p_n) = 0 \qquad (6.11)$$

Since none of the individual terms in the sum can be negative because of inequalities (6.5), all of them are zero. Equation (6.8) then entails that only one p_n is equal to one, the others being zero. *Q.E.D*

(viii) When ρ is written as in Eq. (6.1), a necessary and sufficient condition for ρ to be a projection operator is that all the $|\phi_\alpha >$ be identical up to phase factors, or, in other words, that the sum in Eq. (6.1) reduces to one term.

The condition is obviously sufficient. To show that it is necessary let us note that—as is easily shown from Eq. (6.1)—

$$\text{Tr}(\rho^2) = \sum_{\alpha,\beta} \frac{N_\alpha N_\beta}{N^2} | < \phi_\alpha | \phi_\beta > |^2$$

The condition $\text{Tr}(\rho^2) = 1$ thus gives

$$\sum_{\alpha,\beta} N_\alpha N_\beta (1 - | < \phi_\alpha | \phi_\beta > |^2) = 0$$

Since none of the terms in the sum can be negative (the $|\phi_\alpha >$ are normalized), all of them must be zero. Thus, since both N_α and N_β must be assumed different from zero,

$$| < \phi_\alpha | \phi_\beta > |^2 = 1 = < \phi_\alpha | \phi_\alpha > < \phi_\beta | \phi_\beta >$$

This is the case in which the Cauchy-Schwarz inequality reduces to an equality. The foregoing equality therefore implies that for any α, β

$$|\phi_\alpha >= \lambda_{\alpha,\beta} |\phi_\beta > \tag{6.12}$$

where $\lambda_{\alpha,\beta}$ is a number that, moreover, has modulus 1 since the two kets are normalized. *Q.E.D*

6.2 Pure Cases and Mixtures

Pure Cases

As already noted, an ensemble E that can be described by a state vector $|\psi >$ is called a pure case.

Instead of describing E by means of the (normalized) ket $|\psi>$, we can equally well describe it by means of the statistical operator

$$\rho = |\psi><\psi| \tag{6.13}$$

which is a projection operator. Let A be an observable. Then

$$\text{Tr}(\rho A) = \sum_n <n|\psi><\psi|A|n> = \sum_n <\psi|A|n><n|\psi> = <\psi|A|\psi>$$

In this formalism, therefore, the mean value of an observable A is given by the expression

$$<A> = \text{Tr}(\rho A) \tag{6.14}$$

Similarly let

$$P(a_k) = \sum_r |k,r><k,r|$$

be the projection operator onto the subspace H_k spanned by the eigenvectors of A corresponding to the eigenvalue a_k. Then

$$\text{Tr}[\rho P(a_k)] = \sum_r |<k,r|\psi>|^2$$

Thus the statistical frequency with which a measurement of A is predicted to yield outcome a_k is just

$$w_k = \text{Tr}[\rho P(a_k)] \tag{6.15}$$

Mixtures

An ensemble $\hat{E}$ obtained by combining all the elements of several subensembles E_α is a *mixture*, although, as will be made clear, this statement is not a general definition of the concept. If every E_α can be described by a state vector $|\phi_\alpha>$ and if these $|\phi_\alpha>$ are not all identical up to phase factors, the mixture obtained is not a pure case. Let us use the word "mixture" in the restrictive sense that it excludes pure cases.

It is very convenient to describe $\hat{E}$ by means of the statistical operator ρ defined by expression (6.1). Then the mean value of any observable A of $\hat{E}$ is

$$< A >= \frac{1}{N} \sum_{\alpha} N_{\alpha} < A >_{\alpha}$$

where $< A >_{\alpha}$ is the mean value of A on E_{α}. Equation (6.14) then shows that

$$< A >= \frac{1}{N} \sum_{\alpha} N_{\alpha} \mathrm{Tr}(|\phi_{\alpha} >< \phi_{\alpha}|A) = \mathrm{Tr}(\rho A) \qquad (6.14a)$$

Thus the expression (6.14) for the mean values, derived above for pure cases, applies equally well to the mixtures considered here. A similar argument shows that the same is true in regard to expression (6.15), which gives the probabilities of observation. Mathematically, what differentiates such mixtures from pure states is that the statistical operators that describe mixtures are not projection operators:

$$\rho^2 \neq \rho$$

This follows from proposition (viii), which shows that, if the statistical operator ρ expressed by (6.1) were a projection operator all the operators $|\phi_{\alpha} >< \phi_{\alpha}|$ would be identical to each other. Under such conditions $\hat{E}$ would be the pure case described by $|\phi_{\alpha} >< \phi_{\alpha}|$ or by the corresponding $|\phi_{\alpha} >$

Finally, it is important to note that, if an ensemble $\hat{E}$ is not a pure case, there are several ways (in fact an infinity of them) in which its statistical operator ρ can be expressed in the form (6.1). If, for instance, the $|\phi_{\alpha} >$ are not mutually orthogonal, it is always possible to consider a representation $\{|n, r >\}$ in which the matrix ρ (see Section 6.4) is diagonal. Then ρ can also be expressed as

$$\rho = \sum_{n,r} |n, r > q_n < n, r| \qquad (6.16)$$

Note that the q_n then have an interpretation in terms of probabilities. For example, if A is an observable with eigenvalue equation

$$A|n,r> = a_n|n,r>$$

the probability that, on the ensemble described by ρ, a measurement of A would yield outcome a_m is, according to formula (6.15)

$$w_m = \text{Tr}(\rho \sum_s |m,s><m,s|) = k_m q_m$$

where k_m is the dimension of the eigensubspace of A corresponding to eigenvalue a_m (i.e., the number of possible values of the index r).

• *Remark 1.* Let an interesting property of statistical operators be mentioned here. It concerns the case in which we know with certainty that if we perform a measurement of an observable A the outcome will be some definite value a_n. Formulas (6.1) and (6.15) then yield

$$\sum_{\alpha,r} \frac{N_\alpha}{N} | <k,r|\phi_\alpha>|^2 = \delta_{n,k}$$

Since the N_α are all different from zero by assumption, this shows that

$$<k,r|\phi_\alpha> = 0$$

for any k different from n; in other words, it shows that every $|\phi_\alpha>$ belongs to the subspace H_n spanned by the $|n,r>$ In the case in which we know that simultaneous measurements of the compatible observables $A, B, C,$ would yield the definite outcomes $a_m, b_n, c_q,$ the foregoing argument shows that the $|\phi_\alpha>$ are all contained in the intersection of the subspaces $H_m, H_n, H_q,$ When this intersection is one-dimensional, $\hat{E}$ therefore reduces to the pure case described by the eigenvectors common to $a_m, b_n, c_q,$ Because of the possibility of writing down any statistical operator in the form (6.16), the latter result holds good for any statistical operator, irrespective of whether or not it is given explicitly by a formula such as (6.1).

• *Remark 2:* Remark 1 applies in particular to the case in which the considered measurement M of A is ideal and our knowledge of what its outcome will be comes from a previous ideal measurement M_0 of A, made just before M. This shows that as a consequence of measurement M_0, the statistical operator ρ describing the ensemble of systems must be changed abruptly from whatever it was before M_0 took place to a form involving only kets $|\phi_\alpha>$ belonging to H_n. The notion of collapse is thus extended to mixtures.

Whenever H_n is not one-dimensional, Remark 8 of Chapter 3 also applies. In other words, the exact composition of the collapsed state operator depends on the nature of the measuring instrument. However, the observation made at the end of this Remark 8 also holds good here. If it is requested that Rule 8 of Chapter 3 applies strictly (and that not only the probability of such and such outcomes but also the nature of the collapsed statistical operator should not depend on the order in which the measurements are performed), then the precise structure of the collapsed ρ is determined [1]. The relevant formulas appear below.

6.3 Time Dependence of Statistical Operators

Let ρ be a statistical operator, that describes a pure case or a mixture of the kind we have just defined; ρ is given by the general formula (6.1). The time evolution of the $|\phi_\alpha>$ in the Schrödinger representation is given by Eq. (5.8) so that

$$\rho(t) = U(t, t_0)\rho(t_0)U^\dagger(t, t_0) \qquad (6.17)$$

Because of Eq. (5.11) for the time variation of $U(t, t_0)$ and the corresponding equation concerning U^+, (6.17) yields

$$i\hbar \frac{d\rho(t)}{dt} = [H, \rho] \qquad (6.18)$$

Formula (6.18) is called the "Schrödinger equation for the statistical operator." In spite of a formal similarity (up to a sign) with Eq. (5.19), it holds in the Schrödinger picture.

It is important to observe that

$$\rho^2(t) = U(t,t_0)\rho(t_0)U^\dagger(t,t_0)U(t,t_0)\rho(t_0)U^\dagger(t,t_0)$$
$$= U(t,t_0)\rho^2(t_0)U^\dagger(t,t_0)$$

so that if $\rho^2(t_0) = \rho(t_0)$, then

$$\rho^2(t) = \rho(t)$$

In other words, if an ensemble of systems is a pure case, it cannot evolve into a mixture if the time evolution is governed by the Schrödinger equation (3.2).

6.4 Density Matrices

$\{ \ , |n >, \ \}$ being, as above, an orthonormal basis of the Hilbert space H of the system under study, any operator A operating in H can be described in this basis by the (finite or infinite) matrix $\mathbf{A}$, the elements of which are[1]

$$A_{m,n} = < m|A|n >$$

$\mathbf{A}$ is said to be the matrix corresponding to A in the *representation* $\{ \ , |n >, \ \}$ In particular a *density matrix*, ρ, (a matrix in the proper sense) can in this way be associated with any statistical operator ρ In view of Eq. (6.1) its matrix elements are

$$\rho_{m,n} = \sum_{\alpha=1}^{\mu} < m|\phi_\alpha > \frac{N_\alpha}{N} < \phi_\alpha|n > \qquad (6.19)$$

$$= \sum_{\alpha=1}^{\mu} < m|\phi_\alpha > \frac{N_\alpha}{N} < n|\phi_\alpha >^* \qquad (6.20)$$

Formulas (6.14) and (6.15) are then easily transcribed to

$$< A > = \mathrm{Tr}[\rho\mathbf{A}] \qquad (6.21)$$

and

$$w_k = \mathrm{Tr}[\rho\mathbf{P}(a_k)] \qquad (6.22)$$

---

1. Some more information about matrices in quantum mechanics are to be found in Appendix 1.

respectively, where the symbol Tr[**M**] just means the trace of matrix **M** in the usual sense, that is the sum of its diagonal elements.

Although the $|x' >$ defined by means of formula (3.3b) and made use of in Section 5.3 are not normalizable in a strict sense, still they also can be used for associating a "matrix" to an operator, and in particular to a statistical operator ρ The lines and columns of such a generalized matrix are labeled by the continuous variables x' and x'' Its general element is

$$\rho(x',x'') = < x'|\rho|x'' > \qquad (6.23)$$

If the ensemble is a pure case, described by $|\psi >$, this formula leads, with the help of Eq. (6.13), to

$$\rho(x',x'') = \psi(x')\psi^*(x'') \qquad (6.24)$$

This formula is easily generalized to the case of mixtures as well as to the case of systems composed of more than one particle.

6.5 Alternative Formulation of the Quantum Rules

The rules given in Chapter 3 apply only to ensembles that can be described by state vectors. As shown in Section 6.2, these rules entail as consequences new expressions for quantum probabilities and mean values, which can be called the—derived— "statistical operator rules." However, it is also possible to formulate the statistical operator rules directly instead. These rules are then chosen as basic axioms, and the rules described in Chapter 3 become but special consequences of these axioms. In the new formulation, ensembles (be they pure cases or mixtures) are directly described by statistical operators, observables are still described by Hermitean operators, and the rules yielding the statistical distribution and the mean values of the predicted measurement outcomes are, respectively, expressions (6.15) and (6.14).

When stated in this manner, the rules of quantum mechanics apply directly, not only to pure cases but also to mixtures. Formally, this is a distinct advantage. In the case of mixtures defined by their statistical operators, the new formalism also has another merit. Since such a mathematically defined mixture is, as shown, equivalent not to one but to several physical mixtures such as $\hat{E}$, the old formulation, in terms of state vectors, would imply, in some cases, some arbitrariness in the choice of the latter. The new formulation has no such defect.

Finally, it should be mentioned that such considerations have opened the way to other axiomatic formulations of quantum mechanics that are both concise and very general. One of them goes back to Birkhoff and von Neumann [2] and was further developed by von Weissäcker [3], Mackey [4], Jauch and Piron [5,6], and others. The basic structure in this approach is the set of all the "yes-no" questions. Mathematically, this constitutes a lattice. Supplementary axioms are necessary in order to recover all the usual principles of quantum mechanics, so that this approach has a certain degree of adaptability that may offer some advantage. Starting from it Jauch [7], for instance, has suggested some modifications of quantum mechanics that would dispose of a few of the difficulties described below. At the same time, these modifications change some of the predictions of quantum theory. They thus originate a *new* theory and hence fall outside the scope of this study.

Another axiomatic formulation was initiated by von Neumann, Jordan, and Wigner [8] and developed by Segal [9], Haag and Kastler [10], Primas [11], and others. Its basic idea is to give, axiomatically, the structure of an algebra to the set of all bounded observables, and to relate formalism and experiment by means of a mean value rule that reads as follows: *The mean value of the outcomes of measurements of any observable A on an ensemble of systems of the same type is a real, positive linear functional m(A) of the element A of the algebra that corresponds to this observable.* Again, it can be shown that the usual formulation in terms of Hilbert spaces, statistical operators, and so forth can be recovered at the price of some supplementary assumptions of a quite general nature, the rejection of which would not, as it seems, open interesting vistas.

These axiomatics constitute a very inspiring field of study. However it should be remembered that, at least in the domain in which usual quantum mechanics is undoubtedly applicable—that is, in the domain of the systems endowed with a finite, even if immensely large, number of degrees of liberty—they are entirely equivalent to the latter, which, for this reason, remains the main subject of our investigations.

Henceforth we therefore understand the expression "the quantum rules" as meaning "the rules dealing with statistical operators and/or state vectors." We can therefore make Assumption Q more precise by restating it as follows.

Assumption Q' For the purpose of making statistical predictions about the outcomes of measurements bearing on observables that belong properly to the physical systems S composing an ensemble E, where E is specified without reference to future observations on these S, E can always be described by some statistical operator. By "belong properly" is meant that the said observables can be measured by operating exclusively on *S*.

The study of quantum mechanics thus understood—and the investigation of the types of epistemological descriptions consistent with it—is the main subject of this book.

6.6 The State Operator

As the content of Section 6.4 has shown, when the quantum rules are expressed by means of statistical operators, these operators play, in this new formulation, the role the state vectors play in the set of rules stated in Chapter 3. In this sense it may therefore be said that the notion of a statistical operator generalizes the notion of a state vector. At any rate, this certainly holds true as long as these concepts are merely used as tools for making statistical predictions concerning what will be observed; which, after all, is one of the main roles of state vectors. The idea that such operators may *generally* be used to represent states of systems is therefore a natural one. And in fact it is one the literature on quantum physics uses quite extensively. In this context the expression "statistical operator" is accordingly often replaced by the expression "state operator."

A word of caution is in order, however. The word "state" being usually attributed a meaning that extends far beyond the strictly operational one just mentioned, the expression "state operator" may well convey, in the mind of its users, some too hasty physical interpretations of the formalism it is a part of; and this, as we shall see, can lead to erroneous views. At the present stage, however, let us simply mention this fact. Detailed comments are deferred to Chapter 8.

6.7 | Measurement and the "Density Matrix" Formalism

In this formalism, the probability that, upon a measurement of an observable A, result a_k is obtained is expressed by Eq (6 15), as we saw This formula thus replaces and generalizes Rule 6 of the state-vector formalism Let us now explain by what formulas Rules 9, 10, and 10a, which state how the ensemble is to be described *after* measurement, are to be replaced (and generalized) in the density matrix formalism

Let an ideal measurement be performed on an ensemble of systems S, described by the state operator ρ That is, let an observable A be measured on each system S (this implies of course an interaction with an instrument, but we refrain at this stage from analyzing this interaction) Then we can either

(a) pick out the systems S on which outcome a_k has been obtained, or

(b) consider again the ensemble of *all* systems S—that is, mix up all the systems again

In both cases the question is What state operator is to be attributed to the resulting ensemble? In fact, the operator looked for turns out, in both cases, to be quite easily expressible in terms of ρ Let us derive the corresponding formulas [12, 13], first in the simple case in which all the eigenvalues of A are nondegenerate and then in the case in which some of them are degenerate

Nondegenerate Eigenvalues

Case (a): The state operator in question, ρ_k, is a projection operator on a one-dimensional vector space $|\psi_k >$ It is therefore

$$\rho_k = P_k = |\psi_k >< \psi_k| \qquad (6\ 25)$$

Because of Eq (6 4a),

$$\mathrm{Tr}(P_k\rho) = < \psi_k|\rho|\psi_k > \qquad (6\ 26)$$

(Note that this expression is just the probability p_k that the value a_k is observed) Therefore Eq (6 25) can be rewritten as

$$\rho_k = \frac{|\psi_k >< \psi_k|\rho|\psi_k >< \psi_k|}{\mathrm{Tr}(P_k\rho)} = \frac{P_k\rho P_k}{\mathrm{Tr}(P_k\rho)} \qquad (6\ 27)$$

Case (b): The state operator ρ' we are looking for is a weighted sum of the state operators (6 27), the weights being the relative numbers of outcomes a_k, which are given by Eq (6 26) This immediately yields

$$\rho' = \sum_k P_k \rho P_k \tag{6 28}$$

Degenerate Eigenvalues

In this case simple expressions are obtained only if it is assumed that the measurement process obeys Rule 10a For the sake of calculation it is then convenient to express ρ as a linear combination

$$\rho = \sum_i w_i \rho_i = \sum_i w_i |\phi_i><\phi_i| \tag{6 29}$$

of projection operators on mutually orthogonal states $|\phi_i>$ Such an expansion is, as already mentioned, always possible, and the ρ_i are the statistical operators of subensembles whose union constitutes a mixture (a *proper* mixture in the sense that will be defined in Section 7 2) having ρ as its statistical operator Of course, the w_i are the relative numbers of individual systems in all of these subensembles

Case (a): Let P_k be the projection operator onto the subspace pertaining to eigenvalue a_k Each subensemble described by a $|\phi_i>$ is transformed by the measurement into its projection on this subspace renormalized to unity, that is, into

$$|\chi_{i,k}> = \frac{P_k|\phi_i>}{(<\phi_i|P_k|\phi_i>)^{1/2}} \tag{6 30}$$

The subensemble labeled i is thus transformed into a subensemble that has

$$|\chi_{i,k}><\chi_{i,k}| \tag{6 30a}$$

as its statistical operator We must now recombine all of these statistical operators that correspond to a given value of k into a weighted sum, the weights being the relative population numbers within this "k" set; that is, $n_{i,k}/n_k$, where (N being the total population):

$$n_{i,k} = Nw_i < \phi_i|P_k|\phi_i > \qquad n_k = \sum_i n_{i,k} \qquad (6.30b,c)$$

When Eq. (6.30) is carried over into the product $n_{i,k}|\chi_{i,k}><\chi_{i,k}|$, the factors $<\phi_iP_k|\phi_i>$ in the numerator and the denominator cancel one another, so that, on the one hand,

$$n_{i,k}|\chi_{i,k}><\chi_{i,k}| = Nw_iP_k|\phi_i><\phi_i|P_k$$

and thus

$$\sum_i n_{i,k}|\chi_{i,k}><\chi_{i,k}| = NP_k\rho P_k \qquad (6.31)$$

and on the other hand,

$$n_k = N\sum_i w_i < \phi_i|P_k|\phi_i >= N\,\mathrm{Tr}(P_k\rho)$$

Finally, therefore,

$$\rho_k = \frac{P_k\rho P_k}{\mathrm{Tr}(P_k\rho)} \qquad (6.32)$$

Case (b): The total probability that outcome a_k is obtained is obviously given by formula

$$\sum_i w_i\mathrm{Tr}(P_k|\phi_i><\phi_i|) = \mathrm{Tr}(P_k\rho) \qquad (6.32a)$$

Hence the statistical operator ρ' we are looking for, obtained by combining the subensembles labeled k with weights proportional to their population number, is again

$$\rho' = \sum_k P_k\rho P_k \qquad (6.33)$$

Admittedly, not every measurement procedure leads to Rule 10a, as already noted. For example, the measurement procedures corresponding to von Neumann's proposal cited earlier lead to

Rule 10a′ For such procedures formulas (6.32) and (6.33) are, of course, not valid; but nor is, for them, the rule (Rule 8, in Chapter 3) that the time order of immediately consecutive measurements of two compatible observables is totally irrelevant.

6.8 Wigner's Formula for Probabilities

Again, let us consider an ensemble E of N systems S described by the statistical operator—or "state operator"—ρ Let us now assume that on these systems a measurement is made of an observable A that has eigenvalues $\{ \ , a_j, \ \}$ and that this first measurement is immediately followed by a measurement of another observable B with eigenvalues $\{ \ , b_k, \ \}$ Let P_j^A and P_k^B be the projection operators onto the subspaces pertaining to a_j and b_k respectively; and let us ask for the probability $p(a_j, b_k)$ that the outcomes of these two measurements should be a_j and b_k respectively.

Among the N systems S there are

$$N_j' = N \, \text{Tr}(P_j^A \rho) \tag{6.34}$$

of them on which the first measurement yields outcome a_j and immediately afterward the ensemble of these N_j' systems is described by the statistical operator [see Eq. (6.32)]:

$$\rho_j' = \frac{P_j^A \rho P_j^A}{\text{Tr}(P_j^A \rho)} \tag{6.35}$$

so that each of them has a probability $\text{Tr}(P_k^B \rho_j')$ of yielding outcome b_k. On the whole, then, the number of systems yielding outcomes a_j and b_k is

$$N_j' \, \text{Tr}(P_k^B \rho_j') = N \, \text{Tr}(P_k^B P_j^A \rho P_j^A) \tag{6.36}$$

The probability we are looking for is therefore

$$p(a_j, b_k) = \text{Tr}(P_k^B P_j^A \rho P_j^A) \tag{6.37}$$

This formula can easily be generalized to successive measurements of more than two observables $A, B, \ , F, G$ and to the case in which the measurements are separated by finite time intervals. The final result is the Wigner formula [14]

$$p(a_j, b_k, \ , f_r, g_s) =$$
$$\text{Tr}(\mathcal{P}_g^G \mathcal{P}_f^F \quad \mathcal{P}_k^B \mathcal{P}_j^A \rho \, \mathcal{P}_j^A \mathcal{P}_k^B \quad \mathcal{P}_f^F) \qquad (6.38)$$

where the $\mathcal{P}_j^A$, $, \mathcal{P}_g^G$ are now the Heisenberg-picture projectors onto the corresponding subspaces, the Heisenberg-picture projectors being derived from the Schrödinger-picture projectors by means of the formula [compare Eq. (5.37)]

$$\mathcal{P} = U^{-1}(t'', t') P U(t'', t') \qquad (6.39)$$

where t'' is the time of the considered measurement and t' the time of the preceding measurement (the initial time in the case of P^A). Of course, in the case in which the Hamiltonian is zero the $\mathcal{P}$'s coincide with the P's.

6.9 Wigner's Phase-Space Distribution Function

Let ρ be a statistical operator and A a Hermitean operator, and let us define two functions of the real variables x and p as follows:

$$f(x,p) = (2\pi)^{-1} \int ds\, e^{-isp} < x + \hbar s/2|\rho|x - \hbar s/2 > \qquad (6.40)$$
$$a(x,p) = \hbar \int ds'\, e^{-is'p} < x + \hbar s'/2|A|x - \hbar s'/2 > \qquad (6.41)$$

Following Wigner [15] let us consider the quantity K defined by the formula

$$K = \iint dx\ dp f(x,p)\, a^*(x,p) \qquad (6.42)$$

with the help of formulas (6.40), (6.41), and the well-known formulas

$$(2\pi)^{-1} \int e^{i(s'-s)p} dp = \delta(s'-s) \qquad (6.43)$$

and

$$< u|A|v >^* = < v|A|u > \qquad (6.44)$$

K can be written

$$K = \iint dx\, ds < x + \hbar s/2 |\rho| x - \hbar s/2 >$$
$$< x - \hbar s/2 |A| x + \hbar s/2 > \qquad (6.45)$$

Let us then perform the change of summation variables defined by

$$X = x + \hbar s/2$$
$$X' = x - \hbar s/2$$

whose Jacobian is $\hbar^{-1}$ K then reads

$$K = \iint dX dX' \; < X |\rho| X' >< X' |A| X >$$

which, in virtue of Eq. (3.17) takes the form

$$K = \int dX < X |\rho A| X > \equiv \mathrm{Tr}(\rho A) = < A >$$

This shows that with the help of formulas (6.40) and (6.41) the mean value of an observable A on an ensemble described by the statistical operator ρ can be expressed as the right-hand side of Eq. (6.42). This formula is formally quite similar to the one that, in classical mechanics, yields the mean value of a physical quantity $a(x,p)$ depending on the position x and the momentum p of a particle when the ensemble over which this mean value is defined is described by the joint probability distribution function $f(x,p)$, which for this reason, and because these developments are due to Wigner, is called the Wigner distribution function.

The Heisenberg "uncertainty" relations imply that in general $f(x,p)$ cannot be interpreted as really being a joint probability distribution. As we know, since x and p do not commute, they cannot be measured simultaneously and if, say, p were exactly measured, the ensuing wave function would be a plane wave, so that the probability of getting some well-specified p value upon a measurement of p and then some well-specified x value upon a subsequent measurement of x could not be equal to $f(x,p)$ In fact, $f(x,p)$ is not even necessarily nonnegative for all the values of its arguments, as it should be, of course, if it were a probability distribution. But still, this function is useful in some contexts (see Section 14.10).

References

1. R. Balian, *Am. J. Phys.* **57**, 1019 (1989).
2. G. Birkhoff and J. von Neumann, *Ann. Math.* **37**, 823 (1936).
3. C.F. von Weissäcker, *Naturwissenschaften* **42**, 521 (1935).
4. G. Mackey, *Mathematical Foundations of Quantum Mechanics*, W.A. Benjamin, New York, 1963.
5. C. Piron, *Helv. Phys. Acta* **37**, 439 (1963).
6. J.M. Jauch and C. Piron, *Helv. Phys. Acta* **42**, 842 (1969).
7. J.M. Jauch in *Foundations of Quantum Mechanics, Proceedings of the Enrico Fermi International Summer School, Course IL*, B. d'Espagnat (Ed.), Academic Press, New York, 1971.
8. J. von Neumann, P. Jordan, and E.P. Wigner, *Ann. Math.* **35**, 29 (1934).
9. I.B. Segal, *Ann. Math.* **48**, 930 (1947).
10. R. Haag and D. Kastler, *J. Math. Phys.* **5**, 848 (1964). See also A.M. Gleason, *Math. Mech.* **6**, 885; H. Margenau and J.L. Park, *Int. J. Theor. Phys.* **1**, 211 (1968).
11. H. Primas, *Chemistry, Quantum Mechanics and Reductionism*, Springer-Verlag, Berlin, 1981.
12. G. Lüders, *Ann. Phys.* **8**, 322 (1951).
13. L. Furry, *Boulder Lectures in Theoretical Physics*, Vol. 8A (1965), U. of Colorado Press, Boulder, 1966.
14. E.P. Wigner, *Am. Journ. Phys.* **31**, 6 (1963). E.P. Wigner in *Foundations of Quantum Mechanics, Proceedings of the Enrico Fermi International Summer School, Course IL*, B. d'Espagnat (Ed.), Academic Press, New York, 1971.
15. E.P. Wigner, *Phys. Rev.* **40**, 749 (1932).

Proper and Improper Mixtures

7 In regard to mixtures and their description, a few rather elementary features of the relevant mathematical formalism were explained in Chapter 6. Other, more elaborate, aspects of this formalism could of course be reported on. However, an abundant literature already exists on this subject. What is of greater interest to us (and has, also, been explored in less detail by the various authors who wrote in the field) is the question of the possible physical interpretations of the formalism in question. This chapter is an approach to such problems.

7.1 Operators and Observables

Observable dynamical quantities are described by Hermitean operators, and in the early days of quantum mechanics the correspondence was believed to be one-to-one. It was assumed that (i) every observable can be related to some Hermitean operator and (ii) every Hermitean operator corresponds to one observable.

It seems that no convincing evidence against assumption (i) has yet been found. On the other hand, it is now known that assumption (ii) cannot be true in complete generality. Indeed it has been shown (by Wick, Wightman, and Wigner [1]) that *superselection rules* exist in physics, according to which some linear combinations of well-known states (such as the states of an isotopic multiplet having different total charges) can never be physically realized. As a consequence, the associated projection operators correspond to no observables. Thus, the superselection rules provide a counterexample to assumption (ii). The observables whose eigenvectors corresponding to different eigenvalues cannot be physically superposed are often called "superselection charges "

Although there is no general recipe for determining which operators correspond to observables and which do not, it is not without interest to note that the question whether or not, on a given type of systems, the Hermitean operators all correspond to observables is tightly linked to the the one of knowing whether or not a given density matrix is measurable (in the sense that all its elements are experimentally knowable)

To show this, let us assume that all the Hermitean operators Q of a Hilbert space H corresponding to systems of a certain type S are observable This implies that on a given ensemble E, described by ρ, of systems S we can measure the quantities

$$\text{Tr}(\rho Q)$$

corresponding to any Hermitean operator Q, indeed this is simply done by splitting E into a sufficient number of subensembles the elements of which are randomly selected, and by measuring on each one of these subensembles the mean value of one of the observables in question In particular, we can measure this way the quantities $\text{Tr}(\rho Q)$ corresponding to the Q operators given by

$$|i><j| + |j><i| \quad \text{and} \quad i(|i><j| - |j><i|)$$

where $\{|i>\}$ define the representation But by addition and subtraction these quantities immediately yield the elements $\rho_{i,j}$ of the density matrix, so that these elements can be measured

7.2 Density Matrices Describing Subsystems

Let the normalized ket

$$|\psi> = \sum_{i,j} c_{i,j} |v_i> \otimes |u_j>$$ (7 1)

describe, at a given time t, an ensemble E of N physical systems Σ, each composed of two subsystems U and V A Hilbert space $H^{(U)}(H^{(V)})$ is attached to the $U(V)$ systems, and $\{|u_j>\}$ ($\{|v_i>\}$) is a complete system of orthonormal kets in $H^{(U)}(H^{(V)})$ Such a situation frequently occurs (e g , as a result of the Schrödinger-like time evolution when U and V have interacted in the past). Here we are interested *in the ensemble of the V systems,* which we call E_V

Let A be an observable that pertains to systems V The mean value of A on E is

$$< A > = < \psi | A | \psi > = \sum_{i,j,r,s} c_{i,j}^* c_{r,s} < v_i | \otimes < u_j | A | v_r > \otimes | u_s > \quad (7.2)$$

Of course, $< A >$ is also the mean value of A on E_V The fact that A belongs to system V means that the operator A has no effect on the vectors in $H^{(U)}$ so that Eq. (7.2) can be written as

$$< A > = \sum_{i,j,r} c_{i,j}^* c_{r,j} < v_i | A | v_r > \quad (7.3)$$

$$= \sum_{i,r} \rho'_{r,i} A_{i,r} = \mathrm{Tr}(\rho' A) \quad (7.3a)$$

with

$$A_{i,r} = < v_i | A | v_r > \quad (7.4)$$

$$\rho'_{r,i} = \sum_{j} c_{r,j} c_{i,j}^* \quad (7.5)$$

and

$$\rho' = \sum_{r,i} | v_r > \rho'_{r,i} < v_i | \quad (7.6)$$

Since the quantities $\rho_{r,s;i,j}$ defined by

$$\rho_{r,s;i,j} = c_{r,s} c_{i,j}^* \quad (7.7)$$

are the elements of the density matrix ρ that describes E in the representation $\{ | v_i > \otimes | u_j > \}$, Eq. (7.5) can also be written as

$$\rho'_{r,i} = \sum_{j} \rho_{r,j;i,j} \quad (7.8)$$

and Formula (7.6) as

$$\rho' = \mathrm{Tr}^{(U)} \rho \quad (7.9)$$

where, as is apparent from Eqs. (7.1) and (7.7),

$$\rho \equiv \sum_{r,s;i,j} |v_r> \otimes |u_s> \rho_{r,s;i,j} <v_i| \otimes <u_j| \qquad (7.10)$$

is the state operator of E, and where the symbol $\mathrm{Tr}^{(U)}$ (partial trace) means that the usual operation of taking the trace is carried over only in $H^{(U)}$

$$\mathrm{Tr}^{(U)}\rho = \sum_t <u_t|\rho|u_t> \qquad (7.11)$$

Equation (7.3a) shows that expression (6.14)—which gives the mean value of an observable when E is a mixture of ensembles E_α that can be described by state vectors—can formally be extended to the ensemble E_V now under consideration. In a similar way, expression (6.15) for the statistical frequencies can be extended to the ensemble of the V systems. These extensions simply require that the statistical operator of the ensemble of the V systems be defined by means of formulas (7.5) and (7.6). It is moreover easily verified that the operator ρ' defined by these expressions has properties (i–iii) of Section 6.1. It therefore also has properties (iv–vii). Finally, it can be verified that

$$\mathrm{Tr}(\rho'^2) \leq 1 \qquad (7.12)$$

Hence ρ' has all the formal properties shown by statistical operators describing mixtures.

Under these conditions it seems appropriate to extend the meaning of the expression "state operator" as defined in Section 6.6 so that it also covers entities such as the here considered ρ' and say that ρ' is "the state operator describing E_V"

7.3 Proper and Improper Mixtures

Concerning all the points considered in the previous section the defined ensembles E_V do not differ from the mixtures $\hat{E}$ already encountered in Sections 6.1 and 6.2. This fact is quite remarkable. It motivates the aforementioned convention, extending the generic name "mixture" to ensembles of type E_V Such a convention cannot however be fully accepted—without reservations or qualifications—until it has been verified that any ensemble of type E_V can be identified with an ensemble of type $\hat{E}$. Let us therefore investigate whether or not this is the case, by considering every available piece of information.

At first sight the arguments for identifying the two concepts look impressive. Indeed it is well known that any measurement on an ensemble of systems can always be reduced to measurements of mean values on that ensemble. As a consequence, if E_V is an ensemble of N subsystems such as V and if p_n and $|n,r>$ are, respectively, the eigenvalues and eigenvectors of the statistical operator (7.6) attached to E_V, a mixture $\hat{E}$ exists that is composed of subensembles of the E_α type (i.e., each E_α is describable by a ket) and that cannot be distinguished from E_V by any measurement that bears on systems V alone.[1] This $\hat{E}$ is made up of the ensembles $\hat{E}_n$ that are described by the kets $|n,r>$ and have the populations Np_n. If we could *only* make measurements on the systems V and could not make any on systems U, we should not be able to differentiate such an ensemble from a mixture of type E_V The same is true, as is easily verified, if we are able to measure, along with the quantities pertaining to V, only a limited number of suitably chosen correlations between the U and V systems. Since such cases often occur in practice, the use of the generic name "mixture" for describing ensembles that are either of the $\hat{E}$ type or of the E_V type is quite convenient for applications.

When fundamental questions are at stake, however, the situation is altogether different for, at least in respect to systems $\Sigma = U + V$ for which every Hermitean operator corresponds to an observable, it is possible (d'Espagnat [2]) to falsify what, in view of the above, would seem quite a natural assumption. This is the assumption that the ensemble of the systems U and the ensemble of the systems V are both, separately, mixtures of the type already encountered in Section 6.2 (type $\hat{E}$). More precisely, it is the assumption that both ensembles are mixtures of N systems, on N_α of which a physical quantity A_α (the A_α with different values of α not being necessarily all different) *has* some definite value a_α, corresponding to the eigenvalue equation

$$A_\alpha|\phi_\alpha> = a_\alpha|\phi_\alpha>\qquad(7.13)$$

---

1. We exclude also measurements of correlations between elements of the same ensemble; see Section 8.3

(on the meaning of the verb "to have," see Section 4.3).

To falsify this assumption it is sufficient to establish its inconsistency in a simple case, so let us consider an ensemble E of N spin $1/2$ particle pairs $U + V$, lying in a singlet spin state.[2] In such a case Eq. (7.1) reads

$$|\psi> = 2^{-1/2}(|u_+> \otimes |v_-> - |u_-> \otimes |v_+>) \qquad (7.14)$$

where (in units $\hbar/2$)

$$S_z^U |u_\pm> \pm |u_\pm> \qquad (7.14')$$

and

$$S_z^V |v_\pm> = \pm |v_\pm> \qquad (7.14'')$$

and S_z^U and S_z^V are—as is often said—"strictly anticorrelated" since a straightforward application of Rule 3 shows the outcomes $(+,+)$ and $(-,-)$ of simultaneous measurements of S_z^U and S_z^V have zero probabilities.

According to Eqs. (7.5), (7.6), and (7.14), the ensemble E_V of the N particles V is described by the statistical operator

$$\rho_V = 2^{-1}(|v_+> <v_+| + |v_-> <v_-|) \qquad (7.15)$$

a set of eigenvectors of which is $\{|v_+>, |v_->\}$ And a similar formula holds good concerning the statistical operator ρ_U describing the ensemble E_U of the N particles U. Now, unquestionably ρ_V adequately describes (to be sure: together with an infinite number of other mixture) a mixture $\hat{E}_V$ composed of two pure cases $\hat{E}_+^V$ and $\hat{E}_-^V$, where $\hat{E}_+^V$—with eigenvector $|v_+>$—is an ensemble of $N/2$ particles having their z spin component S_z^V equal to 1 (still in $\hbar/2$ units) and $\hat{E}_-^V$—with eigenvector $|v_->$—similarly is an ensemble of $N/2$ particles V having their S_z^V equal to -1. And of course the same holds true with U everywhere substituted to V, because of the obvious U, V symmetry. But in spite of all this, it is still inconsistent to physically identify E_V to the thus defined $\hat{E}_V$, and E_U to a similarly defined $\hat{E}_U$.

The proof of this inconsistency is that if a pair $U + V$ of E were really constituted of one element of $\hat{E}_U$ and one element

---

2. See Appendix 1, if necessary, for definitions and details.

of $\hat{E}_V$ it would (as a consequence of the very definition of $\hat{E}_U$ and $\hat{E}_V$) necessarily be the case that either its U part belonged to $\hat{E}_+^U$ and its V part to $\hat{E}_-^V$ or its U part belonged to $\hat{E}_-^U$ and its V part to $\hat{E}_+^V$ (the other two possible cases being trivially ruled out in view of their false consequences on feasible correlation measurements on S_z^U and S_z^V). Let us then consider the ensemble E_+ of the pairs whose U parts belong to $\hat{E}_+^U$ (and whose V parts therefore belong to $\hat{E}_-^V$). Since (a) by assumption there are no hidden variables (completeness hypothesis) and (b) the set $\{S_z^U, S_z^V\}$ constitutes a complete set of compatible observables (CSCO) on the four-dimensional Hilbert space $H^U \otimes H^V$ of the two spins, such an E_+ cannot differ from the ensemble $\hat{E}_+$ we could build up by measuring simultaneously S_z^U and S_z^V on a very large ensemble of (arbitrarily prepared) U, V pairs, selecting the pairs for which the outcomes are $+1$ and -1, respectively, and picking out $N/2$ of these [detailed proof: (i) because of the no-hidden-variables hypothesis the quantum rules are known to apply to both $\hat{E}_+$ and E_+ (if any doubt arises concerning their applicability to E_+ see the latter part of Remark 1 in Section 4.4), (ii) the set $\{S_z^U, S_z^V\}$ being a CSCO, both $\hat{E}_+$ and E_+ can, in view of (i) and of Rule 9 of Chapter 3 be described by a ket in $H^U \otimes H^V$, and (iii) because of Proposition A in Chapter 3 this ket is unavoidably the same for both, namely $|u_+ > \otimes |v_- >$, the no-hidden-variables assumption then guarantees that $\hat{E}_+$ and E_+ coincide]. The same argument holds of course concerning the (complementary) $\hat{E}_-$ and E_- ensembles. This shows that to identify E_V to $\hat{E}_V$ and E_U to $\hat{E}_U$ would unavoidably imply identifying E itself—the ensemble of all the N pairs—with a mixture $\hat{E}$ in proportions $1/2, 1/2$ of ensembles $\hat{E}_+$ and $\hat{E}_-$ However, in $H^U \otimes H^V$ E and $\hat{E}$ are described by statistical operators that not only formally differ but also are *testably* different (the tests are through correlation measurements of spin components in some direction other than O_z). To sum up: The assumption that E_V is physically identical to $\hat{E}_V$ and E_U to $\hat{E}_U$ has been shown to imply a false consequence. It is thereby falsified.

Since mixtures of the $\hat{E}_V$ type and mixtures of the E_V type are physically different concepts, it is necessary, at least when fundamental problems are discussed, to differentiate them also in the language. They are called *proper* and *improper* mixtures, respectively [2].[3]

• *Remark 1.* In the simple example used above the result arrived at may, a posteriori, seem rather obvious since, because of the spherical symmetry of the overall system, the O_z axis plays no privileged role in the picture. Consequently ρ_V [Eq. (7.15)] is just a multiple of the unit operator, and such a statistical operator is also the one that describes any proper mixture $\hat{E}_n^V$, in proportions $1/2, 1/2$, of spin $1/2$ particles whose spin component along a direction n is well defined (and equal to ± 1). Obviously E_V cannot be physically identical at the same time with all these $\hat{E}_n^V$, which, as *proper* mixtures, are physically different from one another (think of the possible ways of preparing them, by mixing two pure cases with $S_n = +1$ and -1). But this symmetry argument applies only to such very simple special cases. It is not made use of in the foregoing proof, which is therefore more general.

• *Remark 2:* In view of the content of Remark 1 we could at first sight wonder whether, by any chance E_V and E_U could be physically identified with properly specified mixtures of several $\hat{E}_n^V$ and several $\hat{E}_n^U$ respectively. By generalizing the argument developed above (see, e.g., Ref. [3: sec. 7.2]) it can however be shown in a straightforward way that this, in fact, is impossible. Thus the following theorem holds true.

Theorem. An improper mixture cannot in general be identified with any proper one.

• *Remark 3.* This result can be straightforwardly extended to the case in which the index α in the mixture-defining Eq. (6.1) varies in a continuous manner (so that the number of pure cases E_α composing $\hat{E}_U$ and $\hat{E}_V$ is infinite), as long as the $|\psi_\alpha >$ are normalizable to unity. The case in which the $|\psi_\alpha >$ are normalizable to delta functions does not, in such matters, create a real problem either, because it is an idealized case anyway and when dealing, as here, with conceptual problems we are at liberty to refer to the well-known mathematical procedures that make it

---

3. The names "mixture of the *first* kind" and "mixture of the *second* kind" are also used.

possible to treat this case as a limit of one in which the $|\psi_\alpha>$ are normalizable.

• *Remark 4:* It should be clear that the above impossibility proof crucially depends on the no-hidden-variables assumption. If this assumption is relaxed, the pure case E *can* be considered as composed of two different subensembles E_+ and E_- defined as in the foregoing. But these ensembles are not quantum.

• *Remark 5:* The relevance of the distinction between proper and improper mixtures has been questioned [4] essentially on the ground that, strictly speaking, if the existence of some systems having classical properties and utilizable as ensemble-generating apparatuses is not postulated at the start, proper mixtures cannot really be created. The answer to this is that, admittedly, if the concept of a proper mixture refers to nonexisting entities, there is no point in introducing it. But on the other hand, if such a standpoint is taken up, this entails that the statement according to which, immediately after an observable has been subjected to an ideal measurement, it *has* the observed value is either false or meaningless. The arguments concerning the difference between proper and improper mixtures are conceived of for the benefit of the physicists with a "realist" turn of mind who, prima facie, would tend to consider that, quite on the contrary, the statement in question is meaningful and true (see Sections 4.3 and 8.3). The related problem of possibly weakening the meaning of the verb "to have" (which might conceivably blur the said difference) will be taken up in Chapter 15.

7.4 The Homogeneity of Ensembles

This section deals with the following question, first investigated systematically by John von Neumann [5]: When is it possible to split an ensemble E into two (or more) different ensembles?

It is necessary that the concepts involved in the formulation of this question should first be made nonambiguous. For this purpose, let us make the two following statements.

Statement (a) bears on the notion of splitting. When we say of a given ensemble E of systems of type S that it can be split into two ensembles E_1 and E_2, we do not mean that the operation can actually be performed. What we have in mind is just the abstract possibility of considering ensembles E_1 and E_2 the sum of the numbers of elements of which is the number N of elements in E and which are such that, *for all the observables pertaining to systems of type S*, the statistical frequencies (i.e., the probabilities) of the prospective measurement outcomes are exactly the same on E as they are on the mixture of E_1 and E_2 (by definition the mixture of E_1 and E_2 is just, in this context, the ensemble composed of all the elements of E_1 *plus* all the elements of E_2).

Statement (b) is simply a definition of the word "different." Two ensembles are said to be different with respect to some observable A if the statistical frequencies of the prospective measurement outcomes on observable A are different on the two ensembles. Also, two ensembles are said to be different if there exists at least one observable A with respect to which they are different (needless to say that, here as elsewhere, the word "observable" refers exclusively to the dynamical quantities that conventional quantum mechanics considers as such, not to anything resembling "hidden variables" of any sort).

From a combination of statements (a) and (b), it follows that an ensemble E can be split into two *different* ensembles E_1 and E_2 when and only when (i) the prospective statistical frequencies (or, equivalently, the mean values) of all the observables are the same on E and on the mixture of E_1 and E_2, and (ii) some of these statistical frequencies or mean values are nevertheless different on E_2 from what they are on E_1 (and therefore also from what they are on E).

If it is assumed that every Hermitean operator is the representative of an observable, then, as shown above, the density matrices of the considered ensembles are measurable. If an ensemble E can be split into two ensembles, E_1 and E_2, this then implies that the corresponding statistical operators ρ, ρ_1, and ρ_2 must satisfy

$$N\rho = N_1\rho_1 + N_2\rho_2 \qquad\qquad (7.16)$$

where $N, N_1,$ and N_2 are the number of elements in $E, E_1,$ and E_2 respectively, since otherwise the prospective statistical frequencies $\mathrm{Tr}(\rho\,P_k^A)$ of some observable A would be different on E from what they are on the mixture of E_1 and E_2 (P_k^A is here the projection operator that projects onto the eigenspace relative to the eigenvalue a_k of A). Under these conditions E can be split into *different* subensembles if and only if two different statistical operators ρ_1 and ρ_2 exist that obey Eq. (7.16). By taking into account the conditions that $\text{Tr}(\rho^2)$ must obey, it is easily verified, however, that such statistical operators can be found if and only if ρ is *not* a one-dimensional projection operator.

Thus, if, by definition, we call *elementary* an ensemble that cannot be split into two different subensembles, we have the following proposition:

Proposition. The elementary ensembles are pure cases (i.e., describable by state vectors), and the pure cases are elementary.

This proposition calls for a few remarks.

• *Remark 1.* It should be remembered that the validity of the proposition above hinges on the assumption that all the Hermitean operators correspond to some observables (or, alternatively, on the weaker assumption that the density matrices are measurable). If neither of these assumptions holds, the proposition is false. For example, if the world were such that, in some fixed reference frame, only the third component S_z of the spin were measurable, then a pure case described, for instance, by an eigenvector of S_x would not be elementary. This is so because, according to statement (a), it could be split into several subensembles, each described by one of the eigenvectors of S_z.

• *Remark 2:* Conversely, let a statistical operator ρ satisfying $\rho^2 \neq \rho$ be given. The corresponding ensemble, E, is a mixture, and the mean values of all the observables on this ensemble are of course given by ρ It is possible to show that all these data can be reproduced by introducing a larger Hilbert space H than the one in which ρ operates and by representing E by a ket (i.e., by a pure case) in this space H. This is an application of the well-known Gelfand–Naimark–Segal construction. What happens here is, of course, that several operators in H correspond to no observable, so that the pure case in question is not elementary.

• *Remark 3.* When we say that an ensemble E can be split into two different ensembles E_1 and E_2, this should be understood in the quite specific, operational sense defined by statement (a). In particular, we do not mean to imply that, if the elements of E have interacted with other systems, the correlations between the elements of E and these systems are correctly reproduced when the mixture of E_1 and E_2 is substituted for E. Indeed such a statement would be erroneous. This is the reason why there is no contradiction between the content of this section and some points made in the foregoing one. When we have to do with an improper mixture, such as E_V in Section 7.3, we can say it can be split, only if we disregard (as, to repeat, we do in this section) all the possible correlation measurements that would involve the spin components of system U. For then and only then can we, without running the risk of making false predictions, formally replace this improper mixture by some corresponding proper one: and, of course, only the proper mixtures can be thought of as composed of several subensembles that are physically different.

References

1. G. Wick, A. Wightman, and E. Wigner, *Phys. Rev.* **88**, 101 (1952).
2. B. d'Espagnat, "Conceptions de la physique contemporaine," Hermann, Paris, 1965; contribution to *Preludes in Theoretical Physics: In Honour of V.F. Weisskopf*, A. De Shalit, H. Feshbach, and L. Van Hove (Eds.), North-Holland, Amsterdam, 1966, p. 185.
3. B. d'Espagnat, *Conceptual Foundations of Quantum Mechanics*, Addison-Wesley, Reading, Mass., 1971, 2nd ed., 1976.
4. G.N. Fleming in *The Mini-Course and Workshop on Fundamental Physics*, held at the Colegio Universitario de Humacao, Universidad de Puerto-Rico, 1985.
5. J. von Neumann, *Mathematical Foundations of Quantum Mechanics*, Princeton U. Press, Princeton, N.J., 1955.

Quantum States and the Nonseparability Problem

8 The notion of "state of a physical system" is so much a part of our basic way of thinking and reasoning that we are at pains to imagine how we could possibly dispense with it. On the other hand, some remarkable features of the general quantum formalism, often grouped under the general name "nonseparability," rather drastically limit the domain of validity of the idea. The two notions referred to in the title of this chapter are therefore, in a way, antagonistic. This is just the reason why it is appropriate to put in close relationship the investigations concerning them.

Let it be stressed once more that, in this chapter as in the foregoing ones, we keep—unless explicitly otherwise mentioned—to what may be called *conventional,* or *standard textbook* quantum mechanics, that is, to the working hypothesis that quantum mechanics is *complete,* in the sense defined in Section 4.2. This is not to say that the results we shall get are valid only if this working assumption is made. On the contrary, we shall eventually find out that the main ones have a greater generality. But for methodological reasons it is appropriate to proceed step by step and this is why we keep, for the time being, to standard textbook quantum mechanics.

8.1 Entanglement and Nondivisibility by Thought ("Holistic Aspects"of Quantum Mechanics)

As a first schematic characterization of what, broadly speaking, may be called the nonseparability of quantum physics, it may be noted that this physics disproves a conception of the world that served for centuries as a general guideline to scientists. We refer to the conception that an extended physical system always can—and should—be analyzed in parts. While the more primitive modes of thought had strongly holistic features, right from the beginning (in the late Renaissance), modern scientific thought made analysis one of its most fundamental guiding principles. And, as we all know, this departure from the older ways of thinking proved extraordinarily successful. More specifically, it may be said that the (Descartes inspired) rule according to which an extended physical system always can, and should, be divided by thought into more or less localized elements (linked by forces) is one of the implicit but most basic rules of the whole classical physics.

Of course, this is not to deny that classical physics makes use of many physical quantities that are "functions of n points," with n larger than one. Potential energy is one of the simplest examples. The potential energy of a system of n material points linked by forces is a function of these n points and clearly it would make no sense to speak of its value at one point. But potential energy is merely a derived quantity. If, being informed of the general, mathematical expression of potential energy, we know the position of each material point composing a system, we can *calculate* the potential energy of this system. In other words we do not need to—and indeed we cannot—fix its value as an additional item of description. In fact, in classical Newtonian mechanics and classical electromagnetic theory the *basic* physical quantities (such as mass densities, velocities, electric and magnetic field strengths) from which we can calculate the derived ones are one-point functions. It is meaningful to speak of the values they have at each space-time point and when we know this plus the general equations we can calculate everything, including of course the correlations between all the different parts of the system. This can be expressed by saying that, in classical physics, given the "laws of forces" (i.e., the full Hamiltonian function) a complete knowledge of all the localized parts of a system (including of course the values of the fields at every point) is always at least conceivable and, if possessed, should yield ipso facto a knowledge of all the derived quantities. In other words, classical physics obeys—or at least does not blatantly contradict as a consequence of its axioms—a rule of "divisibility by thought" that can schematically be expressed as follows.

Divisibility by thought (classical physics). Any extended physical system—be it particlelike or fieldlike or partly both—can be thought of as composed of elements or parts localized in different regions of space, an exhaustive knowledge of which is conceivable; and if the Hamilton function for systems of the same general type as this particular one is known, exact complete knowledge of the values of the physical quantities attached to each one of these parts constitutes all by itself an exhaustive knowledge of the whole composite system.

There is no denying the fact that belief in the strict validity of this general view prompts efforts at analyzing, and as such served, and still serves, the advancement of the sciences. But of course this is not a sufficient reason for raising it to the level of an undisputable truth. And in fact, the point that will be made in this section is that quantum physics definitely shows it to be untrue.

To prove this statement, let us first observe that it must be practically obvious for whoever considers that the only truly exhaustive descriptions of physical systems are descriptions by means of state-vectors. This is simply due to the fact that the state vector describing an extended composite system is not, in general, a product of state vectors, each of which would belong to the Hilbert space of one part of the system (in terms of wave functions the argument is that in general the wave function of the composite system is not a product of wave functions, each one having as variables those of one part only). Indeed, consider for instance a system Σ composed of two physical systems U and V (call them its "parts") interacting by means of (time-independent or time-dependent) forces (more precisely· consider an ensemble of such Σ's). Even if this system happens at some time t_0 to be describable by a product state vector (or wave function) its natural time evolution, as determined by the Schrödinger equation, in general results in that, at times $t > t_0$, its state vector is not a product any more. Under these conditions it is no more possible to speak, at time t, of "the state vector of U" or "the state vector of V," since these two state vectors simply do not exist (the same, of course, for the wave functions). At time t, only Σ has a state vector, hence, according to our working hypothesis, an exhaustive knowledge of U is not even conceivable (and same for V). Such a feature, which is peculiar to quantum physics and has no counterpart in classical physics, was called "entanglement" by Schrödinger. It is said that Σ lies in an "entangled" (or, equivalently, "mathematically nonseparable") state.

In order to complete the argument we must of course take into consideration the view according to which not only state vectors but also state operators (density matrices) may constitute adequate, nay even (in the views of some) "exhaustive" descriptions of systems. Even though, at time t, neither U nor V "have" a state vector, still each one has a state operator attached, namely the partial trace of the state operator of Σ taken over the Hilbert space of the partner, as explained here in Section 7.2. It is true that the ensemble of all the U parts, as well as that of all the V parts, is a mixture rather than a pure case. But still the idea has been upheld that in some sense and in some contexts the description of an ensemble by means of the corresponding state operator is "exhaustive" (the state-operator concept is analyzed in more detail in Section 8.3). Clearly the foregoing argumentation cannot be just simply taken over when the presently considered viewpoint is adopted.

Does this remark concerning state operators allow us to consider that with regard to the question of "divisibility by thought" standard textbook quantum mechanics stands—when all is said and done—on the same footing as classical physics? Not in the least, for remember that, as noted, in classical physics, once the "laws of forces" are known, exhaustive knowledge of the parts of a system gives ipso facto an exhaustive knowledge of the whole they constitute. In the quantum description considered here this is obviously not the case [1]. Even if we assume we have full knowledge of the Hamiltonian of the system at all times, a knowledge of the state operator of both U and V does not, by itself, provide us with an exhaustive knowledge of Σ. Indeed it does not even allow us to know whether or not the outcomes of prospective measurements to be performed separately on the U and V parts will turn out to be correlated. If, for example, U and V are spin 1/2 particles and if Σ lies in a singlet spin state, the state operators of U and V are both represented—as far as spin is concerned—by density matrices that are (one-half the) unit two-by-two matrix. Quite obviously, knowing these two state operators does not inform us of the fact that, if the same spin component is measured on both U and V, a strict (negative) correlation will be observed (indeed these matrices are also those that would describe U and V if these systems were totally uncorrelated). Hence we must say that, concerning Σ, such a knowledge is not exhaustive. In other words, the rule of divisibility by thought, as formulated earlier, remains violated even when the possibility of describing systems by state operators is taken into consideration.

If we cared to express this result in some sort of a "realist" language, we might say that it puts us on the horns of a dilemma. Either we consider that the reality of a system is described by its wave function—and then we must grant that when a system such as Σ is in an entangled state it, strictly speaking, has no "parts"—or we impart this descriptive role to the more general "state operator" (or "density matrix") concept—and we apparently have then to concede that the "whole" (here system Σ) is more than its "parts" (here the U and V systems), this being the case even at times when no interaction between these parts exists any more.

To sum up, standard quantum mechanics unquestionably violates the rule of "divisibility by thought" that classical physics accustomed us to consider as valid. In other words it implies *nondivisibility by thought*, which then appears as a basic feature of this theory.

• *Remark:* Moreover, as we know (Section 7.3), the ensemble E_V of, in the foregoing example, the V parts of Σ is but an *improper* mixture (and the same, of course, concerning U and E_U). This implies that it cannot be thought of as composed of pure-state subensembles, the elements of each of which would "have" a definite value of some observable(s) (the one, or those, the pure state in question would be an eigenstate of). For example, contrary to our intuition we cannot think of E_V as composed of a finite or infinite number of subensembles E_n^V in each of which the component S_n^V of the spin of V along some direction n would have a definite value. Correlatively, concerning the ensemble of the Σ pairs, an intuitively attractive picture is thereby shown to be erroneous. This is the idea that the ensemble in question is composed of pairs of particles each of which individually has a spin pointing in some definite direction, the directions of the two spins within a given pair being opposite and the distribution of these directions within the ensemble being isotropic. Within the—here entertained—assumption of quantum mechanical completeness, it is easily checked that applying the quantum rules of Chapter 3 to such an ensemble leads to predictions of observations that contradict the (singlet state predicted) strict negative correlation between the outcomes—within every single pair—of spin components measurements made on U and V along one and the same direction, whatever that direction is.

8.2   Parameter Independence Versus Outcome Dependence

When a composite extended system is in an entangled state, as is the case concerning the Σ pairs considered in the foregoing section, nondivisibility by thought, shown (in Section 8.1) to be true within the realm of standard textbook quantum mechanics, bars—within this realm—any attempt at picturing such a system, however great its extension in space may be, as "really" composed of two (or more) distinct, noninteracting, and possibly distant systems or "parts." It somehow conveys the idea that conceiving of them as "extended wholes" (as the very structure of their wave function suggests) is, when all is said and done, a more adequate picture. But, of course, the fact must then be taken into account that normally, when a measurement on one of its parts, U say, is made, the wave function of the pair must be "reduced," so that the other part, V, is suddenly attributed a wave function or a state vector that it did not previously have. Within some interpretations of the formalism this may be seen as implying that an action at a distance of some sort is suddenly exerted by the measurement made on U on the, possibly quite distant, system V Whether the concept of such an action at distance is acceptable in general will be the subject of a further scrutiny; but we cannot postpone inquiring whether the just recalled fact entails observable, nay, perhaps even usable, effects of action at a distance, unknown from classical physics.

To this end let us here establish two strictly uncontroversial points (Jarrett [2], Bastide [3]) that are useful landmarks in the field. They have been given the names "parameter independence" and "outcome dependence"(Shimony [4]). These notions will be defined quite generally in Section 8.5. Here a less general but more elementary procedure is used for introducing them. It consists in considering a pair of measurements performed on an extended composite system lying in an entangled state (perhaps because of some previous interaction between its parts) such as one of the $\Sigma = U + V$ pairs mentioned in Sections 7.3 and 8.1. Both notions have to do with the possible dependence of one measurement outcome—the one, say, relative to V—on data concerning the other measurement (the one performed on U). Conceivably the probability of the outcome in question may depend or not depend on the *outcome* of the U measurement. Conceivably also, it may depend or not depend on the *parameters* that determine which physical quantity is actually measured on U. The expressions *outcome* and *parameter dependence* and *independence* refer to the four thus defined possibilities.

Outcome Dependence

Entangled states normally entail outcome dependence. Within the realm of the here-assumed completeness hypothesis, this may be considered as a trivial remark. Take for example the already considered ensemble of singlet state pairs Σ of spin 1/2 particles U and V and suppose that a measurement of the z component S_z^U of the spin of U is performed, with outcome +1 (in $\hbar/2$ units). The a priori (as opposed to conditional) probability that the outcome of a measurement of the z component S_z^V of the spin of V is, say, −1 (in the same units) is of course 1/2 since it is by definition (the limit $N \to \infty$ of) the ratio of the number of favorable cases to the total number N of elements in the considered ensemble, and, for symmetry reasons, the probabilities of the two possible outcomes must be equal. But the a posteriori (or conditional) probability for this same outcome, that is, the probability that it is −1 given that the one on U is +1 is different. Indeed this conditional probability is obviously equal to 1 because of the strict correlation between S_z^U and S_z^V [That this correlation is implied by the formalism can be seen in two ways. Either (as in Section 7.3) we calculate the joint probability of outcomes $S_z^U = +1, S_z^V = -1$, using the state vector of the pair and formula (3.4), and divide it by the probability of outcome $S_z^U = +1$ to get the conditional probability looked for, or we apply Rule (10a) of Chapter 3 to describe the state vector after the S_z^U measurement; both methods lead, of course, to the same result.] The fact that a priori and a posteriori probabilities are different constitutes what is called the "outcome dependence" feature of quantum mechanics.

Note that outcome dependence does not yield new means, more powerful than those of classical physics, of signaling at a distance (e g , superluminal signaling) This is essentially due to the fact that the physicist who performs measurements on U, while he can choose what measurements he will perform, obviously cannot decide their outcomes himself

Parameter Independence

A definitely less trivial remark is that entanglement does not lead to parameter dependence Instead, quite the contrary, parameter *independence* can be proved to hold in all cases

That parameter independence holds true in the foregoing, standard, $U + V$ pairs example is fairly obvious and can easily be checked by explicit calculation *whatever* spin component is measured on U, for somebody who does not know the outcome of this measurement the probability that a measurement of S_z^V has outcome +1 (say) remains $1/2$ But the important point is that parameter independence is generally true This simple feature of the standard quantum formalism, mentioned without proof in early works [5], was for a long time a part of what may be called the "unspoken general knowledge" of quantum physicists But, for reasons that will become clear below, the desirability of an explicit proof increased in the last decades Such proofs were given by Eberhard [6] and by Ghirardi, Rimini, and Weber [7] The following is a simple one [8] based on the Wigner formula

Again let the composite system Σ under study be composed of two subsystems U and V Let A and B be observables measured respectively on U at time t_A and V at time t_B, and let a_i and b_m be the corresponding outcomes We assume $t_A < t_B$ Let $|\psi>$ be the (entangled) state vector of the composite system and let, moreover, $P_i^A(t)$ and $P_m^B(t)$ be the projection operators, in the Heisenberg picture, corresponding respectively to a measurement of A yielding outcome a_i and one of B yielding outcome b_m (in the standard example U and V are assumed to propagate freely so that the Heisenberg-picture spin operators just coincide with the more familiar Schrödinger-picture ones)

For the probability $w_{i,m}$ that the measurement results are a_i and b_m the Wigner formula (6 38) yields

$$w_{i,m} = \text{Tr}[P_m^B \, P_i^A \, \rho \, P_i^A] \qquad (8\ 1)$$

What we are interested in is the probability w_m that the measurement of B yields b_m, irrespective of the prior outcome of the (prior) measurement of A, knowing only that this measurement was made. The A measurement outcomes being mutually exclusive, w_m is just the sum of $w_{i,m}$ over all i's; that is,

$$w_m = \Sigma_i \text{Tr}[P_m^B \, P_i^A \, \rho \, P_i^A] \tag{8.2}$$

This formula is a very general one in that it would apply irrespective of the nature of the considered system and the observables A and B. But here we have to do with a composite system in which A belongs to component U and B to component V In other words, operator $A(B)$ operates in the Hilbert space $H^U(H^V)$ of $U(V)$ The same, of course, holds true concerning the projectors $P_i^A(P_m^B)$, so that:

$$[P_i^A, P_m^B] = 0 \tag{8.3}$$

Hence, due to a well-known property of the trace (see Appendix 1):

$$w_m = \sum_i \text{Tr}[P_i^A \, P_m^B \, \rho \, P_i^A] = \sum_i \text{Tr}[P_i^A \, P_i^A \, P_m^B \, \rho]$$

But $P_i^A \, P_i^A = P_i^A$ since P_i^A is a projector, and therefore:

$$w_m = \sum_i \text{Tr}[P_i^A \, P_m^B \, \rho] = \text{Tr}[\sum_i P_i^A \, P_m^B \, \rho] \text{ (linearity of the trace)}$$

$$= Tr[P_m^B \, \rho] \tag{8.4}$$

since $\sum_i P_i^A = 1$.

Formula (8.4), however, is just the one that expresses the probability that result b_m be obtained when B *alone* is measured on Σ. This shows that the fact of making or of not making some previous measurement on U changes nothing to the probabilities of observing such and such result when measuring something on V This completes the promised proof.

Contrary to outcome dependence, parameter dependence could, if it existed, be used for a kind of Morse signaling by choosing the measurement made on U according to some pre-established code. The fact that it does *not* exist bars this possibility, and therefore the question asked at the beginning of this section must be answered in the negative. The here analyzed correlation at a distance entails none of the directly observable experimental effects that would unambiguously show action at a distance.

• *Remark 1.* The preceding proof is based on Wigner's formula, that is, in general, on Rule (10a). Since, as we saw (Chapter 3: Remark 8) the measurement of an incomplete set of commuting observables does not necessarily obey Rule (10a), a hope might remain of (in cases more complex than the standard example) circumventing the foregoing proof. The proofs of Refs. [6] and [7] are more powerful in that they do not rely on any projection postulate whatsoever. They therefore show that the just mentioned hope has to be given up.

• *Remark 2:* Of course, both outcome dependence and parameter independence are just the features we quite trivially expect ensembles of *classical* composite systems to exhibit whenever the component parts are correlated. But in the classical case the correlation found between the outcomes of measurements performed on these component parts is trivially accounted for by referring to the preestablished physical differences that must then exist between the systems in question. In the case of quantum systems lying in a *pure* entangled state—the one here investigated—such an explanation is, in standard texbook quantum mechanics, quite impossible since, by assumption, such physical differences (which would then be hidden variables) do not exist. So that the observed (and quantum mechanically predicted) correlation implies nondivisibility by thought, a highly nontrivial conclusion. Correlatively, parameter independence is, in the quantum case, not intuitive.

• *Remark 3. Outcome dependence versus objective local theories.* The notion of objective local theories was introduced by Clauser and Horne [9] in a more general context than the one here treated. Their definition states that a theory is locally objective if it obeys parameter independence and if, when the *objective state* λ of an extended composite system such as the Σ pair in Section 8.1 is fully specified, the equality

$$(B|\lambda, A) = (B|\lambda) \tag{8.5}$$

holds—where $(X|Y)$ denotes the probability of X given Y and $A(B)$ is the outcome of a measurement performed on the $U(V)$ part of Σ—provided that these two measurements are made in regions sufficiently far apart that no action at a distance takes place between them (e.g., they may be spacelike separated). Here "objective state" means, within a realist philosophy, the individual reality of the physical system "as it really is." When such a philosophy is combined with the working assumption that quantum mechanics is "complete" (as defined in Section 4.2), this objective state can only be the quantum state, which in the present case is described by the state vector of Σ, so that λ must be identified with this state vector and Eq.(8.5) then coincides with the requirement of outcome independence of measurable probabilities. Since, as we saw, outcome independence is trivially violated by standard quantum mechanics, it follows that, as Clauser and Horne put it, standard quantum mechanics is not an objective local theory.

8.3   States and Statistical Operators

As noted in the introduction to this chapter, the questions of nonseparability and of an adequate description of the quantum state are rather tightly intermingled. We shall come back to the former, but before that we must briefly consider the latter.

Statistical operators were defined in Section 6.6, where we saw that the notion of a quantum state is often generalized so as to cover not only pure cases but also mixtures. Then, since mixtures are described by statistical operators it is convenient, as already mentioned there, to designate the latter by the name *state operators*.

When this convention is made it is considered that states of systems are in general described by state operators and that those that can be described by a state vector are just the particular ones the state operator of which is a one-dimensional projector $|\psi><\psi|$. As noted in Section 6.6 this approach makes it possible to reformulate and generalize the basic rules of quantum mechanics, which for a great many practical problems is a considerable advantage. In quantum statistical mechanics, for example, the statistical operator—or the corresponding density matrix—is an entity that in many aspects closely corresponds to the classical phase-space density, and as such it is one that plays a basic role in the description of the time evolution of the systems that quantum statistical mechanics considers.

As already mentioned, the state operator concept is of some use for investigating the properties of ensembles of subsystems of larger systems, although, as we also saw, the sometimes uttered claim that its introduction makes it possible to remove "nondivisibility by thought" from quantum physics is definitely erroneous. But apart from this, other questions, constituting the subject of this section, also have to be considered when we decide to associate statistical operators to states. One of them is related to the fact that if a statistical operator ρ is not a projection operator there exists not one but an infinity of ways in which it can be expressed as a linear combination of projection operators. To give just one example, let us consider the case when in Eq. (6.1) the $|\phi>$'s are not all mutually orthogonal. As pointed out in Section 6.2, we can then of course diagonalize ρ and, since it is Hermitean, the eigenkets $|k>$ we thus get are necessarily orthogonal to one another, so that they cannot coincide with the $|\phi>$'s. Nevertheless ρ, which by assumption can be written in the form (6.1), can just as well be written

$$\rho = \Sigma_k p_k |k><k| \qquad (8.6)$$

The question then is: Can the right-hand sides of Eqs. (6.1) and (8.6) describe one and the same ensemble, which is a logically necessary condition if ρ is really to describe a (well-defined) state?

The answer is that yes, this is conceivable, but only at a certain price—a price that may well be considered as forbidding by those who believe physics should describe "what really exists" rather than just merely abstract constructs. The point can be made as follows [10]. Let us consider two methods that can be used to prepare an unpolarized beam of spin 1/2 particles. Method I consists in mixing, by means of suitable magnets, two fully polarized beams of equal intensity, one polarized along the Oz direction, the other one along the opposite direction. Method II is identical to method I except that the $\pm Ox$ directions replace the $\pm Oz$ directions. Such procedures are only slightly more complicated than the usual one, which just consists in preparing the beam without subjecting the particles to the influence of any magnetic field whatsoever. Since, in the latter case, we usually do not hesitate to say the beam is the material realization of an ensemble, and attribute to it a density matrix, namely

$$\rho = \frac{1}{2}\begin{pmatrix} 1 & 0 \\ 0 & 1 \end{pmatrix} \qquad (8.7)$$

there seems to be no good reason not to say the same concerning the beams prepared by methods I and II. However the density matrix that must then be attributed to any one of these two beams turns out to be this same ρ Eq. [8.7] again (and this is why the beams prepared by methods I and II are said to be unpolarized).

So, are the ensembles prepared by methods I and II identical? If we just say· "No, quite obviously they are not, since we know they are composed differently" this argument may well be labeled unconvincing—or even metaphysical—by the more cautious or (what amounts to the same) the more operationally minded physicists. For they will claim that only an experimentally testable difference between the predictions we can derive from such pieces of knowledge can convincingly establish the two ensembles are different. Can we produce such a difference? Yes we can, provided we treat these "ensembles" as what, after all, they physically are, that is, as systems of (noninteracting) particles. For it then becomes clear that they should be viewed as obeying the same laws as any other quantum system, namely the quantum mechanical ones, which, we must remember, were described in Chapter 3 as applying to ensembles of systems. Ensembles of these beams should therefore be considered. Such ensembles may in principle be prepared and can therefore be subjected to statistical measurements. For example, the fluctuations of the quantity

$$\Sigma_z = \sum_{n=1}^{N} \sigma_{z,n} \qquad (8.8)$$

where $\sigma_{z,n}$ is twice the z component of the spin of the nth particle in the beam and N is the number of such particles, can be experimentally measured on ensembles of beams prepared by any of the two methods. We then observe that these fluctuations are different in the two cases. In fact they are characterized by the standard deviations

$$\sigma_I = 0, \qquad \sigma_{II} = \sqrt{N} \qquad\qquad (8.9)$$

So that the beams prepared by methods I and II can in principle be distinguished from each other. In the same way, the beams prepared by method I can be distinguished, at least in principle, from the unpolarized beams produced by a third method, the one consisting in preparing N pairs of spin $1/2$ particles U and V in a singlet state and selecting particles V (thus building up an improper mixture of these particles, see Section 7.3). This is due to the fact that, again, the standard deviation of the Σ_z of such beams is $\sqrt{N}$ (of course a similar fluctuation measurement on Σ_x would make it possible to discriminate such beams from those produced by method II). It is worth noticing that this possibility of experimentally distinguishing the beams prepared by method I from the other ones remains available, at least in principle, even in cases in which the number N of constituent particles is not totally under control and hence may vary from one beam to the next. It suffices that these variations (as well as those between the numbers of $S_z = +1/2$ and $S_z = -1/2$ particles in each beam) be appreciably smaller than $\sqrt{N}$, which seems a technically reachable goal.

Returning now to the idea of defining a state by means of a statistical operator, we must first wonder whether or not we consider it as acceptable that two physical systems S and T should be said to be in the same state in cases in which a set of identical replicas of S and one of identical replicas of T lead to different experimental predictions with respect to some observables. If our answer to this question is no, as it is most likely to be, the preceding shows that we can *define* a state by means of a genuine (i.e., non-pure-case) statistical operator, only at the price of deciding that the ensembles concerning which the quantum rules do make predictions are essentially nothing else than abstract constructs. It is true that such abstract constructs (where we can take N infinite; they are usually called Gibbsian ensembles) are operationally quite useful. But, as we see, since we cannot consider them as composite physical systems, their meaning is not extensible beyond the purely operational (i.e., predictive) domain. Consequently it must be borne in mind that when a "state" is defined by reference to a non-pure-case statistical operator, the term is to be taken in a strictly limited sense, a purely operational one that conveys only a small part of the meaning we normally give to the word "state." But of course the question remains open as to whether or not some similar limitation also applies to pure cases. In Section 10.12 we shall meet with relativistic problems showing that identifying state vectors with descriptions of physical states also leads to quite serious difficulties.

8.4 Local Causality and Its Violation

Within any intuitive conception of locality the measurement-induced wave function collapse described by Rules 10 and 10a (Chapter 3) seems to entail some nonlocality since, according to these rules, a measurement performed on the U part of an extended $\Sigma = U + V$ system such as the one considered in the standard two-spin example entails the "creation" of a (previously nonexisting) ket specifically attached to the (possibly far away) V But the rules in question, being essentially operational in nature do not, strictly speaking, have any implication concerning the "ontological" status of the wave function or state vector. These entities could just be regarded as mathematical devices for predicting observable correlations between events, so that their "reduction" at a distance might conceivably imply no *physical effects* at a distance. If we want not to take sides prematurely on such a delicate point—all ontological questions have to be handled with great care and this one more than most others!—we cannot therefore consider that the mere presence of Rules 10 and 10a among the basic rules of quantum mechanics makes this theory indisputably nonlocal. In this section and the next we therefore inquire whether some nonlocality aspects can be found to quantum mechanics even within a more cautious approach, not unwarrantedly bestowing "reality" to mathematically defined concepts just because they proved efficient.

To this effect let us first observe, following John Bell [11] that in Maxwell's theory the fields in a space-time region R are determined by those in any spacelike slice K of the backward light cone of R (Fig. 8.1). Since K is limited, hence, in some sense, localized, the foregoing remark leads to the statement that Maxwell's theory exhibits *local* determinism. Now, quantum mechanics is

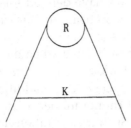

Figure 8.1. A space-time region R with its backward light cone and a spacelike slice K of this backward light cone.

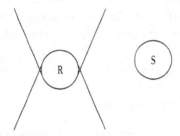

Figure 8.2. Two spatially separated space-time regions, R and S.

not a deterministic theory, but even of undeterministic theories we might a priori expect that they should exhibit a feature of locality generalizing the one just described. More precisely, we would intuitively expect this feature to be such that if S is a space-time region spatially separated from R—that is, lying outside the envelope lightcone of R (see Fig. 8.2)—events in S cannot be "causes" of events in R. This of course would not mean that the two sets of events would be uncorrelated since, just as in Maxwell's theory, they could have "common causes," lying in the overlap of the backward light cones of R and S. But it would mean that somehow causality is propagated from one event to another by sequences of nearby intermediate events separated by timelike intervals, each acting as a direct cause to the immediately succeeding one so that even indirect causes of an event are not further away than permitted by the velocity of light.

To some extent these ideas are still intuitive and to give them a precise formulation it is good to remember that—as Hume pointed out a long time ago—the apparent obviousness of meaning of the word "cause" is to a great extent illusory, so that unless we have defined this word (which we did not) we had better avoid using it. This can be achieved by noticing that when a view such as the one described above is adhered to, it is natural to assume that when the events within some space-like slice of the backward light cone of R (such as K in Fig. 8.1) are sufficiently specified, any additional information about S is redundant (it cannot influence the probabilities of events taking place in R). Such considerations led John Bell to formulate "local causality" as follows.

Local causality. A theory is locally causal if the probabilities of events taking place in a space-time region R_V are unaltered by specification of what events take place in a space-time region R_U spatially separated from R_V *when what happens in the backward light cone of R_V is already sufficiently specified:* for example, by a full specification of the events in a space-time region R' (Fig. 8.3) that completely shields from R_V the overlap of the backward light cones of R_U and R_V

It is clear from what we mentioned that local causality is a natural generalization of local determinism to theories in which determinism is given up. In fact, it seems it is the most natural one that can be thought of. Under these conditions it is a philosophically quite significant point that *standard quantum mechanics violates local causality* and that this is true even of its relativistic form, known as quantum field theory. Imagine, for example, again following Bell [12], a β-radioactive nucleus surrounded at a great distance by Geiger counters, the whole set-up being shielded from cosmic rays, etc. The probability is nonzero that one specific counter registers within some given time interval, whereas the probability is zero that it registers *given* that during this time interval another distant counter registers. Could it be that this difference is due to an incomplete specification of the dynamical quantities describing the circumstantial state of affairs within some region such as R' (Fig. 8.3)? Not if we stick to the hypothesis that quantum theory is complete in the "standard textbook" sense specified in Chapter 4. For indeed, there are then no physical parameters bearing on the outcome, other than those that specify the state vector. Regarding the definition of the probability that a counter in some spacetime region R_V yields some given outcome, there can then be no question of

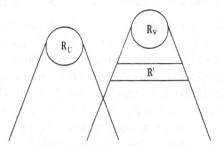

Figure 8.3. A space-time region R' completely shielding off from region R_V the overlap of the backward light cones of R_U and R_V.

making the result displayed by a counter in some other region R_U redundant by referring to values of additional parameters related to some region such as R' Since in such a setup the distance between the counters can in principle be chosen so that the space-time regions R_U and R_V in which the measurements take place are spatially separated, it is then clear that the difference between the aforementioned a priori and a posteriori probabilities is a violation of local causality.

• *Remark:* As is well known, the main reason for taking interest in the local causality notion is not that local causality is violated by standard textbook quantum mechanics. It is that it is also violated by *any* theory that leads, relatively to quite feasible experiments (involving such simple extended composite systems as the Σ pairs considered in foregoing sections), to the same experimental predictions as quantum mechanics and in which the experimentalists are assumed free to choose at whim, what measurement they are going to perform. This is the import of the Bell theorem (more precisely, "Bell 2"), to which we shall turn in the next chapter.

8.5 "Nonlocality," "Nonseparability," and All That

A few miscellaneous points have still to be added. Some of them are mainly of a semantic nature. They concern the relationships between concepts designated by various names in the literature as well as some differences between the meanings attributed to one and the same expression by various authors and in various contexts. Others are more fundamental. Let us begin with these.

Parameter and Outcome Independence Redefined

The foregoing formulation of the local causality assumption has the nice feature of being compact. On the other hand, for epistemological purposes it may be useful in some cases to make it *less* compact by splitting it into a *parameter independence* assumption and an *outcome independence* assumption. This, of course, is easily done [13]. The outcome independence assumption is obtained merely through a slight change in the local causality assumption, this change being just to replace the phrase "what events take place in ( ) R_U" by the phrase: "the outcome of a measurement performed in R_U." Similarly, the parameter independence assumption is obtained by replacing the same phrase by "what is done in R_U (e.g., choice of a measurement)." These new formulations of parameter and outcome independence may be viewed as precise and general formulations of the notions already considered in Section 8.2.[1] It is because standard textbook quantum mechanics violates *outcome* independence that it violates local causality.

A Word on Semantics

When Bell's theorem appeared it was immediately clear that some expressions had to be coined in order to express in words its bearing and significance. As is usual in such cases there was some wavering in the process, so that words such as *locality* and *separability* and their negative counterparts do not always have exactly the same meaning in the writings of different authors, while more than one are also sometimes used, by different authors, to designate the same concept. Under such circumstances the best that can be done is to define here the semantic conventions that are to be made use of in *this* book.

First, since local causality appears violated, a name for this violation is needed. To this end, two words are available in the literature: *nonseparability* and *nonlocality*. Nonseparability looks to be the more appropriate one within contexts where the completeness assumption is made, since the violation in question then essentially amounts in practice to nondivisibility by thought and,

---

1. Note however that, being more precise, they are also somewhat less comprehensive. In particular, while the definition of parameter independence sketched in Section 8.2 seems somehow to bar out dependence on the values the instrument parameters had at *any* time, this is clearly not the case concerning the one just given. Hence the latter does not suffice to bar out the kind of indirect dependence considered in Appendix 4.

as pointed out by Zeh [14], is kinematical rather than dynamical. On the other hand, when the completeness assumption is *not* made, as is the case in most developments bearing on the Bell theorem(s), the word "nonlocality" is the one that best pictures the existing state of affairs since in the theories with supplementary variables the particles composing an extended system are attributed some kind of individual existence although they *interact* nonlocally. In the following, both words are therefore used but with "technically" the same meaning, which we choose to be a rather broad one: they refer to the violation of *either* the compound premise "local causality and free will"—that is, the premises of Bell 2 (see Chapter 9)—*or* the premises of Bell 1 *or* Bell 3 (same chapter).

With regard to the positive counterparts, "locality" and "separability" of these two words, because of the risks of ambiguity we shall, as a general rule, avoid using them in this book, except however in the composite expression "Einsteinian separability," which, on the contrary, will be frequently made use of. The meaning of this expression is defined in Section 9.1 and—let attention be drawn to this—is *not* the opposite of "nonseparability" as just defined.

• *Remark 1.* In other texts the word "separability" is used without a qualificative. It then has a meaning not to be confused with that of Einsteinian separability. In Ref. [8], in particular, I defined a "principle of separability" that has more similarity with local causality than with Einsteinian separability. It bears on pairs of particles (or systems) similar to the $U + V$ pairs we considered earlier, assumed to be created at a source S in an "objective state" λ. Essentially it consists in postulating that when two space-time regions R_U and R_V are "sufficiently" far apart, the intrinsic probability, with fixed λ, that a measurement made on U in R_U has some given outcome is independent of—or depends only very little on—both the nature and the outcome of any measurement made on V in R_V The main difference from local causality is that no reference is here made to relativity theory and that the role of the "complete specification of the events in R'" is played by a complete specification of the (hidden *plus* nonhidden) variables λ at the source. Needless to say, this principle of separability is very similar to (and strongly inspired by) the notion of "objective locality" considered by Clauser and

Horne [9] In conjunction with an assumption that the instrument settings do not influence the objective state λ at the source, it also leads to the Bell inequalities and is therefore disproved, if this assumption is maintained, by means of an argument very similar to the one made use of in Bell 2

• *Remark 2* In order to prevent possible confusions let it be here observed that, clearly, the wave-like or field-like character of a theory has nothing to do with the locality problem As noted previously, a classical field theory such as Maxwell's electromagnetic theory, in which all the basic fields are one-point functions, obeys local causality Hence, statements such as "a wave is a nonlocal entity" are not compatible with the here adopted definition of nonlocality

References

1   E Schrödinger, Proc Camb Phil Sci **31**, 555 (1935); B d'Espagnat, *Conceptions de la physique contemporaine*, Hermann, Paris, 1965; contribution to *Preludes in Theoretical Physics: In Honour of V F Weisskopf*, A De Shalit, H Feshbach, and L Van Hove (Eds), NorthHolland, Amsterdam, 1966, p 185

2   J P Jarrett, *Noûs*, **18**, 569 (1984)

3   C Bastide, *Phys Letters* **103 A**, 305 (1984)

4   A Shimony, "Events and Processes in a Quantum World," in *Quantum Concepts of Space and Time*, C Isham and R Penrose (Eds), Cambridge U Press, 1986

5   B d'Espagnat, *Phys Rev* **D 11**, 1424 (1975), fn 30

6   P H Eberhard, *Nuovo Cim* **46 B**, 392 (1978)

7   G C Ghirardi, A Rimini, and T Weber, *Lett Nuovo Cim* **27**, 293 (1980)

8   B d'Espagnat, "Nonseparability and the Tentative Descriptions of Reality," Physics Reports **110**, 203 (1984)

9   J F Clauser and M A Horne, *Phys Rev* **D 10**, 526 (1974)

10   B d'Espagnat, *Conceptual Foundations of Quantum Mechanics*, Addison-Wesley, Reading, Mass 2nd ed 1976, sec 10 2

11   J S Bell, *Speakable and Unspeakable in Quantum Mechanics*, Cambridge U Press, 1987, ch 7

12   J S Bell, "La nouvelle cuisine," in *Between Science and Technology*, A Sarlemijn and P Kroes (Eds), Elsevier Science, North-Holland, 1990 (ch 6)

13   B d'Espagnat, "One or Two Bell Theorems?" in *Bell's Theorem and the Foundations of Physics, Proceedings of the International Conference in memory of John Bell [held in Cesena, 7–10 October 1991]*, World Scientific, 1993

14   H D Zeh, *The Physical Basis of the Direction of Time*, Springer-Verlag, Berlin, 1989

The EPR Problem and Nonseparability

9 In this chapter we temporarily depart from conventional "orthodoxy" in that we set aside the completeness hypothesis and its generalizations (Assumptions Q and Q', Sections 4.4 and 6.5). In other words, for the time being we give up the hypothesis that supplementary (or "hidden") variables do not exist (we do not assume they exist either; we just leave the matter open). But of course we hold fast to the hitherto never contradicted assumption that the experimentally verifiable predictions of quantum mechanics are all valid. The obvious reason for doing so is that, if considered as a set of predictive rules, this theory has, as we all know, been tested with considerable preciseness in a huge variety of different experiments. Let it be stressed once again that it is quite possible to give up completeness without thereby violating any of the experimental predictions of quantum physics.

9.1 Incompleteness as Derived from the Einstein Assumptions

In 1935 Einstein and his two younger collaborators, B. Podolsky and N. Rosen (EPR) issued a paper [1] in which they showed that, assuming the validity of but very general and highly plausible assumptions, the incompleteness of quantum mechanics could be demonstrated. The opinion that quantum mechanics is a complete description of the atomic and subatomic phenomena was, already at that time, so much engrained in the mind of the physicists that this EPR result was soon referred to as a "paradox." Of course, there is nothing intrinsically paradoxical in the assertion that a theory is incomplete, and therefore such an appellation is surprising. A circumstantial explanation of why it became popular among physicists may be found in the conjunction of two facts: (1) knowledge of the Heisenberg indeterminacy relationships and (2) the prevailing philosophical view that what, as a matter of principle, cannot be known has, by definition, no existence. The latter view barred on a priori philosophical grounds any possibility of interpreting the Heisenberg relationships as merely being limitations on our knowledge, devoid of bearings on the very nature of things. It therefore suggested interpreting these relationships as indicating that the concept of a particle having both quite a definite position and quite a definite velocity (just the concept that the EPR argument leads to) is meaningless: hence the "paradox." But, to this, realists of course responded, and still respond, by rejecting the very philosophical conception on which this argument is based. Thus, when all is said and done, the EPR result is a real paradox only in the eyes of a special bunch of people. It will not be called one here.

The EPR Assumptions

The EPR argument was based on assumptions, or premises, that, prima facie look most plausible. One of them (called Einsteinian separability) was more or less kept implicit in the original EPR paper but was made explicit in Einstein's later writings on the subject (and anyhow it falls in nicely with relativity theory). It is also known as the no-faster-than-light-influences assumption. The other one (they are two in number) is the EPR criterion proper.

Einsteinian separability. Let R_U and R_V be two spatially separated space-time regions (Fig. 9.1). The real factual situation within R_V is independent of what is done in R_U.

The words "the real factual situation" and "what is done in R_U" are borrowed from Einstein himself [2]. The name "Einsteinian separability" is just a convenient semantic convention that we adopt (other authors use the name "Einsteinian locality").

EPR reality criterion. [1]. If, without in any way disturbing a system we can predict with certainty (i.e., with probability equal to unity) the value of a physical quantity, then there exists an element of physical reality corresponding to this physical quantity.

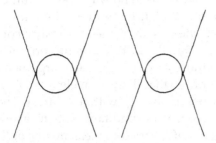

Figure 9.1. Two spatially separated space-time regions.

The "Proof of Incompleteness"

It is from these premises that EPR derived the incompleteness of quantum mechanics. This they did on an example of their own but to describe the logical structure of their argument it is, as David Bohm [3] showed, simpler to use another one that, for short, will henceforth be referred to just as "the standard Bohm example." It concerns such Σ pairs of spin $1/2$ particles U and V lying in a singlet spin state as already considered in Sections 7.3 and 8.2. As an easy calculation shows, the state vector of such a pair can be written:

$$2^{-1/2}(|u_n^+ > |v_n^- > - |u_n^- > |v_n^+ >) \qquad (9.1)$$

where $|u_n^\pm >$ ($|v_n^\pm >$) are the two eigenvectors of the component S_n^U (S_n^V) of the spin of U (V) along a direction n that we can choose as we like. (In other words the state in question is totally isotropic). It follows from this that in fact the strict (negative) correlation mentioned in Section 8.2 holds concerning not only the spin components of U and V along Oz but also their spin components along any direction n. In particular, suppose we consider spin component measurements performed in spatially separated space-time regions R_U and R_V (Fig 9.1). For any n, if, in R_U, we measure S_n^U we thereby know with certainty, because of the strict correlation in question, the value that would be found for S_n^V if that observable were measured in R_V And because of the spatial separation between R_U and R_V and Einsteinian separability[1] we

---

1. Note that, in some descriptions of the argument, including the original EPR one, no reference is made to relativity and Einsteinian separability. The additional assumption is then made instead that at the time of the measurement the two particles no longer interact.

know that V has in no way been disturbed by the measurement made on U. Hence the EPR criterion can be applied and tells us that there is an element of reality corresponding to observable S_n^V. But, the EPR reasoning continues, since the direction n is arbitrary, precisely the same argument goes through for any direction, so that there is an element of reality corresponding to every observable S_n^V, with n arbitrary. These physical quantities are of course not simultaneously measurable (we have no guarantee that if we tried to measure one of them, then another one, and then again the one measured first, the same result would be obtained on the first and last measurements) and correspondingly, at best one of them is described by a state vector. But still, the argument shows that they all are elements of physical reality so that the quantum mechanical description of physical reality by means of a state vector is, at best, an incomplete one. Since these elements of reality are not simultaneously observable, it is clear that their existence in no way contradicts the experimentally verifiable predictions of standard quantum mechanics.

9.2 | Discussion and Comments

As Sections 9.3 and 9.5 point out, the set of the EPR assumptions has now been disproved on a scientific (as opposed to a priori, or philosophical) basis. Nevertheless, and be it only for the purpose of appreciating the true bearing of this disproof, it is important to scrutinize some features of the argument reviewed above.

(i) The Status of Einsteinian Separability

A comparison between Einsteinian separability and local causality (Section 8.4) shows the definitions of these two notions are similar. However, it is necessary to stress that Einsteinian separability must be distinguished not only from local causality but also from *both* the ideas of parameter independence and outcome independence that compose the latter. Some of these differences are somewhat subtle, but all of them are nevertheless quite definite.

With local causality the difference is in the phrase "what is done." Local causality essentially tells us that suitably defined probabilities concerning events in R_V are independent of everything that takes place in region R_U, which is spatially separated from R_V. The expression "everything that takes place" obviously refers not only to the events we can create, or decide upon, in R_U but also to those that we can*not* decide upon, and in particular to *measurement outcomes* as displayed by instruments inside R_U (clearly we cannot dispose of these). On the contrary, Einsteinian separability refers exclusively to what can be *done*, that is to the events *we* can induce, such as choosing which physical quantity we shall measure. In this respect it is thus less restrictive than local causality.

Does the preceding mean that Einsteinian separability expresses only—but faithfully—a *part* of the local causality assumption? That it just drops the outcome independence aspect of this assumption and keeps unchanged its parameter independence part? No, it does not. The difference here is that Einsteinian separability refers to "the real factual situation" in detail. Indeed, if the EPR "proof of incompleteness" (Section 9.1) is to succeed, this must include the real factual situation of the *microsystem V* On the contrary, parameter independence refers only to the probability of getting this or that "measurement result" when *V* is made to interact with some measurement apparatus. Here in other words, if it refers to some "real factual situation" it is only to that of the macrosystem this apparatus is. In this respect Einsteinian separability is therefore *more* restrictive than just parameter independence. But of course it makes sense only to the extent that the notion of "real factual situation" makes sense also. This induces a comment on the EPR criterion.

(ii) The Status of the EPR Criterion

In ordinary language (and also in scientific language when speaking of macroobjects) when we assert of some quantity *A* that it "has value *a*" on some system, we tacitly imply, not only that when we, or other people, measure *A*, value *a will* be found but also, much more generally, that if we or anybody else measured *A*, value *a would* be found. In other words, our assertion implicitly extends to the counterfactual domain. There is no denying that the fact of making such assertions and attributing to them the meaning in question is at least one essential feature of the (in the philosophical sense) realist conception of things. And clearly the EPR criterion of reality must be considered an endeavor at capturing the said feature, in view of applying it.

(iii) Delicate Points in the EPR Formulation

While the reasoning in Section 9.1 seems at first sight quite elementary and clear, closer inspection shows it involves delicate points linked in part with this reference to realism.

One of these delicate points has to do with the meaning of the expression "the value of a physical quantity." Since the EPR reasoning bears on microsystems, this expression must apply to physical quantities (such as spin components) pertaining to microsystems. If we assume (as we then must!) that the notion of a real factual situation of a microsystem makes sense, it would seem, prima facie, that we have a choice: either (i) we consider the expression in question refers to the value the quantity "really has" or (ii) we turn to an operational (and counterfactual) definition, and state that "the value of a physical quantity" means "the outcome a measurement of this quantity *would* yield if it were performed." But choice (i) is open to criticism: since the very purpose of the criterion is to specify conditions under which the notion of (physical) reality can be attached to a physical quantity, it would be circular to make use of the notion of a value really possessed for stating the experimental conditions under which the criterion can be applied. Hence we are left with choice (ii).

On the other hand, with choice (ii), questions arise as to the justification of using strict correlations for predicting, in the way this was done above when applying the criterion. And this, in its turn, has some bearing on which definition of the word "predict" may consistently be retained in the criterion. To see why this is so, consider the following example. Imagine an opaque bottle containing a solution that, when brought in contact with air, precipitates once and for all, in the form either of pink crystals or green powder, consequent on minute details lying beyond human control. It is clear that if two or more "measurements," that is, temporary opening-up of the bottle to look inside it, are performed suc̲ ̄sively their "outcomes," that is, the observed color, are always the same, that is, show strict correlation. Suppose then that the bottle is opened at time t_1 by some observer O That observer O then knows with certainty what color another observer, O' *would* observe *if* she, in turn, opened the bottle at a time t_2 later than t_1 But, of course the criterion does not apply in this case since it is by disturbing the system (opening the bottle) that O gets the piece of knowledge that allows him to predict. In order to get rid of such possible disturbances we might think of considering the case in which the counterfactual (O') measurement—the idea of which serves to define the "value of the physical quantity" [remember we are in "choice (ii)"]—is imagined as taking place at some time t_0 *earlier* than t_1 In this case, nondisturbance (by what O does on what O' observes) is guaranteed to hold, simply because of the universal law that the present cannot disturb the past. On the other hand, as we saw, generally speaking a strict correlation necessarily always exists between the two measurement outcomes. Can we then *formally* apply the criterion, and infer from it the existence, at time t_0, of a physical quantity corresponding to the color observed by O at t_1? Obviously not, since at t_0 the bottle has not yet been opened so that its content is colorless. And the reason why the argument fails is that with counterfactual assertions we must be careful. More precisely· in the just considered inference the strict correlation inherent in the model is made an incorrect use of. It is true that it holds good concerning any pair of actually performed measurements, but this does not imply that we are allowed to draw from an actually performed measurement (at t_1) information on a measurement counterfactually imagined performed at an earlier time.

The same is true concerning an elementary "realistic" interpretation of nonrelativistic quantum mechanics: the interpretation in which the wave function is assumed to be "real" and completeness is assumed as well. In the standard Bohm example ket (9.1), then, is the only reality as long as no measurement has reduced it. Obviously therefore, if a measurement of S_n^u is performed at t_1, with outcome, say, +1, this cannot imply that at an earlier time t_0 there existed an element of reality corresponding to value -1 of S_n^v Here, again, it is true that a measurement made at t_1 cannot disturb the real factual situation at the earlier time t_0 and it is also true that a strict correlation holds between the outcomes of *actually* performed measurements on S_n^u and S_n^v But, as we see, such pieces of knowledge do *not* imply the type of inference that the EPR criterion would be expected to allow for.

What must be kept in mind concerning all this questioning is that to predict with certainty means to predict with a probability unity, and the very notion of probability is dependent on that of the set of the a priori possible cases. It is on the—usually implicit—specification of the latter that the difference between past and future has quite a decisive—though also, usually, quite implicit—role. In questions bearing on causality, predictions, and so on, the set of the a priori possible cases is normally built up taking the past as fixed and given. Otherwise said, such probabilities are conditional, and conditioned upon the past. Basically, it is for this reason that the foregoing fancy descriptions of uses of the criterion to try to infer something on the past are aberrant. And this means that, in the criterion, the verb "to predict" cannot be given a domain of meaning exceeding the limits it has in common language in that it would extend to "prediction of past events." Recourse to strict correlation effects in order to predict actual, or even counterfactual events (what *would* take place *if*) is justifiable only if these events do not take place in the past. Or, to express the same idea differently, no counterfactual assumption (such as that of an imagined measurement) can be made concerning the past.

All this clearly shows that the use of the EPR criterion in the proof of incompleteness cannot be justified by merely referring to the circumstance that, in some reference frames, R_U is later than R_V and the observation that the future never changes the past. Fortunately, relativity theory helps us here. When R_U and R_V are spatially separated, as assumed in the preceding incompleteness proof, there are some reference frames in which R_U is later than R_V—and this may be considered as justifying the nondisturbance assumption in the proof—but there are also reference frames in which R_U is earlier than R_V, and in which, therefore, the word "predict," in the criterion, applies in its literal sense: and then, referring to the strict correlation between the S_n^U and S_n^V measurements for justifying the "prediction" does not raise the aforementioned difficulty (to put it differently, outside the past envelope light cone of R_U, counterfactual assumptions are admissible).

To sum up, these considerations lead to the views: that in order to appreciate the convincing power of the EPR criterion the latter cannot be dissociated from the way it is employed; that there are some delicate points there relatively to the modes of justification of the non-disturbance premise; but that, on the whole, these delicate points are not of such a nature as to block the previously described "proof of incompleteness." Hence the use of the criterion is not, when all is said and done, necessarily conditioned by an assumption—additional to Einsteinian separability or replacing it—of no interaction at a distance.

Concerning this criterion and its use, there is a second delicate point however. It is relative to the word "can" in the phrase "can predict." It must be granted (see also Ref. [4: fn 10]) that in the above-reported original formulation of the EPR criterion this phrase is ambiguous. It may be understood either (i) in the sense that data are actually available for making the prediction or (ii) in the broader sense that a measurement *could* provide the data for the prediction. For the argument to succeed, the criterion must be true in this second, broader sense, as is apparent from the fact that the foregoing "proof of incompleteness" proceeds in two steps. In the first one it is noted that if we actually measure S_n^U in R_U we know with certainty what the outcome of a measurement of S_n^V would be if performed in R_V, and, from this and the criterion, it is inferred that an element of reality exists, corresponding to observable S_n^V In the second step this measurement in R_U is assumed not to be actually performed (some $S_{n'}^U$ with $n' \neq n$ is measured instead) and the reasoning proceeds on the basis that it *could* be performed.

Now, it might be argued that, because of Einsteinian separability, it makes no difference whether we actually perform this measurement or not, so that, in the problem at hand, applicability of the criterion in the second sense actually follows from its applicability in the first sense. It is this point that is worth analyzing. To this end we may imagine that, at the start, an S_n^U measuring instrument—a Stern-Gerlach device, say—is in place and that, within R_U, it is reoriented in direction n' The question then is: Taking Einsteinian separability into account, can we say that this operation in R_U does not modify the real, factual situation in R_V and therefore changes nothing to the fact that an element of reality corresponding to S_n^V exists? The answer is yes or no, depending on whether the reorientation in question is, in the instrument proper time, made after or before the interaction of the Stern–Gerlach device with particle U. If it takes place *after*, it obviously has no disturbing effect: data are *actually* available for the prediction concerning S_n^V on V, and the inference is therefore correct. But, of course, this leads us nowhere since at the time of the reorientation a measurement has already been made of one definite spin component, S_n^U of U and there is no question that another, subsequent, measurement, also performed in R_U, of some other spin component of U, should then reveal anything concerning V In the case that the reorientation takes place before the interaction of the device with U this objection fails. But in this case no datum concerning S_n^U is *actually* available for making the prediction concerning S_n^V, and therefore the criterion taken in sense (i) cannot be made use of.

All this shows that there is no easy path from the criterion taken in sense (i) to the criterion taken in sense (ii) and that therefore, if we want not to take hasty commitments on such points, we must distinguish from one another two conceivable reality criteria. For later reference let us call them, Reality Criterion 1 and Reality Criterion 2.

Reality Criterion 1. If data are available that make it possible, without in any way disturbing a system, to predict with certainty the value of a physical quantity pertaining to this system, then there exists an element of reality corresponding to this physical quantity.

Reality Criterion 2. If the situation is such that without in any way disturbing a system, data could be made available that would make it possible to predict with certainty the value of a physical quantity pertaining to this system, then there exists an element of reality corresponding to this physical quantity.

As we saw, it is Reality Criterion 2 that must be assumed for the EPR reasoning to succeed. It obviously involves an element of counterfactuality and, to some extent, Bohr's famous criticism of EPR [5] can be considered as a rejection of precisely this element. On the other hand, it would not be proper to state that there is no itinerary at all leading from Reality Criterion 1 to Reality Criterion 2. The point is that the second one can be replaced by the first together with a postulate whose only intuitively simple expression unfortunately takes up a somewhat "metaphysical" or "ontological" form and that, for this reason, will be called "Postulate R" (R for "real"). Postulate R (linked with counterfactuality, of course) essentially is that when we have some cogent reasons to assert that something is real, then we must consider the fact that it *is* real as being independent of our information. The special feature Einsteinian Separability was noted to have near the end of paragraph **(i)** (where we commented on the status of this concept) amounts to taking Postulate R quite seriously into account In the standard example, assuming we have measured S_n^U in R_U, applied Reality Criterion 1 and inferred from it the existence of an element of reality associated with S_n^V, applying postulate R we can argue that this element of reality exists quite independently of our knowledge concerning it, so that it would exist even if, contrary to what is the case, we had not performed, in the space-like separated region R_U, the S_n^U measurement that made it possible for us to assert its existence We can then keep this knowledge in store when we think of experimental conditions in which some other component, S_n^U, would be measured, and so on

On this issue it is worth noticing that the somewhat imprecise notion of "reality" on which Postulate R is based can be replaced by the use of more precise concepts pertaining to formal logic not—admittedly—to elementary, conventional logic but rather to the refined extension of conventional logic known as *modal logic* The key to this (see [6,7]) is the notion of *counterfactual implication*, considered as capturing the essentials of the notion of reality, and the related one of *counterfactual definition* of dynamical properties of systems Details are to be found in the references above (see also Section 11 1)

• *Remark* In connection with the Bell theorems (see below) and the various ways of deriving the Bell inequalities, proposals for actually dispensing with the notion of reality were made by Stapp [8], who, however, used in its place a notion of counterfactual definiteness somewhat different from the just-mentioned type of counterfactuality Also in connection with the Bell theorems, it was shown (d'Espagnat [9, 6]) that, formally, not only the notion of reality but even the notion of counterfactuality can be dispensed with This, however, is at the price of replacing the latter by a new—and somewhat roundabout—postulate, called *inductive causality*, see Ref [6]

9.3 The Bell Theorems

The well-known Bell's inequalities, written down by John Bell [10] in 1964 and generalized by Clauser, Horne, Shimony, and Holt [CHSH, 11] in 1969 are statements, the essential characteristics of which are that (1) they involve but experimentally measurable dynamical quantities for which the quantum mechanical rules of calculation (recorded here in Chapter 3) make unambiguous statistical predictions, (2) they are *violated* by the predictions in question, and (3) on the other hand they *follow* from sets of premises all of which seem prima facie so reasonable that the falsification of such sets provides very important pieces of information concerning the nature of physical knowledge.

This section bears a title in which the Bell theorems are mentioned in the plural. This is meant to emphasize a fact I consider [12] as shedding additional light on the subject, and which is that there exist not just one but *several* inequivalent sets of premises from any one of which the Bell inequalities can be derived. A consequence of this is that we have to do with several distinct theorems, each of which disproves a specific set of premises on the basis of the fact that the Bell inequalities are violated. In fact there are three such theorems. Let them be called *Bell 1*, *Bell 2*, and *Bell 3*, for short. The technicalities of the proof are about the same for all three theorems and have appeared in a number of books and articles (see, e.g., [13]). They are not to be reproduced here. Let it merely be stressed that, contrary to what is repeatedly asserted by some physicists the existence of supplementary (or "hidden") variables does not have to be postulated for deriving any of these theorems.

Bell 1

Bell 1 is essentially the outcome of Bell's original 1964 paper. Its proof, which proceeds in two steps, is based on the use of the standard Bohm example considered in Section 9.1 in conjunction with the EPR derivation of incompleteness. In fact, the set of premises disproved by Bell 1 is just the set of the two premises used in the derivation in question, namely Einsteinian separability (or alternatively the idea that U and V no longer interact at the times when they undergo the considered spin component measurements) and the EPR criterion of reality; and the first step in the proof of Bell 1 is nothing else than this EPR derivation. As may be remembered (see Section 9.1) the outcome of the latter is that there exist elements of reality corresponding to the spin component of V (and also of U for symmetry reasons) along any direction n in space. These elements of reality, characterizing what may be called the *objective state*, λ, of the pair, are of course not described by the state vector of the pair, which means they are supplementary (also conventionally called *hidden*) variables, and of course they are *local* such variables since they all are one-point functions and more precisely functions of the point in space where it is planned to subject the corresponding particle to a measurement. It is worth stressing once again that the existence of these supplementary variables is here a *consequence* of the premises—following from the strict correlation between the two particle spins—so that this existence does *not* have to be postulated.

The second step in the proof of Bell 1 is to derive the Bell inequalities from the existence of these local hidden variables, just as Bell did.

Bell 2

Bell 2 (see, e.g., Ref. [14]) is a disproof (through a derivation of the Bell–CHSH inequalities) of the one premise of local causality (Section 8.4) (or, what is the same, of the two premises of parameter independence and outcome independence as stated in Section 8.5), together—it must be added—with the supplementary hypothesis (also implicitly assumed in Bell 1 and Bell 3) that the experimentalist is really free to choose at whim what component to measure.[2] In Section 8.4 it was pointed out that standard textbook quantum mechanics violates local causality. Bell could show that in fact *any* theory compatible with local causality (and in which the experimentalist can, at the last moment, choose at whim the observable to be measured) entails Bell's inequalities and therefore violates experimentally verifiable (and verified), quantum predictions (see, e.g., the experiments of Aspect, et al. [15]).

From the point of view of the experimentalist one great difference between Bell 2 and both Bell 1 and Bell 3 is that Bell 2 also applies to particle pairs lying in a quantum state other than the $S = 0$ one and therefore exhibiting no strict correlation. For

---

2. This assumption rules out any possibility that the orientations **a,b,**. of the instruments should have a functional link with the probability distribution $\rho(\lambda)$, and this is indeed a condition for the theorem to hold good. Appendix 4 describes a model in which the Bell inequalities are violated and the quantum mechanical predictions recovered just because of the violation of *this* particular condition.

this reason Bell 2 is often seen as a generalization of Bell 1 or 3. In Section 9 4 it is, however, shown that they are distinct theorems.

Bell 3

Bell 3 resembles Bell 1 in that it exclusively concerns the case of particle pairs in which a *strict* correlation is predicted between the outcomes of two distant measurements of the same spin component. But the argumentations of the two are different. Formally Bell 3 is based on the idea, first put forward by Wigner [16], of considering both measurements in a symmetric way and using the fact that probabilities cannot be negative. In Refs. [17] and [18] I proceeded in the same way but pointed out in addition that the thus derived Bell inequality is related to a much more general one (valid within an infinite variety of qualitatively different situations most of which have nothing to do with particles and correlations) and gave a detailed account of the premises that the theorem (here called) Bell 3 shows cannot be true together. Essentially they are that (a), when faced with a phenomenon of correlation at a distance, particularly a strict one, we normally are not content with just a rule-of-the-game predicting it: we demand a genuine "explanation," that is, an explanation referring to strongly objective general and individual properties possessed by the involved systems; (b) there is no action at a distance between the particles involved; and (c) inductive inference is a legitimate reasoning procedure. The latter idea more precisely is that if, in some way or other, it has been proven that all the elements, chosen at random, composing an unbiased sample of a given population have some property, inductive inference can be applied, to the effect of asserting that *all* the elements of this population have this property. On this basis the first step then consists in imagining that on four representative, equally populated samples of randomly selected Σ pairs of the Bohm standard example, measurements are made of the spin components of U and V along the same, arbitrary direction (the four directions at stake differing from one another). In view of the (theoretically predicted) strict correlation between the outcomes, premises (a) and (b) may then be used to show that in each sample the U and V particles individually had, before they were subjected to the measurement in question, definite spin components along the chosen direction (or equivalent built-in local parameters) and that these values determined the outcome of the measurement. Premise (c) then tells us that the result thus obtained on each sample can be generalized to the whole population of U and V particles, including those on which no measurement is made. This completes the first step in the argument. The second step then consists in deriving Bell's inequalities from this result, using no abstract theoretical concept whatsoever but merely the plain, obvious fact that the number of elements of a statistical ensemble of systems cannot be smaller than zero. A few additional details must be carefully considered for ascertaining that the argument sketched does succeed. They are made explicit in ref. [18].

• *Remark 1.* As pointed out in Section 8.2, parameter independence rules out the possibility of superluminal signaling. This induced some physicists to conjecture that local causality violation is nothing new. To make their point they proposed an analogy. They suggested we should imagine a very powerful lighthouse located at the center of a circular wall the radius of which is so large that the rotating spot created on the wall by the rotating lamp in the lighthouse travels faster than light. Of course, the existence of this phenomenon does not make faster-than-light signaling possible. The argument of these physicists was that since also the violation of local causality does not allow for superluminal signaling, the two facts should basically be of the same nature. This, however, does not follow from the premises and is not true. In the case of the spot, the correlation observed between the lighting of two distant points on the wall is due to a set of common causes (the parameters that determine at every time the orientation of the lamp). And the nature of the whole setup implies that, for anybody who, at any time, would know *exactly* the values of these parameters, the probability (equal to either 1 or 0 in this deterministic case) that a given point on the wall is lighted up at some given time would be totally independent of what takes place at any other point on the wall. Local causality is therefore *not* violated in the considered phenomenon, whereas, as recalled above, it is violated in the phenomena in which the Bell inequalities are violated.

• *Remark 2:* One aspect in which Bell 2 obviously generalizes Bell 1 is that, as we noted, Bell 1 applies only to strictly correlated pairs and its premises then imply that the supplementary variables shown to exist in the derivation first step *determine* the outcomes of the measurements. In contrast, nondeterministic supplementary variable models can be considered that obey local causality and that Bell 2 therefore shows are at variance with the quantum mechanical predictions. In this connection it is, however, worthwhile to mention the fact that—as shown by Fine [19]—to any such model a deterministic supplementary variable model yielding the same measurement outcomes can be associated.

9.4 Comparing the Bell Theorems

Since the word *theorem* was used above in the plural, it is necessary to check that the premises of Bell 1, Bell 2, and Bell 3 are indeed different.

Let us first compare the premises of Bell 1 and Bell 2 [12]. A first difference is that, contrary to what is the case concerning Bell 2, outcome independence is definitely not a premise of Bell 1. This is because, as pointed out in Section 9.2, Einsteinian separability refers only to *what is done* in R_U. In particular, if "what is done in R_U" is a measurement, it refers to the choice the experimentalist makes of what measurement he or she does. It does not refer to the outcome of this measurement. Moreover, in the strict correlation case on which Bell 1 bears, this outcome, according to the argumentation above, turns out to be determined by the objective state λ of the pair and the just mentioned choice, which here is that of **n**. Hence it is not a free variable, on which the outcome of a measurement performed in R_V could depend in addition to its dependence on λ and **n**. For this reason, outcome independence had not to be introduced as an additional premise in Bell's original 1964 proof.

However, there is a counterpart to this. Namely, if Bell 1 spares the "outcome independence" premise, in compensation, so to speak, it relies on premises that are stricter than those of Bell 2 on another point for, as also pointed out in Section 9.2, contrary to parameter independence, Einsteinian separability refers to "the real, factual situation" of individual systems. Hence it is, in this respect, more stringent than parameter independence, which merely refers to a priori probabilities which, in cases of interest, are not unity (not even in the cases of strictly correlated pairs).

As for Bell 3, its premises obviously differ from those of Bell 2 since, as do those of Bell 1, they involve the notion of properties possessed by microsystems. But they also differ from those of Bell 1. A noteworthy difference is that, while in Bell 1 only one measurement, the one on U, is thought of as being actually performed, and the fact V possesses a given property is—via the EPR criterion—inferred without V being subjected to any actual measurement, in Bell 3, on the contrary, the measurements on both U and V are considered as being performed, so that the fact that U and V possess given properties is inferred through a reference to something else than the EPR criterion proper. This "something else" is, as mentioned, just the idea that a strict correlation logically cries out for an explanation going beyond a mere computational recipe in that it should be couched in strongly objective terms; an idea that may be seen as another of our (rightly or wrongly) deeply ingrained rationales for attributing properties to objects.

9.5 The GHZ Proof

A proof of Bell 1 that involves no inequality, and therefore no basic reference to statistics, was given by Greenberger, Horne, and Zeilinger [GHZ, 20]. Its premises are the EPR criterion of reality and a natural extension of Einsteinian separability to several, mutually spatially separated, space-time regions. It goes as follows [4] (see also Mermin [21]). Let a spin 1 particle with z spin component $M = 0$ decay into two spin 1 particles I and II, each of which then decays into a pair of spin 1/2 particles. It is assumed that orbital angular momentum L and spin S are both separately conserved during the whole process, so that the z component of the total spin of the system remains zero. The spin state is $|S, M > = |1, 0 >$ and after the first decay it can be expressed as

$$|1, 0 > = 2^{-1/2}(|1 >_I | -1 >_{II} - | -1 >_I |1 >_{II}) \qquad (9.2)$$

where $|m >_X$ $(X = I, II)$ is the spin state of particle X that corresponds to eigenvalue m of the z component of the spin of X (the coefficients in Eq. (9.2) are the familiar Clebsch–Gordan coefficients for this combination of spins; see Appendix 1). Concerning the second step decays, conservation of the z spin component gives, for each particle I and II,

$$|1> \to |+,+>$$
$$|-1> \to |-,->$$

where $|+,+>$ $(|-,->)$ denotes the spin state of the decay product particle pair corresponding to both components having their z spin component equal to $+1/2$ $(-1/2)$; so that the final spin state of the four particles system is

$$|\psi> = 2^{-1/2}(|+>_1 |+>_2 |->_3 |->_4$$
$$-|->_1 |->_2 |+>_3 |+>_4) \qquad (9.3)$$

where numbers 1 and 2 label the decay products of I and numbers 3 and 4 those of II.

Let the experimental setup be so arranged that the decay particles 1, 2, 3, and 4 move freely and can be subjected to spin-component measurements in spatially separated space-time regions R_1, R_2, R_3, and R_4 respectively. In order to have a simply visualizable picture not contradicting the assumed orbital angular momentum conservation we may for example assume [4] that particles 1 and 2 move in the positive z direction and particles 3 and 4 in in the negative z direction, and that moreover the beams bearing particles 1 and 2 as well as those bearing particles 3 and 4 are spatially displaced so as to be extremely distant from one another. We then consider making measurements of either S_x or S_y (the spin components along x and y) on each one of the four particles, within the just defined corresponding space-time regions, so that, according to the EPR general argumentation, measurements performed on any of these particles do not disturb the others. Let these spin components be measured in units $\hbar/2$ so that the corresponding operators are just the Pauli σ matrices, let the outcomes be designated by the letter $m(m_x$ or $m_y)$ and let the symbols σ, S, and m pertaining to particle $i(i = 1,2,3,4)$ be labeled by the upper index i. The following argumentation is essentially based on the fact that ket (9.3) happens to be an eigenket of quite a number of commuting Hermitean operators involving these σ's In particular, it is easily verified that

$$\sigma_x^1 \sigma_x^2 \sigma_x^3 \sigma_x^4 |\psi> = -|\psi> \tag{9 4a}$$

$$\sigma_y^1 \sigma_y^2 \sigma_x^3 \sigma_x^4 |\psi> = +|\psi> \tag{9 4b}$$

$$\sigma_y^1 \sigma_x^2 \sigma_y^3 \sigma_x^4 |\psi> = -|\psi> \tag{9 4c}$$

$$\sigma_x^1 \sigma_y^2 \sigma_y^3 \sigma_x^4 |\psi> = -|\psi> \tag{9 4d}$$

Now, quite generally, the lemma holds that when a system is in an eigenstate $|\psi>$ of a product of commuting Hermitean operators ABC , if g denotes the corresponding eigenvalue the product of the outcomes of separated measurements of observables A, B, C must be g for if the eigenvalue equations for A, B, C are

$$A|\phi_{k,l,m, \, ,r}> = a_k|\phi_{k,l,m, \, ,r} >$$
$$B|\phi_{k,l,m, \, ,r} > = b_l|\phi_{k,l,m, \, ,r} > \tag{9 5}$$
$$C|\phi_{k,l,m, \, ,r} > = c_m|\phi_{k,l,m, \, ,r} >$$

then

$$< \psi|\phi_{k,l,m, \, ,r} > = g^{-1} < \psi|ABC \quad |\phi_{k,l,m, \, ,r} >$$
$$= g^{-1} a_k b_l c_m \quad < \psi|\phi_{k,l,m, \, ,r} >$$

so that

$$< \psi|\phi_{k,l,m, \, ,r} > (1 - a_k b_l c_m \, /g) = 0 \tag{9 6}$$

hence the probability

$$\sum_r | < \phi_{k,l,m, \, ,r}|\psi > |^2$$

that the outcomes of the measurements be $a_k, b_l, c_m,$ respectively is zero for all $a_k, b_l, c_m,$ the product of which differs from g, **Q.E.D.** An obvious consequence is that, g being given, knowledge of $b_l, c_m,$ (complete list except a_k) makes it possible to unambiguously predict a_k

Now, since S_x^2, S_x^3, and S_x^4 can be measured, as we saw, without disturbing particle 1 in R_1 (remember that, in appropriate reference frames, these measurements take place later), it follows from Eq. (9 4a), from the above and from the EPR criterion of reality that S_x^1 corresponds to an element of reality and therefore has, in region R_1, a definite value m_x^1 independent of what is done and observed in other regions. Naturally the argument applies to Eqs. (9 4b), (9 4c), and (9 4d) as well, and in the same way to the physical quantities S_y. To sum up, then, if valid, the aforementioned premises entail, in this particular example, that all the eight quantities S_x^1, S_x^2, S_x^3, S_x^4, S_y^1, S_y^2, S_y^3, S_y^4 have *definite* numerical values $m_x^1, m_x^2, m_x^3, m_x^4,\ m_y^1, m_y^2, m_y^3, m_y^4$ even if they are not measured. Of course these values can only be +1 or −1 since these are the only possible outcomes.

Equation (9 4a) then shows that if, for example, S_x^2, S_x^3, and S_x^4 have values m_x^2, m_x^3, and m_x^4 all equal to +1 S_x^1 has value $m_x^1 = -1$, and more generally that the product of m_x^1, m_x^2, m_x^3, and m_x^4 is −1. Eqs. (9 4b–d) have similar implications so that what has finally been shown is that the EPR assumptions entail the four equalities:

$$m_x^1 m_x^2 m_x^3 m_x^4 = -1$$

$$m_y^1 m_y^2 m_x^3 m_x^4 = +1$$

$$m_y^1 m_x^2 m_y^3 m_x^4 = -1$$

$$m_x^1 m_y^2 m_y^3 m_x^4 = -1$$

$$(9.7)$$

But these equalities are not mutually compatible, as is immediately seen by comparing the product of all the left-hand sides, which is +1, with that of all the right-hand sides, which is −1. This shows that the EPR assumptions cannot be reconciled with the observable predictions of quantum mechanics.

It will be observed that particle 4 plays no role in the argument, which shows that equivalent results can be obtained by considering three particles only. The reason why a four-particle system has been used here instead is just that the principle of its preparation could be described more easily.

9.6 A GHZ-Like Proof of Bell 2

As we just saw, the proof by Greenberger, Horne, and Zeilinger is based on essentially the same two premises as those of Bell 1 and shows, same as Bell 1, that these premises cannot both be true if the experimentally verifiable predictions of quantum mechanics are true. In this sense the GHZ proof may be viewed as a "proof of Bell 1." Let us show here that an alternative "proof of Bell 2" following these lines can also be given.

To this end, let us consider the four space-time regions $R_1,$ $, R_4$ in which it is contemplated making measurements on particles 1, , 4 respectively. Let $\mathbf{a}_i$ and A_i be respectively the orientation of the ith measuring apparatus (determining which spin component is measured) and the outcome (+ or −) of this measurement. Let R_i' ($i = 1,$, 4) be a space-time region that completely shields off from R_i the overlaps of the backward light cone of R_i with the backward light cones of all the three other regions and let λ collectively designate a set of variables that specify events in the four backward light cones and in particular *fully* specify those in all four R_i''s as well as the complete objective state of the source (redundancy will occur here but is harmless). With the notation

$$(X|YZ) \tag{9.8}$$

meaning "the conditional probability for X *if* Y and Z," local causality (plus the hypothesis that the experimentalist is free; see above) implies that when λ is fully specified the probability for a given outcome A_i is unaltered by specification of $\mathbf{a}_k$ and/or A_k, with $k \neq i$; therefore this probability can be written

$$(A_i|\lambda, \mathbf{a}_i) \tag{9.9}$$

By the same token the similarly defined probability for the measurements on particles 2, 3, 4 say, to yield outcomes A_2, A_3, and A_4 is unaltered by specification of $\mathbf{a}_1$ and A_1 and can therefore be written

$$(A_2, A_3, A_4|\lambda, \mathbf{a}_2, \mathbf{a}_3, \mathbf{a}_4) \tag{9.10}$$

Let then

$$p_\pm = p_\pm(\lambda, \mathbf{a}_1) = (A_1 = \pm|\lambda, \mathbf{a}_1) \qquad (9.11)$$

$$q_+ = (A_2 = +, A_3 = +, A_4 = +|\lambda, \mathbf{a}_2, \mathbf{a}_3, \mathbf{a}_4)$$
$$+(A_2 = +, A_3 = -, A_4 = -|\lambda, \mathbf{a}_2, \mathbf{a}_3, \mathbf{a}_4)$$
$$+(A_2 = -, A_3 = +, A_4 = -|\lambda, \mathbf{a}_2, \mathbf{a}_3, \mathbf{a}_4)$$
$$+(A_2 = -, A_3 = -, A_4 = +|\lambda, \mathbf{a}_2, \mathbf{a}_3, \mathbf{a}_4) \qquad (9.12)$$

and let q_- be obtained from q_+ by exchanging the + and − signs. Let now the vectors $\mathbf{a}_1$, , $\mathbf{a}_4$ all be chosen along O_x so that the A_i are the outcomes of measurements of the S_x^i. From Section 9.5 we know that the probability is zero for the product of these outcomes to be +1 [Eqs. (9 4a), and the ensuing lemma]. On the other hand this probability can be expressed as a weighted average of the corresponding probability $P(\lambda, \mathbf{a}_1, , \mathbf{a}_4)$ relative to one particular set of λ's over all possible such sets. This latter probability is of course just the sum of the probabilities of the eight composite possible outcomes in which the number of elementary outcomes −1 is even. And these probabilities are themselves just the products of the probabilities of these elementary outcomes, that is, of probabilities such as $(A_i|\lambda, \mathbf{a}_i)$ (in general, such probabilities of composite events are the products of the a priori probability of one of them by the conditional probability of the "next" one, and so on; but here these conditional probabilities are equal to the a priori ones because of local causality). This last remark also holds, of course, for each component term in $q_\pm$ and hence, when applied to $P(\lambda, \mathbf{a}_1, , \mathbf{a}_4)$ and to these $q_\pm$ components this argumentation easily yields:

$$P(\lambda, \mathbf{a}_1, \mathbf{a}_2, \mathbf{a}_3, \mathbf{a}_4) = p_+q_+ + p_-q_- \qquad (9.13)$$

so that, in view of the above:

$$\int [p_+q_+ + p_-q_-]\rho(\lambda)\, d\lambda = 0 \qquad (9.14)$$

where $\rho(\lambda)$ is the probability density of the objective states of the composite system. Since all the quantities in the sum are nonnegative, this in turn implies

$$p_+q_+ = 0 \qquad (9.15a)$$

$$p_- q_- = 0 \tag{9.15b}$$

for any actually existing objective state ($\rho \neq 0$) But since

$$p_+ + p_- = q_+ + q_- = 1 \tag{9.16}$$

Eq. (9.15b) also reads

$$(1 - p_+)(1 - q_+) = 0. \tag{9.17}$$

Together with Eq. (9.15a), Eq. (9.17) yields

$$p_+ + q_+ = 1 \tag{9.18}$$

which, together with Eq. (9.15a) again, yields in turn that either

$$p_+ = 1, \qquad q_+ = 0$$

or

$$p_+ = 0, \qquad q_+ = 1$$

In view of the definition (9.11) of $p_\pm$, the preceding shows that in the GHZ case local causality entails that the outcome of any individual measurement of S_x on particle 1 is predetermined *by the set of the variables* λ. Hence the outcome of any, yet to be performed, measurement of S_x on particle 1 can be labeled m_x^1, without any further indexes specifying, for example, what other measurements are or will be performed on the other three particles and what their outcome will be. The same of course holds, on symmetry grounds, concerning the three other particles 2, 3, and 4 and, since the argument also applies to the other combinations of measurements of S_x and S_y considered in Eqs. (9 4b–d), also concerning the measurements of the quantities S_y. Hence Eqs. (9 4) entails the set of Eqs. (9.7), which is self-contradictory as we saw. This proves Bell 2 in the GHZ way, that is, without inequalities.

• *Remark:* Compared to the proofs of Bell theorems based on the Bell inequalities, those using the GHZ approach have the obvious advantage of not being statistical, so that when we use them we do not have to worry about such things as the law of large number, the notions of representative samples, standard or nonstandard fluctuations, etc. Since statistics is, after all, an applied science (the experimentally available ensembles not being infinite), on matters of principle such as those the Bell's theorems deal with it is gratifying that a nonstatistical type of proof should be available: And it therefore does not come as a surprise that a number of physicists who beforehand had kept aloof from and refused to commit themselves on such questions became convinced of their significance when they read of the GHZ proof. On the other hand, it must be granted that the standard proofs of the Bell theorems also have some definite advantages in other respects.

The main advantage is of course that the standard proof of Bell 2 holds also concerning systems exhibiting no strict correlations, which, as mentioned, makes experimental testing easier (this whole experimental work, the highlights of which are the contributions of Freedman and Clauser [22], Fry and Thompson [23], and Aspect, et al. [15], is very extensive and important but distinctly falls outside the subject proper of this book and will therefore not be reviewed).

Another advantage of the standard proofs is that the Bell inequalities can be derived from highly plausible general assumptions (essentially local causality) that borrow nothing from quantum mechanics. Hence the experimental result that these inequalities are violated constitutes in principle a direct disproof of these general assumptions, where "direct" means that it is valid in its own right, quite independently of whether quantum mechanics is right or wrong. The GHZ derivation does not achieve quite as much, for, in it, local causality is not used alone, for making any definite prediction that quantum mechanics would then falsify. A consequence of this is that a "pre-1920" physicist could in principle have had the (improbable) idea of checking local causality by showing first that it entails the Bell inequalities and by then performing the corresponding experiments in a variety of cases. He would have found out that while these inequalities are experimentally verified in most instances, they are, surprisingly enough, not obeyed in all cases. Thus the universal local causality hypothesis would have been invalidated in much the same way as that of the ether was, by the unexpected outcome of the Michelson, Morley experiment. In contrast, no such reconstruction of history can be imagined along the GHZ line, for, there, it seems that no general consequence can be derived—independently of quantum mechanics and from just undeniably existing correlations *plus* local causality—that quantum mechanics would falsify.

The Bell Theorems Without Spins and Without Polarizations

On the theoretical side it has been shown [4] that the GHZ argument can be developed without involving any spin, essentially by replacing the spin-component measurements by presence measurements performed by means of counters. Since the GHZ argument is the counterpart of Bell 1 and implies a violation of the Einsteinian premises, this makes the violation of the latter by quantum mechanics even more striking. As an exercise in these matters the interested reader may attempt to show that by combining the argument described in Ref. [4] with the one presented in the foregoing section, it is also possible to prove the GHZ counterpart of Bell 2 without considering spins.

On the experimental side important developments have taken place in the 1990s due to the discovery of the so-called parametric down conversion technique, with the help of which it is now technically possible to produce photon pairs the elements of which have sharply correlated momenta. These experiments essentially consist in splitting each photon of the pair and recombining it with a variable phase delay. It is well known that in the ordinary Young two-slit experiment, instead of observing fringes on a screen one can operate with a fixed counter, vary the phase in one of the beams by inserting an adjustable phase plate, and observe the ensuing intensity variations on the counter. Here the same procedure is used but on *both* particles and with the difference that it is the dependence on such adjustable phases of some *coincidences* between the counts of the *distant* counters triggered by each one of the two particles that is investigated. Quantum theory predicts a sinusoidal dependence whereas local causality, if applicable, entails the validity of a Bell-CHSH inequality that is incompatible with such a dependence. Experiments of this type were suggested independently by several authors, notably Horne, Shimony, and Zeilinger [24] and Franson [25] and performed notably by Rarity and Tapster [26], who found a violation of the Bell inequality by 10 standard deviations (more on this in Section 13.2).

References

1. A. Einstein, B. Podolsky, and N. Rosen, *Phys. Rev.* **47**, 777 (1935).
2. A. Einstein, in *Albert Einstein: Philosopher, Scientist*, P.A. Schilpp (Ed.), Library of Living Philosophers, Evanston, Ill., 1949.
3. D. Bohm, *Quantum Theory*, Prentice-Hall, Englewood Cliffs, N.J., 1951.
4. D. Greenberger, M. Horne, A. Shimony, and A. Zeilinger, *Am. J. Phys.* **58**, 1131 (1990).
5. N. Bohr, *Phys.Rev.* **48**, 696 (1935).
6. B. d'Espagnat, "Nonseparability and the Tentative Descriptions of Reality," *Physics Reports* **110**, 203 (1984).
7. B. d'Espagnat, *Reality and the Physicist*, Cambridge U. Press, 1989.
8. H.P. Stapp, *Nuovo Cimento* **40 B**, 191 (1977); *Foundations of Physics* **10**, 767 (1980).
9. B. d'Espagnat, *Phys. Rev.* **D 18**, 349 (1978).
10. J.S. Bell, *Physics* **1**, 195 (1964).
11. J.F. Clauser, M.A. Horne, A. Shimony, and R.A. Holt, *Phys. Rev. Lett.* **23**, 880 (1969).
12. B. d'Espagnat, "One or Two Bell Theorems?" in *Bell's Theorem and the Foundations of Physics, Proceedings of the International Conference in Memory of John Bell Held in Cesena, 7–10 October 1991*, World Scientific Publishing Company, New York, 1993.
13. B. d'Espagnat, *Conceptual Foundations of Quantum Mechanics*, Addison-Wesley, Reading, Mass. 2nd ed. 1976, ch. 12.
14. J.S. Bell, *Speakable and Unspeakable in Quantum Mechanics*, Cambridge U. Press, Cambridge, 1987, "La nouvelle cuisine," in *Between Science and Technology*, A. Sarlemijn and P. Kroes (Eds.), Elsevier Science Publ. B.V., North-Holland, 1990, ch. 6.
15. A. Aspect, P. Grangier, and G. Roger, *Phys. Rev. Lett.* **49**, 91 (1982); A. Aspect, J. Dalibard, and G. Roger, *Phys. Rev. Lett.* **49**, 1804 (1982).
16. E.P. Wigner, *Am. J. Phys.* **38**, 1005 (1970).
17. B. d'Espagnat, *Scientific American* **241**, 158 (1979).
18. B. d'Espagnat, *In Search of Reality*, Springer-Verlag, New York, 1983.
19. A. Fine, *Phys. Rev. Lett.* **48**, 291 (1982).
20. D. Greenberger, M. Horne, and A. Zeilinger in *Bell's Theorem, Quantum Theory and Conceptions of the Universe*, M. Kafatos (Ed.), Kluwer Academics, Dordrecht, The Netherlands, 1989.
21. N.D. Mermin, *Am. J. Phys.* **58**, 731 (1990).
22. S.J. Freedman and J.F. Clauser, *Phys. Rev. Lett.* **28**, 938 (1972).
23. E.S. Fry and R.C. Thompson, *Phys. Rev. Lett.* **37**, 465 (1976).
24. M.A. Horne, A. Shimony, and A. Zeilinger, *Phys. Rev. Lett.* **62**, 2209 (1989).
25. J.D. Franson, *Phys. Rev. Lett.* **62**, 2205 (1989).
26. J.D. Rarity and P.R. Tapster, *Phys. Rev. Lett.* **64**, 2495 (1990).

Other reading

M. Redhead, *Incompleteness, Nonlocality and Realism, a prolegomenon to the philosophy of quantum mechanics*, Clarendon Press, Oxford, 1989.

On Measurement

10 The notions and theorems described in the preceding two chapters raise basic questions to which we shall have to come back. But before that we must consider different, though closely related, sets of problems: those that concern measurement. Unless otherwise specified we investigate them within the realm of conventional textbook quantum mechanics (no hidden variables).

For reasons to be unfolded here, measurement constitutes a riddle, and a great many theories were put forward as attempts to solve this riddle. The number is so considerable indeed that to try and review them all would be an almost impossible undertaking. There is all the less ground for trying to grapple with it, as a survey of a number of approaches to measurement theory was already given in Refs. [1] and [2]. Hence what will be done here is merely to explain the problem, sketch and discuss its current most popular suggested solutions, and have a look at a series of related questions, such as decoherence, formulation within the Heisenberg picture, and lack of relativistic covariance of the wave-packet collapse.

10.1 The Basic Problem

In Section 8.1 it was pointed out that in the case in which two quantum systems temporarily interact, after the interaction is over the composite system is generally left in what is called an "entangled state," which means that its state vector is not just a product of state vectors corresponding to each system.

Macroscopic systems being composed of atoms and fields, there is no reason why they should not obey the quantum laws. It therefore seems quite natural to consider them as quantum systems, albeit extremely complex ones. If this point of view is taken, then it is, of course, to be expected that at least in some cases the interaction that takes place during the measurement time between a measurement apparatus A (a macroscopic system of a certain type) and the quantum system S on which the measurement is made by means of A will bring the composite system $S + A$ into an entangled state. As will presently be explained, it is this fact that raises the essential interpretational difficulty of quantum measurement theory. Although taking the interaction between A and its environment adequately into account lightens the difficulty, it does not remove it altogether as, at a later stage, will be shown.

To describe, then, the basic problem (also known as the "Schrödinger cat problem" because, to picture it more vividly, Schrödinger, in the account he gave of it replaced the instrument pointer by a cat), let us call $|\psi_1 >,$ $, |\psi_n >,$ the eigenvectors of an observable L that pertains to a system S and whose value an experimentalist wants to measure by means of an apparatus A. For the sake of simplicity, let us assume that A can only be in one of a discrete set of eigenstates $\{|n, r >\}$, where n labels the possible positions $G = g_n$ of a pointer, and r is a degeneracy index related to all the other variables in A. Needless to say, such a description of the apparatus is highly schematic; any conceptual difficulty to which it may lead will therefore have to be examined with care, within the framework of a more realistic formalism, before it can be said to be a true one. Nevertheless, this simple picture is useful for illustrating the main features of the problem.

Let us first consider the case in which S is initially in an eigenstate $|\psi_m >$ of L.

$$L|\psi_m >= l_m|\psi_m > \qquad (10.1)$$

and in which A simultaneously lies in one of the states $|0, r >$, so that the coordinate G has the value g_0. A is so devised (this is a part of our assumption) that, under such circumstances, the effect of the interaction is to bring it into one of the states $|m, s >$ without changing the state of S. In other words, the state vector that describes the composite system $S + A$ undergoes the change described by·

$$|\psi_m > \otimes |0, r > \to |\psi_m > \otimes |m, s_{m,r} > \qquad (10.2)$$

If this is the case, after the interaction the pointer of A is in the position $G = g_m$, which is in a one-to-one correspondence with the eigenvalue l_m of L. Thus Eq. (10.2) can indeed be considered as a model describing a process of measurement of L by means of A.

Now, however, let us consider a different initial situation, in which A is still described by $|0, r >$ but S is in a superposition of states $|\psi_m >$

$$|\psi > = \sum_m a_m |\psi_m > \qquad (10.3)$$

where the a_m's are parameters. The superposition principle guarantees that (except in the special cases when superselection rules operate) Eq. (10.3) is indeed a possible initial state. Then, as a result of Eq. (10.2) and the linearity of the time evolution law (Chapter 3: Rule 3), the final state of the compound system is necessarily described by

$$|\Psi_f > = \sum_m a_m |\Psi_m > \qquad (10.4)$$

with

$$|\Psi_m > = |\psi_m > \otimes |m, s_{m,r} > \qquad (10.4a)$$

There is a sense in which Eq. (10.4) can be said to adequately describe the "measurement process" in the considered situation. For this equation shows that if, immediately after the interaction between the system S and the instrument A, an observer happens to measure (by a procedure we do not analyze) both L (on S) and G (on A) on a whole ensemble of $S + A$ systems described by the right-hand side of this equation, he is bound to find a 100 percent correlation between the results. Each time he finds $L = l_m$ he must find $G = g_m$ and conversely (this immediately follows from Rule 8, of Chapter 3). In other words, as soon as he has measured G he knows the value he will find upon measurement of L. Under these conditions he can dispense with the latter operation. We say therefore that by measuring G on A he has performed a measurement of L (obviously this argument can be extended to a whole chain of instruments $A^{(1)}, A^{(2)}, \ldots, A^{(k)}$, each measuring the "pointer coordinate" $G^{(1)}, G^{(2)}, \ldots, G^{(k)}, \ldots$ of the one preceding it in the chain; the latter is called the *von Neumann chain*). Moreover, while the final state operator

$$\rho = |\Psi_f><\Psi_f| = \sum_{m,n} a_m a_n^* |\Psi_m><\Psi_n| \tag{10.5}$$

admittedly involves cross-terms (terms with $m \neq n$), still such terms play, in practice, hardly any role. The reason is that—as formulas (3.4) or (6.22) together with Eqs. (10.4) and (10.5) easily show—the probabilities of getting a given outcome for G *and* a given outcome for any observable B pertaining to S upon simultaneous measurements of G on A and B on S involve no such terms. And the same is true, of course, concerning the probabilities relative to measurements of observables pertaining either to A alone or to S alone (because of the mutual orthogonality of the $\{|\psi_m>\}$ and $\{|m, s_{m,r}>\}$ respectively). In other words, the cross-terms in question can only come in when correlations between observables of S and some observables of A *differing from* G are investigated. More detailed investigations (Jauch [3]; see also [1]) even show that essentially the class of the latter contains none of the observables of A that commute with G. Under these conditions ρ can—for all practical purposes—be replaced by

$$\rho' = \sum_m |a_m|^2 |\Psi_m><\Psi_m| \tag{10.5a}$$

But if we compare this ρ' with expression (6.1) we observe that it is identical to the statistical operator of a (proper) ensemble in which a proportion $|a_m|^2$ of the systems have $L = l_m$ and $G = g_m$. Since, applied to state vector (10.3), probability rule (3.3) would have yielded just this result, we see that, in the just sketched theory of the measurement process, the probability rule in question is recovered.

Operationally all this is rather satisfactory, but does it mean that the physical interpretation of Eq. (10.4) is totally clear? The answer is no, for difficult questions are still to be faced. Let us have a quick view of them.

1. Above, we implicitly took it for granted that the overall, composite quantum system under study is composed of two definite, distinct parts: S, the system on which the observable is measured, and A, the instrument of observation. But

questions may be asked as to what defines the borderline between the two. Is it merely defined by the fact that *we* find
it convenient to consider some parts of the global system as
being "parts of the instrument" or is it determined in a more
physical way, by the very structure of the system at hand? In
other words, if we were just given a complex system—that is,
its Hilbert space, its Hamiltonian, etc.—could we, by mere inspection of these data, determine whether or not this system
is composed of a "measured system" and an instrument? And
could we determine what elements of the global system belong
to each? In a theory such as quantum mechanics, some basic
rules of which refer to measurement, such questions are significant.

   2. Even assuming this question is settled, we must face the
problem of what determines the "pointer basis" of the instrument. To understand what is here at issue let us consider the
simple case in which the index m merely runs from 1 to 2
and let us suppress the degeneracy indexes, so that Eq. (10.4)
reads

$$|\Psi_f> = a_1|\psi_1> \otimes|1> + a_2|\psi_2> \otimes|2> \qquad (10.6)$$

   Let us then assume that the Hermitean operator G' whose
eigenvalue equation is

$$G'|V_\pm> = g'_\pm|V_\pm>$$

with

$$|V_\pm> = 2^{-1/2}(|1> \pm|2>)$$

also represents an observable of the instrument. Equation (10.6)
can then also be written

$$|\Psi_f> = 2^{-1/2}[(a_1|\psi_1> + a_2|\psi_2>) \otimes |V_+>$$
$$+ (a_1|\psi_1> - a_2|\psi_2>) \otimes |V_->]$$

and it is clear that if we somehow get to know the value, g'_+ or g'_-,
of G', we thereby also know the state, $|\chi_+> = a_1|\psi_1> + a_2|\psi_2>$

or $|\chi_- > = a_1|\psi_1 > -a_2|\psi_2 >$ of the measured system.[1] This immediately raises the questions: "What does the instrument actually measure?" and "What is the pointer coordinate—G or G' or ?"

In general the states $|\chi_+ >$ and $|\chi_- >$ are not orthogonal, so that they are not the eigenvectors of any observable and to try and answer the question at issue some physicists argued that a measurement should measure not a state but an observable and correlatively proposed the rule that in an expansion of $|\Psi_f >$ of the type (10.4) both the $\{|\psi_m >\}$ and the $\{|m, s_{m,r} >\}$ should be sets of orthonormal kets, which, at least when $a_1 \neq a_2$, would eliminate any pointer coordinate other than G. The proposal is attractive in the sense that, except for special choices of the a_m's, it does make the admissible pointer basis unique, and thus determines unambiguously what is measured and what is the pointer coordinate. But it meets with some difficulties, in particular when the evolution in time of the global system is taken into account. We shall see (Section 10.6) that the environment theory proposes a different, more flexible and, as it seems, on the whole more satisfactory solution (in this connection see also Section 13.7).

3. Even assuming that difficulties 1 and 2 have been settled, we must apparently bring at some place the von Neumann chain to a close. At some stage (p) it must be said: "And finally the observer observes the quantity $G^{(p)}$ on $A^{(p)}$ and finds $g_n^{(p)}$ " As long as the theory does not analyze this last, crucial point it must be considered as—to say the least—incomplete. This is true even though a positive feature of the von Neumann chain concept must be acknowledged, namely the fact that, at least if a strictly operational standpoint is taken, then, under fairly general conditions, the place in the chain where the "cut" is made is arbitrary. These conditions (to be made more precise below) are tantamount to the requirement that the systems which, in the chain, lie "after" the cut should be classically describable. For this reason the cut in question is often just referred to as the "quantum-classical cut."

---

1. Suppose for example that we actually measure G' and get g'_+ as a result. By applying Rule 10a (Chapter 3) we then find that, just after this measurement, the state vector of the overall system is $|\chi_+ > \otimes |V_+ >$

4. To place this cut at the borderline between "matter" and "consciousness," as was suggested by some authors [4,5], raises difficulties, be it only because of the plurality of consciousnesses. But to place it between the system and some macroscopic physical system called "the instrument" raises the problem that when macroscopic systems such as instrument pointers are considered, we like to think of them in the way both classical physics and commonsense do, that is, as having definite positions in space, definite velocities, and so on. More precisely, even though we willingly grant that their center-of-mass position and velocity should not violate the Heisenberg indeterminacy relationships, still we feel reluctant to admit that an instrument pointer, for example, could sometimes not lie in any definite interval of the instrument scale (nor in any other definite region of space either!). Indeed, one essential idea that the expression "a classical system" is meant to express is that the center-of-mass position of such a system must at any time *have* (perhaps within some tiny uncertainty limits) a definite value. Now, as explained in detail in Section 4.3, this, in turn, means that we can at least imagine somebody who would know with certainty, on any particular such system and within the limits just alluded to, what the outcome would be of a measurement of this center-of-mass position if it were performed. This generates the two difficulties 4a and 4b now to be made explicit. These difficulties are very much linked but for maximal clarity it is suitable to formulate them separately.

4a. This difficulty stems from the just recalled meaning of the verb "to have" together with the fact that in replacing ρ by ρ', that is in going over from Eq. (10.5) to Eq. (10.5a), we had to assume the practical inobservability of some entities. Indeed, anybody who accepts the description in terms of ρ' and interprets ρ' as describing a proper mixture can truly say—in the sense defined above—that a proportion $|a_m|^2$ of the elements of the ensemble do *have* $L = l_m$ and $G = g_m$. But we must remember there is a difference between ρ and ρ', corresponding to the fact that there exist Hermitean operators operating in the Hilbert space of the $S + A$ system whose mean values for the two cases are different. If these operators correspond to physical quantities that conceivably could be measured, if the "commonsense" view is kept that "of course" an experimentalist can choose at whim what observable to measure on $S + A$, and if it is considered that the purely *practical* obstacles making some measurements uneasy are not to be taken into account, then the difficulty cannot be coped with. Most measurement theorists do not, therefore, accept such stringent conditions and are content that under realistic conditions the operators in question do not correspond to practically observable physical quantities. But the reference, in such, allegedly basic, descriptions, to what is practically feasible and what is not introduces a touch of subjectivity that raises questions.

4b. As already noted, difficulty 4b is very much linked to the foregoing one. In a sense, however, it is even worse for it remains even under conditions under which difficulty 4a would be considered as removed (unobservability of some physical quantities). We shall refer to it as the "and-or" difficulty. To state it let us turn once more to ρ' [Eq. (10.5a)] and to its interpretation as a proper mixture. To some extent, this interpretation is backed up, as pointed out, in particular, by Gottfried [6], by the fact that the $|a_m|^2$ appear in ρ' precisely in the same manner as probabilities do in classical statistical physics. But is this a truly convincing argument? As stressed by Bell [7] "if one were not actually on the lookout for probabilities,   the obvious interpretation of even ρ' would be that the system is in a state in which the various $|\Psi_m >$'s somehow *coexist*:

$$|\Psi_1 >< \Psi_1| \quad \text{and} \quad |\Psi_2 >< \Psi_2| \quad \text{and}$$

This is not at all a *probability* interpretation, in which the different terms are seen not as *coexisting* but as *alternatives*." In this quotation Bell, admittedly, makes no reference to ensembles. But it is clear that such a reference can hardly help. For indeed, when considering a measurement process, we are interested in what happens to one individual system S, associated with one individual instrument A. And as long as the completeness assumption is made it seems impossible to reconcile the thereby implicitly assumed uniformity of the ensemble described by ρ with the diversity of the elements of the proper mixture that, allegedly, ρ' is claimed to describe. To assume that it is just the unobservability of some physical quantities that *creates* this diversity just sounds absurd.

Difficulty 4b must somehow have been in the mind of most quantum physicists for a long time. It may well be what implicitly prompted Everett to put forward his relative state theory (see Section 12.1). It was made explicit in Ref. [2] and forcefully stressed by J. Bell as just noted. Surprisingly though, it seems to have attracted little or no explicit consideration on the part of any of the physicists who did put forward quantum measurement theories. For this reason it will not be further taken into account in the review of such theories given below. But of course it will be considered in later chapters (Chapter 15 in particular).

10.2 An Explicit Example

At this stage a preliminary question comes to mind. It has to do with the (idealized) process (10.2). Is at least *this* process free of conceptual difficulties? Can it take place in conformity with quantum mechanics? In other words, can some interaction Hamiltonian be built up that does entail such a process? Fortunately the answer is yes, as was first pointed out by von Neumann in his already quoted book and is now shown on an example.

Of course, in such examples maximal theoretical simplicity should be aimed at. Consequently let the system S on which the measurement is made just be a spin 1/2 particle and let the measured quantity L be the z component S_z of its spin, so that Eq. (10.1) reads (with $\hbar/2$ units)

$$L\,|\phi_m> = m\,|\phi_m> \qquad , m = \pm 1 \qquad (10.7)$$

Further, let ξ be the pointer coordinate and let the time evolution of the system take place in accordance with the scheme:

$$g_0(\xi) \otimes |\phi_m> \to g_m(\xi) \otimes |\phi_m> \qquad (10.8)$$

which, when the $g_m(\xi)$ are mutually orthogonal, is just a special realization of scheme (10.2). Here $g_0(\xi)$ and $g_m(\xi)$ are the (normalized) wave functions of A (more precisely· of the pointer of A) before and after the interaction respectively (for convenience the grossly oversimplified assumption of no degeneracy is made). The problem is to find an interaction Hamiltonian H' operating only during a short time interval (t', t'') and having the effect of inducing transition (10.8).

It is easily verified that a solution exists, in the form

$$H' = \beta(t)\, L \otimes (-i\frac{\partial}{\partial \xi}) \qquad (10.9)$$

$\beta(t)$ being equal to zero except during the short time interval (t', t''), during which it is so large that all the rest of the Hamiltonian can be neglected compared to H' Setting

$$\alpha(t) = \int_{t'}^{t} \beta(\tau)\, d\tau \qquad (10.10)$$

and

$$a_{\pm} = \pm\alpha(t'') \qquad (10.11)$$

it is then easily checked that during the measurement time interval the ket

$$|\psi_{\pm} > = g[\xi \mp \alpha(t)] \otimes |\phi_{\pm} > \qquad (10.12)$$

is, for *any* derivable function $g(\xi)$, a solution of the Schrödinger time-dependent equation. Hence scheme (10.8) does hold, with

$$g_{\pm}(\xi) = g(\xi - a_{\pm}) \qquad (10.13)$$

and if $g(\xi)$ is a peaked function choosen so that it is zero for $|\xi| > |\alpha(t'')|$ the condition that $g_{+}(\xi)$ be orthogonal to $g_{-}(\xi)$ is met, so that scheme (10.2) is indeed realized (if $g(\xi)$ is a Gaussian the condition is only approximately met but the overlap of g_{+} and g_{-} can be made as small as desired by choosing $g(\xi)$ with a small enough width).

Of course, this specific example of an idealized measurement process only shows that the scheme (10.2) can be realized. It sheds no light on how to remove difficulties 1 to 4b listed in Section 10.1. But still, the mere fact that process (10.2) can indeed take place in conformity with the Schrödinger equation, that is, without violating any of the uncontroversial rules of quantum mechanics, is already worth noticing, since a conception incompatible with this result is sometimes put forward and, what is even worse, is often—though wrongly—attributed to Bohr. Indeed, in semipopular books or textbooks it is often asserted that, according to Bohr, any measurement performed on an atomic system unavoidably disturbs this system. Clearly, process (10.2) is a counterexample to such a claim. It is true that, in situations more realistic than the model described, a *small* disturbance is most often unavoidable, due to the existence of additively conserved quantities, such as total angular momentum, shared by the system and the instrument (this was first pointed out by E. Wigner [8], see also Araki and Yanase [9] and Ref. [1]), but by choosing appropriate *macroscopic* instruments such disturbances can be brought below any prescribed level. And anyhow, to attribute such a conception to Bohr is a misinterpretation of his standpoint. His claim was different, for in fact it was that it is impossible even to (conceptually, so to speak) make a clearcut separation between the behavior of atomic objects and their interaction with the measuring instruments. So that he seems to have considered it meaningless even to *mention* the attributes of a quantum system as if they existed per se, and this led him to explicitly criticize the use of such notions as that of a "disturbance of a phenomenon by observation" (see Section 11.1 for more detail).

10.3 An Unsatisfactory Qualitative Suggestion

To remove the aforementioned difficulties many proposals have been made, some quantitative and some qualitative only. In this section, a particularly "sweeping" one of the latter type is considered and found unsatisfactory. Though seldom viewed as attractive by the physicists, this proposal is not unfrequently put forward by contemporary philosophers, and particularly by those in the opinion of whom the growing difficulty to communicate met with by experts in different scientific fields constitutes an indication that the very concept of universal laws is obsolete. These thinkers start by pointing out that the quantum mechanical formalism was developed quite specifically in order to deal with phenomena in the microscopic domain and that it is basically in this domain that it met with all the successes commonly referred to in order to prove its validity. It is therefore—they say—far from obvious that its basic notions are relevant at every level of the physical description of nature. To claim they are amounts—they say—to believe in a special brand of *reductionism*, consisting in arguing that quantum physics is the basic theory *because* everything is composed of atoms. But—these persons go on saying—such an argument is all the more shaky as quantum physics itself, via nonseparability and so on, sets rather drastic limitations on the adequacy of any physical description based on atomistic reductionism. The argument is further backed, in the mind of a number of these persons, by the observation that even chemistry—the science closest to quantum physics—has quite a number of laws and concepts that cannot be derived from those of physics in any convincing manner. The conclusion drawn is that the conceptual difficulties quantum mechanics meets have no substance since they all stem from considering that the quantum laws apply to the instruments in spite of these being macroscopic systems.

Although this argument may sound convincing to some ears, in fact it is not. First of all, the difficulties scientists in different fields have to communicate with one another are of course most easily understood by taking into account the complexity of contemporary scientific research, its rapid growth during the last decades, and the time-consuming efforts each scientist must devote to the essential task of mastering his or her own domain and reading the corresponding literature. The difficulties in question therefore do not testify in any way either for or against the unity-of-science conception; and most scientists know very well that, now as in the past, concepts developed in one branch of science not infrequently prove useful in some apparently quite distinct other branches, which testifies in favor of at least some kind of universality of basic laws.

In particular, a few essentially quantum notions, such as quantum tunneling, have been shown quite indispensable for explaining a number of experimental facts concerning some systems of macroscopic size and complexity and which therefore we have every reason to call "macroscopic" (see in particular the experiments reported by Leggett [10, 11]). This seems to contradict in a rather blatant way the theory that nature has two (or more) different domains, one of them being that of the microscopic systems and another one that of the macroscopic physical ones, the laws of one domain being exclusively valid in this domain. But if this theory is given up it is no more possible to uphold the argument described above, according to which, in situations where instruments are at stake, a possible violation of the quantum laws applied to systems of which these instruments are parts is admissible. More precisely, it seems that then the only way to salvage this view would be to consider that the macroscopic systems on which quantum effects are seen are not "really macroscopic." Now, a claim of this sort is fully consistent within a conception of physics in which this science is considered as describing, not reality "as it really is" but just collective appearances; for then a notion such as that of "macroscopic systems" can meaningfully be *defined* by referring to human possibilities. Indeed this is just the definition of these systems that will be taken up in Chapter 15, in connection with the notion of *empirical reality*. Here, however, the context is quite different, for the proponents of the foregoing argument do not question the view that the purpose of physical research is to describe reality as it really is. Under these conditions, the purely negative and human-centered definition of a whole class of important systems, considered here, is obviously unsatisfactory. In fact, it amounts to pushing forward the frontier of the microscopic domain while making it a matter of principle to carefully keep the instruments outside this frontier. This, however, is very much reminiscent of von Neumann's chain, and raises the same question: Where is the cut?

Finally, are the arguments based on nonseparability and on the relative autonomy of chemistry significant concerning the question at stake? In fact they are not. With regard to chemistry it, admittedly, is true (as Primas [12], for example, has very clearly shown) that strictly speaking the laws used in chemistry are not all deducible merely from the general laws of quantum physics. They follow from these laws *plus* some definite ways of "looking at things," ways that prompt us to disregard the nonseparability features that make the wave function of any molecule entangled with those of all the other molecules in the universe. But clearly the truth of this (quite important) remark does not have the effect of validating any claim to the effect that the quantum laws should be suspended in chemistry. Quite the contrary, we must go on considering that these laws are an essential part of any description of the complex systems chemistry deals with.

What has just been noted is also relevant to understanding why the existence of nonseparability does not constitute a valid argument against the notion upheld here that the quantum laws—though, admittedly, discovered in the atomic realm—are, in fact, universal. The point is that the scarcity of typically quantum phenomena in the macroscopic domain has strictly nothing to do with nonseparability. It is due to other causes, which are very well understood *within* the theory that quantum physics is universal. Correlatively, nonseparability, far from being a *corrective* to atomic (i.e., quantum) physics, is an inherent consequence of it. As a result, it is quite mistaken to claim that it is by disregarding nonseparability effects that we can apply the quantum laws on the small scale, that is, to atoms. For these reasons the plausibility of the hypothesis that the quantum laws and constants are universal—a plausibility due in particular to the past successes of the universality hypothesis applied to other laws and constants—is not in the least diminished by the existence of nonseparability.

To sum up, to the extent that the difficulties pointed out in Section 10.2 are real ones in quantum physics, they cannot just be wiped away by the qualitative general arguments examined in this section.

10.4 Still Another Unsatisfactory Qualitative Suggestion

Contrary to the previous one, the qualitative suggestion here referred to stemmed not from philosophers but from physicists. It may even be considered that it constitutes an at least plausible interpretation of what Heisenberg wrote in Chapter III of his book *Physics and Philosophy* [13] concerning the measurement process. The suggestion in question is based on the speculation that the difficulties 1–4b in Section 10.1 are all due to the fact that, in the there-described schematic measurement theory, the instrument A is, before the interaction, ascribed a pure case $|0, r >$ Being a macroscopic system, A is of course composed of a great many particles and has therefore an extremely large number of degrees of freedom. It is totally impossible to know, at one precise time, the values of all the parameters that would precisely specify the state of A. To this subjective uncertainty in the overall picture the interaction of S with A has but the effect of adding a further—this time "objective"—element of uncertainty. In other words the "probability function" (in Heisenberg's terms) controlling the various possible outcomes contains, along with the objective element due to the quantum nature of the process, an unavoidable, important subjective element due to our ignorance of the microscopic initial conditions The suggestion then is that when this combination of the two—subjective and objective— uncertainty "sources" is taken into account, the uncertainty in the measurement result should be explained in a way in which difficulties 1–4b simply cannot arise

The fact on which this argument is based can certainly not be denied We do not know precisely the initial state of the instrument Indeed, in view of the fact that instrument A previously interacted with other systems it could even be claimed (see Chapter 8) that describing the "initial state" of A by means of a state vector is definitely unrealistic Unquestionably, what should be done in the spirit of Heisenberg's remark is to consider a whole ensemble of $S + A$ systems and describe the initial "state" of A by means of a statistical operator

If this is done, then of course the ensemble of the $S + A$ systems is also described by a statistical operator, the time evolution of which during the measurement time interval is described by known equations But then the question arises Are the difficulties of Section 10 1 really removed? In other words Is the statistical operator describing the $S + A$ system after the interaction is over, of such a form that the ensemble E it describes can be interpreted as one containing, in proportions $|a_m|^2$, $S + A$ systems the A part of which have $G = g_m$?

Unfortunately the answer is no This is by no means obvious, for the ensemble E is of course represented by a statistical operator that is not a projector and when this is the case there exists, as we know (Section 8 3), not just one but many (an infinity of) physically distinct proper mixtures that are represented by it At first sight it could be hoped that among all these proper mixtures there should be one having the required properties, if not exactly, at least approximately But I showed in Ref [14] that in fact this is not the case There would be no point in transcribing here the calculations leading to this result, for they are already described at length in Ref [1], where the ultimate failure of various other attempts at refining the approach sketched in this section is also shown (for a critical review of other, more recent, measurement theories see also Ref [2 , addendum])

10.5 Superselection Rules (They Yield No Direct Solution)

In Sections 3.1 and 7.1 superselection rules were mentioned, but merely in a somewhat cursory way. Since a conjecture has sometimes been expressed that this notion could yield quite a straightforward clue to the measurement problem this is a proper place for explaining in somewhat more detail the superselection rules. This will make it possible to (a) describe the reasons why at first sight the notion in question seems to lead to quite a simple solution of the problem and (b) show why, after all (as has been known for a long time by a majority of theorists) this, strictly speaking, is a lure (although it could inspire more elaborate developments[2]).

As already mentioned, while in quantum mechanics the dynamical, observable quantities are quite generally associated to Hermitean operators the converse is not true: it is impossible to consider that *all* the Hermitean operators operating in the Hilbert space of a certain type of system correspond to observable quantities. It may then happen that the Hermitean operator associated with some definite observable Q commutes with all the operators associated with observable quantities, without being merely a multiple of the unit operator. Such an observable Q is called a *superselection charge*. Let its eigenvalue equation be written:

$$Q|k,r> = q_k\ |k,r> \tag{10.14}$$

With $\{|k,r>\}$ taken as a basis, a general matrix element of the matrix corresponding to an observable described by Hermitean operator A takes the form

$$A_{i,s;j,t} = <i,s|A|j,t> \tag{10.15}$$

---

2. Note added in proofs. See e.g., M. Namiki and S. Pascazio, *Quantum Theory of Measurement Based on the Many-Hilbert-Space Approach*, Physics Reports **232**, p. 303, September 1993 [within the realm of the *empirical reality* approach (see Chapter 15) this nonrelativistic theory may be considered as removing difficulty 4_a].

and with the help of the decomposition (3.11) of the unit operator the general matrix element of the product $Q\,A$ may be written

$$
\begin{aligned}
< \iota, s|Q\,A|\jmath, t > &= \sum_{k,r} < \iota, s|Q|k, r > < k, r|A|\jmath, t >\\
&= \sum_{k,r} \delta_{i,k}\delta_{s,r}q_k < k, r|A|\jmath, t >\\
&= q_i\ < \iota, s|A|\jmath, t >
\end{aligned}
\qquad (10.16a)
$$

similarly·

$$
< \iota, s|A\,Q|\jmath, t >= q_j\ < \iota, s|A|\jmath, t > \qquad (10.16b)
$$

so that, since, by assumption, $Q\,A - A\,Q = 0$:

$$
(q_i - q_j)\ < \iota, s|A|\jmath, t >= 0 \qquad (10.17)
$$

This shows that the matrix elements of all the *observable* physical quantities taken between eigenvectors of Q corresponding to different eigenvalues of Q vanish. In other words, in the $\{|k, r >\}$ basis all the nonvanishing matrix elements of the observable physical quantities lie within squares—called *superselection sectors*—strung together along the main diagonal, each square corresponding to one definite q_k value, as:

The reason why, at first sight, the notion of *superselection charges* seems to provide a way of solving the measurement riddle can now easily be explained. With the help of formula (3.12) any Hermitean operator A representing an observable can be written in the form

$$
A = \sum_{\alpha} a_\alpha P_\alpha \qquad (10.18)
$$

where the a_α are the eigenvalues of A and

$$P_\alpha = \sum_v |\alpha, v > < \alpha, v| \qquad (10\ 19)$$

(here $|\alpha, v >$ are the eigenkets corresponding to eigenvalue a_α) is the projector on the corresponding subspace α If Q is a superselection charge then the commutator $[Q, A] \equiv Q A - A Q$ is equal to zero for *any* A, hence in particular for *any* choice of the a_α, which implies that $[Q, P_\alpha] = 0$ Then, if, just before a measurement of A, the state of the system is $|\psi >$, the probability that this measurement yields outcome a_α is

$$p_\alpha = \mathrm{Tr}[|\psi > < \psi| P_\alpha] = < \psi |P_\alpha|\psi > \qquad (10\ 20)$$

and if $|\psi >$ is expanded on the $\{|k, r >\}$ basis according to

$$|\psi >= \sum_{i,s} c_{i,s}|i, s > \qquad (10\ 21)$$

then

$$p_\alpha = \sum_{i,s,j,t} c_{i,s}^* c_{j,t} < i, s|P_\alpha|j, t > \qquad (10\ 22)$$

But since $[Q, P_\alpha] = 0$, Eq (10 17) also applies with P_α substituted for A, and it shows that all the terms in (10 22) that correspond to $i \neq j$ vanish Within any subspace H_i of the relevant Hilbert space corresponding to a given value of i, that is, of the superselection charge, it is then an easy matter to pick up a basis in which the matrix corresponding to P_α is diagonal (this new basis, $\{|i, r >\}$, is obtained by diagonalizing the matrix whose general element is $< i, s|P_\alpha|i, t >$) Expressed on the basis of H constituted by the union of these new H_i bases, all the cross-terms in Eq (10 22) vanish and p_α takes the form

$$p_\alpha = \sum_{i,r} |c_{i,r}'|^2 < i, r|P_\alpha|i, r > \qquad (10\ 23)$$

This result shows that, for *any* observable A whatsoever, the probabilities of getting this or that outcome are identical to those that would obtain if we had to do, not with a pure case, described by a state vector, but with a proper mixture in proportions $|c_{i,r}'|^2$ of systems in states $|i, r >$ Since the stumbling block, concerning a realistic interpretation of the theory, is constituted by the very presence of such cross-terms, prima facie their vanishing seems to yield the clue to solving the measurement riddle. The idea would just simply be to postulate that in such highly complex systems as the ones, labeled $S + A$ previously, which involve a macroscopic instrument, many Hermitean operators correspond in fact to no observable at all (i.e., are physically meaningless) allowing for superselection charges to be present, and that the pointer position is just one of the latter.

A priori the idea looks quite attractive since of course it is true that for all practical purposes a great majority of all the conceivable dynamical physical quantities belonging to such systems must be considered as lying outside the realm of what can possibly be measured. However, the superselection charges notion does not refer to measurements that in practice cannot be performed. It is based, as we saw, on the assumption that some Hermitean operators can strictly not be interpreted as referring to quantities that would be observable, even in principle—that is, they are mere mathematical abstractions that correspond to no physical property of the systems—and that all the other Hermitean operators commute with the superselection charges. The assumption that the pointer coordinate, G, really *is* a superselection charge therefore implies in particular that the Hamiltonian of the $S + A$ system either corresponds to no physical property of this system *or* commutes with G. Since it is quite a basic rule of quantum mechanics that the Hamiltonian corresponds to the energy, which has always been considered as a most important physical property, it seems that the first of these two options would imply quite a momentous change in our understanding of the quantum mechanical formalism, and it is far from clear that such a change could be carried through in a consistent way. We are thus left with the idea that the pointer coordinate G commutes with the Hamiltonian. But, as is well known (and is easily verified), a dynamical quantity that commutes with the Hamiltonian is a constant of the motion. The idea thus implies that, during the measurement process, G should remain the same (or, in other words, that the time evolution of the $S + A$ system leaves it in the one superselection sector, q_i say, where it was at the beginning). Quite obviously this is incompatible with the role of G as the instrument coordinate since if the pointer does not move during the process the measurement cannot yield outcomes differing from one another.

This (negative) result is obviously quite general. It is quite independent of the precise nature of the instrument A and in particular of whether or not the environment of A is considered as being itself a part of A. It is true that the interesting developments reviewed in the next section and which are based on the idea of taking the role of the environment quite seriously into account are sometimes described in a language in which the expression *superselection rules* is made use of. But, strictly speaking, the notion that is there referred to under this name is not the same as the one that the name in question customarily refers to and that has been described above.

10.6 The Role of Environment

A new and promising general line of approach to measurement appeared in the early seventies when physicists such as H. Zeh made the most important remark [15] (to which we shall come back in Chapter 12) that, at the quantum level, macroscopic systems never can be considered isolated; for it was immediately clear that this opened new possibilities concerning the measurement problem. Zeh's essential point is that the quantum energy levels of even quite tiny macroscopic systems are extremely close to one another. From this it follows that transitions between nearby levels can be induced by extremely weak fields. So weak indeed that, as detailed calculations show, even a small dust particle "lost" in interstellar space cannot be considered as remaining quite isolated within any appreciable time interval. In consequence, insofar as theoretical attempts such as those mentioned in the foregoing section picture the $S + A$ systems as if they were isolated from the outside world, such attempts cannot be considered appropriate. They should be replaced by theories explicitly taking into account the interaction of the (macroscopic) instrument A with its environment E. And, hopefully, in these new theories the difficulties 1–4b in Section 10.1 will take more tractable forms. Perhaps, even, they will vanish.

A purposely simplified example that has the advantage of clearly exhibiting both the attractive features and the remaining weak points of such theories is Zurek's 1982 "bit-by-bit" model

[16]. In it, as here in Section 10.2 , the quantity to be measured is the z component $L = s_z$ of the spin of a spin 1/2 particle S. As for the "pointer coordinate" G, for maximal simplicity it is assumed (in an avowedly "unrealistic" way) that it is a dichotomic variable, with *nondegenerate* eigenvalues g_+ and g_- The measurement process is then described by means of an interaction Hamiltonian $H^{A,S}$ operating, as in Section 10.2, for a short time only, and whose effect during this time is symbolically described by

$$|+ > \otimes |V_+ > \to |+ > \otimes |U_+ > \qquad (10.24a)$$

$$|- > \otimes |V_+ > \to |- > \otimes |U_- > \qquad (10.24b)$$

where $|+ >$ ($|- >$) is the eigenket of s_z corresponding to eigenvalue $+1$ (-1) in units $\hbar/2$, $|U_+ >$ ($|U_- >$) is the eigenket corresponding to g_+ (g_-) and

$$|V_\pm >= 2^{-1/2}(|U_+ > \pm |U_- >) \qquad (10.25)$$

(An $H^{A,S}$ having such properties is quite easily constructed, essentially on the same lines as H' in Section 10.2.) In Eqs. (10.24) $|V_+ >$ clearly represents the initial state of the "pointer" while $|U_+ >$ and $|U_- >$ describe its two possible final states, corresponding respectively to $s_z = +1$ and $s_z = -1$. Note that the pointer may be thought of as being itself a spin 1/2 particle and G its spin component S_z, in which case its initial state is just the one corresponding to its spin component S_x, along Ox being definite and equal to $+1$ (still in $\hbar/2$ units).

Of course, at this stage the model involves the difficulties mentioned in Section 10.1, and in particular difficulties 2 and 4a. To make these explicit it is sufficient to assume that initially system S is prepared in a state

$$|\phi_i >= a|+ > +b |- > \qquad (10.26)$$

(that is, in an eigenstate of a component of its spin other than s_z) where a and b are two complex numbers with

$$|a|^2 + |b|^2 = 1. \qquad (10.27)$$

The linearity of the Schrödinger equation then entails that

$$|\phi_i > \otimes |V_+ > \rightarrow |\Phi_f > \tag{10.28}$$

with

$$|\Phi_f >= a|+ > \otimes |U_+ > +b|- > \otimes |U_- > \tag{10.29}$$

so that $|\Phi_f >$ has the general form of the right-hand side of Eq. (10.4), which, of course, gives rise to difficulty 4a. More precisely, the mean value on state $|\Phi_f >$ of an observable such as the product $s_x\, S_x$ then involves cross-terms with coefficients a^*b and b^*a that differentiate it from the mean value this same observable would have if the final ensemble were a mixture of states $|+ > \otimes |U_+ >$ and $|- > \otimes |U_- >$ in proportions $|a|^2$ and $|b|^2$ respectively. At this stage, in other words, the model exhibits difficulty 4a in a very explicit way.

For the purpose of removing such defects, an interaction between A and the environment E is introduced in the model. For computational simplicity the environment in question is assumed to be composed of a large number N of spins $1/2$ that do not interact with one another and start interacting with A just at the time $t = 0$ when the S, A interaction has ceased; and this A, E interaction is described by the interaction Hamiltonian

$$H^{A,E} = \sum_k H_k^{A,E} \tag{10.30}$$

with

$$
\begin{aligned}
H_k^{A,E} &= g_k(|U_+ >< U_+| - |U_- >< U_-|) \otimes (|u_+ >< u_+| \\
&\quad - |u_- >< u_-|)_k \otimes \Pi_{j \ne k} 1_j \\
&\equiv g_k S_z \otimes s_z^{(k)} \otimes \Pi_{j \ne k} 1_j
\end{aligned} \tag{10.31}
$$

where $|u_+ >_k$ and $|u_- >_k$ are the eigenvectors of the z component $s_z^{(k)}$ of the kth environmental spin. For simplicity it is also assumed that the free Hamiltonians of $S, A,$ and E are all zero.

If, at time $t = 0$ the state of the combined $S + A + E$ system is

$$\Psi(0) = |\Phi_f > \otimes \Pi_{k=1}^N (\alpha_k |u_+ >_k + \beta_k |u_- >_k) \tag{10.32}$$

it is then immediately shown that at time t this state becomes

$$
\begin{aligned}
\Psi(t) = a|s_+ > &\otimes \Pi_k [\alpha_k \exp(ig_k t)|u_+>_k \\
&+ \beta_k \exp(-ig_k t)|u_->_k] \\
+ b|s_- > &\otimes \Pi_k [\alpha_k \exp(-ig_k t)|u_+>_k \\
&+ \beta_k \exp(ig_k t)|u_->_k]
\end{aligned}
$$
(10.33)

where

$$
|s_\pm > = |\pm > \otimes |U_\pm >
$$
(10.34)

The statistical operator of the S, A system—describing an *improper mixture* of such systems—is then obtained by taking the partial trace of $|\Psi(t) > < \Psi(t)|$ over the environment Hilbert space and turns out to be

$$
\begin{aligned}
\rho = |a|^2 |s_+ &>< s_+| + |b|^2 |s_- >< s_-| \\
&+ z(t) ab^* |s_+ >< s_-| \\
&+ z^*(t) a^* b |s_- >< s_+|
\end{aligned}
$$
(10.35)

where

$$
z(t) = \Pi_{k=1}^N [\cos 2g_k t + i(|\alpha_k|^2 - |\beta_k|^2) \sin 2g_k t] \quad (10.36)
$$

With N large and the coupling constants g_k chosen at random, $z(t)$ soon becomes quite small and remains small a very long time thereafter (indeed a time comparable to the Poincaré cycle's duration). This effect, which has come to be known as *decoherence*, means that a small time τ after the system-apparatus interaction has ceased the nondiagonal matrix elements of ρ have themselves become (and remain) very small—so small as to be negligible for any conceivable practical purpose. Under such conditions the mean value, on the considered ensemble, of any observable belonging to the $S + A$ system is easily shown to be the same as if this ensemble were a proper mixture, composed, in proportions $|a|^2$ and $|b|^2$, of two subensembles in both of which s_z and S_z would have definite values.

The fact that the two ensembles in question are characterized by definite values of S_z rather than of other "instrument" observables is of course due to the circumstance that $H^{A,E}$ commutes with S_z, and not with these other observables. As a result, it may be considered that taking the interaction between the instrument and its environment into account has removed difficulty 2. At least, this standpoint may be taken up if the distinction between instrument and environment is taken for granted, which, actually, still raises questions. In a way, the model may also be said to meet the demands formulated in Section 10.1 in order to remove difficulty 4a. But it meets them only concerning observables belonging to S, to A, or undividedly to both. Of course, there also exist Hermitean operators that are not restricted to operate either on the Hilbert space $H^{S,A} = H^S \otimes H^A$ of the composite $S + A$ system or only on that, H^E, of the environment, but act on the full $H^{S,A,E} = H^S \otimes H^A \otimes H^E$ Hilbert space corresponding to the overall physical system under study. If some such operators correspond to observables the problem must be reconsidered.

Now, it just happens that in the model here under study operators do exist that are good candidates in this respect. The example I gave [17] is

$$M = s_x \otimes S_x \otimes \Pi_{k=1}^N \otimes s_x^{(k)} \qquad (10.37)$$

To show this, let us remark first that when an operator C is a tensor product of two operators A and B corresponding to observables defined on two different subsystems S_1, S_2 of a larger system, the quantum mechanical mean value $< \psi | C | \psi >$ of C on any $| \psi >$ describing an ensemble of such larger systems is equal, as easily checked, to the value obtained by taking the mean on this ensemble of the products of the outcomes of measurements of A and B, separately performed on each pair S_1, S_2. This shows that C is measurable whenever A and B are and of course this result can be extended to products of an arbitrary number of similarly defined observables. In the case under study it is true that an actual, simultaneous measurement of s_x, S_x, and all the $s_x^{(k)}$ would be "fantastically" difficult to perform as soon as N is large. But this is just a practical difficulty. In theory such a simultaneous measurement is conceivable and in view of what has just been shown it must then be acknowledged that, in theory at least, an experimental procedure exists for measuring M, so that it is difficult not to consider M as an observable.

The reason why this remark is relevant is that the mean value of M on an ensemble described by $\Psi(t)$ [Eq. (10.33)] is, at any time t, altogether different from what it would be if this ensemble were one of those in which each "instrument variable" S_z and each "measured quantity" s_z has some definite value. This is because, as an explicit calculation easily shows, M commutes with any $H_k^{A,E}$ and therefore with the total Hamiltonian, so that its mean value $< M >$ remains the same as what it was at time $t = 0$, namely

$$< M >= (a^*b + b^*a) \, \Pi_k(\alpha_k^* \beta_k + \beta_k^* \alpha_k) \qquad (10.38)$$

whereas on each one of the two ensembles of systems on which s_z and S_z have definite +1 or −1 values $< M >$ is equal to zero. If, technically, an experimentalist were sufficiently well equipped as to be able to apply the above-sketched experimental procedure and if we assume—or take it for granted—that he would be free to apply it, then, if he did (and assuming the experimental quantum mechanical predictions are correct) he would get results that would falsify the commonsense view according to which macroscopic systems always have macroscopically definite positions. M is an example of the observables we called "sensitive" in Section 1.6.

Should it then be said that the idea examined here—that of taking the role of the environment into account in connection with the measurement problem—is of no value whatsoever? Certainly not, for there exists a subtler way of thinking, in which the said idea opens very interesting vistas.

The point is that, in the very sketchy model described, space did not play any role. Indeed it was not even mentioned. As compared to actually existing states of affairs this of course is a gross oversimplification. Physical environments do extend in space and this feature is one of the first that a more realistic model should take into account. For example, the above model could be refined by assuming that the kth environmental spin lies at a distance L_k proportional to k from where the measurement takes place and by making some appropriate corresponding changes in the interaction Hamiltonian. Then a statistical operator ρ_p could be defined by partial tracing over all the environmental spin Hilbert spaces corresponding to $k > p$, where p is a positive integer smaller than N. This ρ_p describes a composite system that is "localized" in the sense that it does not extend in space farther than L_p (more precisely, it describes, of course, an ensemble of such systems). And it is trivially verified that ρ_p can be written according to formula (10.35), with the only difference that $z(t)$, instead of being expressed by Eq. (10.36) is expressed by a similar formula, in which the product only runs from $k = p + 1$ to N but which still entails for z the property of soon becoming extremely small. Under these conditions it is clear that concerning any observable *defined on this composite "localized" system* (let such observables also be called "localized") the predictions from $\Psi(t)$ and those from a mixture, in proportions $|a|^2$ and $|b|^2$, of systems having definite s_z and S_z values are, for all practical purposes, identical. Hence difficulty 4a is essentially removed in such models provided that all the entities that ordinary quantum mechanics refers to as "observables" and that are not "localized" in the above sense should be considered "by fiat" as not being *truly* observable (strictly speaking, the fact that human beings are localized does not suffice to render these quantities unobservable since composite measurements can be performed by several distant observers who later gather to compare notes).

In their 1984 paper [18], to which we shall turn again in Chapter 12, Joos and Zeh suggested the idea that local, classical properties (or at least appearances thereof) have their origin in the nonlocal character of the quantum state contrasted with the local character of conscious beings. "The interference terms"— they wrote—"still exist but they are not *there*!", and further: "the collapse could then be based on the assumption on how a nonlocal reality is experienced subjectively by a local observer." The foregoing remarks may, in a way, be considered as qualitative illustrations of this quite far-reaching idea. We observe here an interesting convergence of views between inferences drawn from quantum measurement theory and conclusions that were drawn mainly from the Bell theorems [1,19] and consisting in distinguishing between a reality per se that can be described only as nonlocal ("independent reality" in my terminology) and what is experienced by a community of essentially local observers (empirical reality). We shall come back to such questions.

• *Remark 1.* The objection centered on Eq. (10.38) to the Zurek-like attempt at reconciling quantum mechanics with the idea that a macroscopic system always has a definite macroscopically defined position rests in fact on two notions. One is that the practical difficulty of performing such and such a measurement does not have to be taken into account when discussing basic questions: concerning these, what is significant is only whether or not some measuring procedure can be *defined*. The other one is that the experimentalist is free to choose at whim what observable to measure. Neither one of these two notions is logically necessary. As for the first one we shall indeed consider in later chapters some possibilities of slightly watering it down. As for the second one, it was already pointed out (in Chapter 1) that it implies in fact a qualitative difference in the ways of considering future (viewed as open) and past (viewed as closed), and this difference may be seen as being unwarrantedly centered on man. Hence their rejection is conceivable. But it seems reasonable to require that if either one is rejected, this should be mentioned instead of being kept implicit.

• *Remark 2:* Unfortunately, however, this demand is not always met. While most of the physicists who developed the theory reviewed here—especially Zeh and Joos—are very careful to use words and phrases that faithfully reflect the epistemological complexity of the matter, there sometimes are exceptions. In his 1982 paper [16], for example, Zurek wrote that when the nondiagonal matrix elements of the density matrix of the $S + A$ system are small "the probabilities on the diagonal of the density matrix are there *because of our* (i.e., the observer's) *ignorance about the outcome of the measurement. It is as yet unknown to us but nevertheless it is definite.*" Elsewhere in the same article he similarly wrote "the interaction with the environment forces the system *to be in one of the eigenstates of the pointer observable rather than in some arbitrary superposition of eigenstates.*" More recently [20] he again wrote that "open quantum systems *are forced into states* described by localized wave-packets." [3]

We would most probably be mistaken if we believed that, on the substance of the matter at hand, there are real differences of opinion between Joos and Zeh on the one hand and the cited author on the other hand. More precisely, when all is said and done there is no serious ground for doubting that the said author subscribes to at least the main lines of the well-balanced standpoint of Joos and Zeh. It is true nevertheless that

---

3. Some controversy ensued. The texts of seven letters commenting on the last quoted article can be found in the April 1993 issue of *Physics Today* (pp. 84–90), together with Zurek's reply.

to the mind of innocent laymen the sentences quoted, likely to be taken at face value, might well convey the quite erroneous idea that the theory does show the pointer (and, more generally, macroscopic systems) to really *be* in such and such macrostates, in the absolute, ontological sense normally attributed to the verb *to be* That the "environmental measurement theory" here under study does not really achieve that much is of course acknowledged by all the physicists who work on the subject [4] What seems most likely is that some of them are considering this aspect of the matter as being "merely philosophical and therefore not worth mentioning " While such a general position is defensible, nay even healthy, in most cases of normal scientific practice, when what is essentially at stake is the agreement between theoretical prediction and experiment, its consistency is questionable in problems in which, as here, this agreement is not what is under study, and what is looked for is basically an understanding of the actual meaning of the formalism

10.7 "Epsilonology" Is of No Help

Even in classical physics completely sharp factual statements cannot in general be made The values of the dynamical variables cannot be given with an infinite number of decimals, no physical system can be said to be "perfectly" isolated, and so on Hence small differences necessarily exist between a system as it really is and its description by means of simple statements Moreover, elaborate experimental setups can in general be conceived of that, if used as measurement devices, would enhance these small differences A trivial example is when we have to do with two localized systems whose mutual distance is large We are then entitled to say that, as far as gravitation is concerned, they may be considered as isolated from each other "to a very good approximation," notwithstanding the fact that we can always conceive of gravitation-measuring devices sensitive enough to detect the small lack of isolation Another,

---

3 *(continued)* Note added *on reprinting*: More recently Zurek responded again (and most positively!) to these critics, by putting forward a new, subtle and thought-provoking world view: W H Zurek, *Phil Transactions of the Royal Society, London* A (1998) 356, 1793–1821

4 This is why the notion of "environment-induced superselection rules" [16] was gradually replaced by that of *decoherence* (see Section 12 3) On such questions the exchange of views between Hepp [21] and Bell [22], is worth looking at

also quite trivial example is the case when we have to do with two strongly peaked, Gaussian-shaped wave packets, the centers of which are separated by a distance much greater than their widths. We are then entitled to say that, to a very good approximation, these wave packets may be considered as separated, notwithstanding the fact that we can always conceive of measuring devices contrived so as to be especially sensitive to amplitudes in the region where the two Gaussian tails overlap. What is read on such a device is an "observable" that takes up very different values in the case under consideration and in that in which the wave packets are *strictly* separated (i.e., when they lie at an infinite distance from each other or when, instead of being Gaussian, they have finite and nonoverlapping supports).

The possibility—well known in classical physics—of conceiving of such devices and thinking of the observables they define is sometimes referred to in attempts at minimizing, or even negating the conceptual difference between classical and quantum mechanics in the domain of measurements. Basically the argument consists first in acknowledging the existence, in the quantum case, of "sensitive" observables such as M [Eq. (10.37)]—observables concerning which the simple sharp statement "the $S + A + E$ ensemble is a mixture of two subensembles" is grossly false—but in pointing out that these observables are extremely difficult, if not impossible, to measure. It is then argued that if we go as far as taking into consideration "far-fetched" observables such as these we must as well take seriously into account the fact that also in classical physics observables can be defined, as we just saw, the measurement of which would falsify elementary "sharp statements" (such as: "The two wave packets are separated") some of which are so useful—and even essential—both in ordinary and in scientific activity. The argument then continues by observing that in classical physics the existence of such far-fetched observables is not viewed as basically invalidating our usual statements about systems being isolated or separated from one another "to a very good approximation." Why then, the argument goes on, should the same standpoint not be taken in quantum physics? This is supposed to mean that the existence of observables such as M should not prevent us from asserting that, in the considered example, the final $S + A + E$ ensemble is composed of subensembles the elements of which have definite s_z and S_z values "to a very good approximation."

As a matter of fact, all the stages in the just-reported argument are quite correct except the last one. The inference it draws concerning the near identity of the classical and the quantum case on this issue is invalid. This is because of one important difference between them, which must now be pointed out.

The difference in question is simply that in classical physics a sharp statement bearing on individual systems and falsifiable by far-fetched measurements only, can in general be replaced by an unsharp (or less sharp) but strictly true one, not falsifiable by any measurement bearing on these same systems; whereas this is not always the case in quantum measurement theory. In the foregoing, "unsharp" means "expressible by an inequality rather than an equality."

More explicitly, the point is as follows. In the example of the two distant, massive objects, the sharp but, as noted, falsifiable statement that these systems are isolated from each other can be replaced by the unsharp but strictly true statement that the gravitational potential energy between them is smaller than such and such a value. There is no observable, far-fetched or not, the measurement of which would falsify this last statement. Similarly, in the classical example of two Gaussian wave packets, the sharp but falsifiable statement that they are separated can be replaced by the unsharp but strictly true statement that the ratios of their "scalar product" (in the familiar sense used in quantum physics) to each one of their "norm squared" is smaller than such and such a value. On the contrary, in the quantum case here under study, and as long as—just as in the classical case—individual systems are considered, such a replacement is impossible. Of course it *is* possible concerning purely mathematical abstractions such as wave functions or kets. Admittedly, in the model considered in Section 10.6 the sharp but falsifiable statement that the final $S + A + E$ ensemble is composed of the two ensembles specified above can be replaced by the abstract, unsharp but strictly true, statement that the ket describing the said ensemble is a quantum superposition of two kets, each one corresponding to definite values of s_z and S_z, and the scalar product of the corresponding environmental kets is small. But can the statement in question in its turn be replaced by some strictly true unsharp one referring to *the individual, physical systems themselves,* so that the inequality expressing it should bear on a physical description of these physical systems? This approach would amount to assuming that the kets describing each subensemble, while not being eigenkets of s_z and S_z in a strict sense, are good approximations to such eigenkets. At time t one subensemble, composed of approximately $N|a|^2$ systems, would then be described, not by

$$|+> \otimes |U_+> \otimes |\chi_+> \qquad (10.39)$$

but by kets such as

$$(1 - \varepsilon)|+> \otimes |U_+> \otimes |\chi'_+>$$
$$+\varepsilon|-> \otimes |U_-> \otimes |\chi'_-> \qquad (10.40)$$

while the other one, composed of approximately $N|b|^2$ systems, would be described, not by

$$|-> \otimes |U_-> \otimes |\chi_-> \qquad (10.41)$$

but by

$$(1 - \eta)|-> \otimes |U_-> \otimes |\chi''_->$$
$$+\eta|+> \otimes |U_+> \otimes |\chi''_+> \qquad (10.42)$$

with ε and η both much smaller than unity (here $|\chi_\pm>$ stand for the Π_k products in the first and second line of Eq. (10.33) and $|\chi'_\pm>, |\chi''_\pm>$ are freely adaptable kets in the environment Hilbert space). The hope of being able to replace the sharp but falsified statement that "the final ensemble is composed of two subensembles the elements of each of which are endowed with definite s_z and S_z values" by an unsharp but nonfalsifiable one could then materialize if, and—apparently—only if, values of ε, η and kets $|\chi'_\pm>$ and $|\chi''_\pm>$ could be found, such that the assertion "the final ensemble is a mixture of the two above-defined subensemble" would not be falsifiable at all. But this is impossible, since the expression for the mean value of the hitherto considered observable M on *any such* ensemble cannot involve any appreciable term in $a^*b + b^*a$ and cannot therefore be identical to expression (10.38).

In private talks theorists sometimes assert that the main difficulty in reconciling a universally valid conception of conventional quantum theory with macroscopic realism stems from our predecessors having demanded too much from "realism" when they requested the instrument pointers to lie in exactly one graduation interval. They consider that the contemporary development of the statistical viewpoint renders such an exactness requirement questionable and they somehow feel that this should give a clue to the measurement riddle. The content of this section shows that such a hope is a delusion. In other words it shows that the conceptual difference existing between classical and quantum mechanics in the domain of measurement is quite a basic one, not to be wiped off by such simple means as having recourse to "epsilons."

10.8 Does the Heisenberg Picture Help?

At this stage it is appropriate to inquire whether, by any chance, the difficulties in understanding the measurement process in realistic terms do not just come from making use of the Schrödinger instead of the Heisenberg picture. As shown in Chapters 3 and 5, state-vector reduction is associated with measurement in both pictures. But in the Schrödinger picture the state vector, or more generally the state operator, has a dual role. On the one hand it evolves in time (according to the time-dependent Schrödinger equation) as the physical system also does, and is therefore quite naturally considered as representing the "physical reality"— whatever this expression means—of the system. On the other hand it also describes our *knowledge* of the system: and it is as such that it has to be abruptly "reduced" ("reduction of the wave packet," Rules 10 or 10a) when a measurement performed on the system changes our information about it. It is prima facie conceivable that this dual role should be at the origin of the difficulties 1–4b listed in Section 10.1. In the Heisenberg picture—as stressed in particular, by Unruh [23]—the two roles in question are, on the contrary, neatly separated and allotted to quite different entities. The operators are time-dependent—in fact they obey the same equations as the corresponding dynamical quantities of classical physics—and they can therefore be thought of as *being* the dynamical physical quantities. As for the state vector, it is time-independent as long as no measurement is made and

changes abruptly when one is performed, which is quite consonant with the idea that its role is but to describe *our knowledge* of
the physical system. By thus separating the behavior of the physical world from the increase in our knowledge, the Heisenberg
picture might well be a good candidate as a framework within
which the difficulties in question may be hoped to disappear.

But is this actually the case? To study the matter it is appropriate to make use of a simplified model (d'Espagnat [24])
in which, as in Section 10.2, the time evolution of the entities
describing both the measured system and the instrument can
be calculated exactly. Such a scheme can be achieved by starting from an explicitly soluble measurement model described,
as is generally the case, within the Schrödinger picture and by
"translating" it into the Heisenberg picture. One very simple
such model was put forward by A. Peres [25]. It consists in the
measurement of the z component S_z of the spin of a spin $1/2$
particle S by means of an instrument A with pointer coordinate
G. The eigenvalue equations

$$S_z|\pm> = \pm|\pm> \tag{10.43}$$

$$G|F_\pm> = g_\pm|F_\pm> \tag{10.44}$$

serve to define the notations. Moreover, let the initial value,
at time 0, of G be g_+ and let the measurement process consist in the fact that if $S_z = +1$ (in units $\hbar/2$) the value of G remains unchanged, whereas if $S_z = -1$ G is flipped to $G = g_-$ An
interaction Hamiltonian that, according to the time-dependent
Schrödinger equation, accounts exactly for such a process is

$$H' = \beta(t)|-><-| \otimes P \tag{10.45}$$

where $\beta(t)$ is a smooth function of time with compact support
$(0, t_1)$ obeying

$$\int_0^{t_1} \beta(t)\, dt = \pi \tag{10.46}$$

and where P is the projector

$$P = \frac{1}{2}\,(|F_+> -|F_->)(< F_+|- < F_-|) \tag{10.47}$$

for indeed with

$$\Pi = |->< -| \otimes P \qquad (10.48)$$

the time evolution operator between times 0 and t_1

$$U = U(t_1, 0) \qquad (10.49)$$

takes [because of equation (5.11)] the form

$$U = \exp(-i\pi\Pi) = 1 - \Pi + \Pi\exp(-i\pi) = 1 - 2\Pi$$
$$= 1 - 2|->< -| \otimes P \qquad (10.50)$$

and it is easily checked that with

$$|\psi_\pm(0) > = |\pm > \otimes|F_+ > \qquad (10.51)$$
$$|\psi_\pm(t_1) > \equiv U|\psi_\pm(0) >= |\pm > \otimes|F_\pm > \qquad (10.52)$$

as requested.

On this model difficulty 4a of Section 10.2 arises of course as soon as we consider shooting on instrument A a particle with a spin pointing toward some direction other than z—that is, whose spin is described by a linear superposition

$$a|+ > +b|- > \qquad (10.53)$$

of states $|+ >$ and $|- >$, for then the linearity of the Schrödinger equation entails that the state vector of $S + A$ at time t_1 must consist of a linear superposition with the same coefficients a and b of the final states $|\psi_+(t_1) >$ and $|\psi_-(t_1) >$, and such a state vector, not being a tensor product of a ket belonging to the Hilbert space of S with a ket belonging to the Hilbert space of A, has no physical interpretation. This is the well-known *entanglement* riddle.

Does the Heisenberg picture remove this riddle? To try and answer this question let us "translate" the model from the Schrödinger to the Heisenberg picture. This is easily done since U is known. The time dependence of the Heisenberg operators is given by the general rule (5.18). In the specific case under study, the relevant operators, which operate within the tensor product of the spin and instrument Hilbert spaces, are, at time zero

$$S_z(0) = s_z \otimes 1 \equiv (K_+ - K_-) \otimes 1 \qquad (10.54)$$

which, with

$$K_\pm = |\pm><\pm| \qquad (10.55)$$

describes the spin of the particle and

$$G(0) = 1 \otimes (g_+ Q_+ + g_- Q_-) \qquad (10.56)$$

which, with

$$Q_\pm = |F_\pm><F_\pm|. \qquad (10.57)$$

describes the instrument "pointer" and can also be written

$$G(0) = g_+ P_+(0) + g_- P_-(0) \qquad (10.58)$$

where

$$P_\pm(0) = 1 \otimes Q_\pm \qquad (10.59)$$

are at time 0 the relevant Heisenberg projection operators.

Rule (5.18) (the use of which is here slightly simplified by the fact that U happens to be Hermitean, $U = U^\dagger$) then yields

$$S_z(t_1) = S_z(0) \qquad (10.60)$$

and

$$G(t_1) = g_+ P_+(t_1) + g_- P_-(t_1) \qquad (10.61)$$

where $P_\pm(t_1)$, the Heisenberg projection operators at time t_1, take up the form

$$P_+(t_1) = K_+ \otimes Q_+ + K_- \otimes Q_- \qquad (10.62a)$$
$$P_-(t_1) = K_+ \otimes Q_- + K_- \otimes Q_+ \qquad (10.62b)$$

Of course, the probabilities that such and such outcomes should be obtained when such and such measurements are performed on both S and A (by means of some superinstrument) are the same when calculated using the Heisenberg picture as when calculated using the Schrödinger picture, the two pictures being completely equivalent in this respect, as is well known. For time t_1 this is easily verified using Eqs. (10.62). But what interests us here are questions such as entanglement—Schrödinger's cat riddle—that have to do with *interpretation* of the theory and to repeat, it is prima facie conceivable that such riddles should take quite a different form in one and in the other picture. It could even be hoped that here—in the Heisenberg picture—they disappear.

Now, one first lesson to be drawn from the model under study is that, yes, entanglement does take here a different form but that, no, it does not disappear. Formally, as we saw, the Schrödinger entanglement is the fact that, in circumstances of interest, the ket that, in the Schrödinger picture, describes the $S + A$ system is not a tensor product. This suggests that a formal definition of entanglement, suitable for being extended to other pictures such as Heisenberg's, should be that the symbol corresponding to the physical $S + A$ system in such pictures is not a tensor product. With such a definition then, clearly, entanglement also occurs in the Heisenberg representation since, as shown by Eqs. (10.61) and (10.62), at time t_1 neither G nor the projection operators $P_\pm$ are tensor products. It is true that this Heisenberg-like entanglement *is* quite different from the Schrödinger-like one since it takes place even in cases in which, as in the model, the initial state of the measured system is an eigenstate of the measured quantity. But to the extent that entanglement is considered a difficulty, this remark can hardly be seen as alleviating it.

A second point is that in view of Eqs. (10.62) an adherent to the interpretation according to which the Heisenberg operators *are* the dynamical quantities is confronted with a conceptual oddity the nature of which can be explained as follows. As we saw, in the model S_z and G both have definite values at t_1 if they had definite values at $t = 0$. This being the case, if we really were entitled to say, as previously, that at time 0 $S_z(0)$ and $G(0)$ [Eqs. (10.54) and (10.56)] are adequate representations of the *physical attributes* of the $S + A$ system, why should somebody coming into our laboratory at time t_1 and *then* being informed of the situation, not be allowed to proceed just as we ourselves did at time 0, that is, build up at time t_1 a description of the $S + A$ system made up of operators S_z and G that would *both* be tensor products? By simultaneously describing our (common) knowledge of the composite system at time t_1 by means of a ket suitably tailored to this description of the system (and different of course from the one *we* use) this person would have a Heisenberg picture of both the system and our knowledge of it just as adequate at all times as the one we ourselves built up above.

Since we have to do with the same system and the same "knowledge" in both cases, this plurality—or at least duality—of descriptions, which has no obvious parallel in the Schrödinger picture, is intriguing. At this stage, the only hope there is of salvaging the view that the operators represent the physical reality of the system and the ket our knowledge of it seems to be to identify in some way the state of affairs at hand with the fact that when we decide to represent a vector by means of its components we are free to choose the coordinate system in the way we like and the set of numbers we get depends on this choice of ours. However, in the present case there is nothing that corresponds to the vector notation. The idea is sometimes entertained that the Heisenberg operators *are* just what would correspond to this coordinate-system-independent notation; but the above shows this to be a delusion.

Moreover, anyhow, the analogy with vector components is basically superficial. In physics, vectors the components of which are numbers are *classical* entities. And the mathematical symbols, ρ, **E, B, r**   that are made use of in classical physics, be they vectors, scalars, or whatever, all have two roles. One of them, call it *Role a*, is that they serve for writing down the general laws, such as the Maxwell equations, considered as good candidates for describing the *general* structure of the world. The other one, call it *Role b*, is that they designate the values that the quantities they refer to actually have in some particular instance (not all planetary systems have the same number of planets nor are their planets exactly at the same places at the same time, and so on). Now unquestionably, when, in classical physics, we claim that a measurement (assumed ideal) "merely lets us know what the values the dynamical quantities ρ, **E, B,** etc., of the particular system under study actually are, *without in any way altering them*," we refer to Role *b* of these symbols, not to Role *a*. Similarly, for the (parallel) assertion that "a quantum measurement, far from changing the dynamical quantities themselves— here represented by the Heisenberg operators such as $S_z(t)$ and $G(t)$—of a particular system, merely modifies the *knowledge* we have of them" to have a meaning, it would be necessary that these operators should be interpretable as having Role b, that is, as representing some dynamical properties that, under such and such circumstances, the system in question contingently has. But this is not the case. The Heisenberg operators only have Role a. On our model this shows up through the fact that in the equations that describe, for example, $G(t)$ [Eqs. (10.58) and (10.61)], the two possible values of G, g_+ and g_-, *both* appear. Apparently the illusion that these operators have Role b as well is entertained by some authors. It presumably stems from the fact that in a case in which two systems, such as S and A in our model, come into interaction it seems extremely natural to describe the overall system they constitute before they interact by means of a tensor product of two operators, each of which operates only in the Hilbert space of *one* system. This partial determination of the involved operators goes over and beyond the mere general formulas derived from the Lagrangian of the theory, and seems to take into account at least some of the contingent properties the particular systems under study happen to have at the considered time. But, as our simple model clearly shows, this is not actually the case. The partial determination in question is in fact nothing more than a convenient choice *we* make. In the model, it can suitably be made either at time 0 or at time t_1 but not at both, and there is no reason whatsoever to consider that the contingent physical features (or "dynamical properties") of the systems at hand make one of these choices more natural than the other.

Finally, therefore, we come to the conclusion that the Heisenberg operators can in no way reflect the contingent dynamical properties of this or that particular system, so that, in case we really want the dynamical properties in question to have correspondents in the theory, we have no other choice than to impart such a role to the wave function (or state vector)—which then becomes something more than a mere description of our knowledge. And under these conditions it is not clear that concerning the interpretational riddles anything is gained by going over from the Schrödinger to the Heisenberg picture. But does this mean that the Heisenberg operators have no correspondence with reality whatsoever? No, it does not. Their role and usefulness (Role a) in the general equations of the theory amply testify to the contrary. However, this is not the contingent reality of this or that particular physical system. It is the structural reality—so to speak—of the world. Hence the existence and usefulness of the Heisenberg picture is highly consonant with the view, to be developed in later chapters, that the one word "reality" covers in fact two different concepts both of which are meaningful. There is, on the one hand, the just mentioned structural reality of the world, which has been called "independent reality," and on the other hand the observable, contingent reality of this or that particular system: the one our wave functions, density matrices, and so on tell us something about. From all that has been seen in the foregoing sections it then seems to follow in a rather convincing way that this "empirical reality," as we called it earlier, is intersubjective rather than strongly objective.

10.9 Measurement-Induced Disturbances; Zeno Effect[5]

To explain the so-called *watch-dog*, or *Zeno*, effect, the easiest way is to use a simple model, and again we shall make use of one Peres put forward [26]. Let a spin 1/2 particle whose spin is initially directed along Ox be immersed within a constant magnetic field B directed along Oz. The particle spin then tends to precess around Oz with an angular velocity ω depending on B, so that its spin state vector at time t is

$$|\psi> = 2^{-1/2}(e^{i\omega t}|+> + |e^{-i\omega t}|->) \qquad (10.63)$$

where $|+>$ and $|->$ are the eigenvectors of S_z corresponding to eigenvalues $+1$ and -1 respectively. Suppose, however, that an ideal measurement of the x component S_x of the spin is repeatedly performed, at times

$$t_1 = T/n, \quad t_2 = 2T/n, \qquad t_n = T \qquad (10.64)$$

We ask for the probability p that all the outcomes of these measurements be $+1$ (in $\hbar/2$ units). Since the eigenvector corresponding to this value of S_x is

$$|\xi> = 2^{-1/2}(|+> + |->) \qquad (10.65)$$

---

5. The rather technical Sections 10.9 to 10.11 may be bypassed on first reading.

the probability p_1 that the first measurement yields +1 is

$$p_1 = 2^{-1}|(<+|+<-|)(e^{i\omega t}|+> +e^{-i\omega t}|->)|^2 \quad (10.66)$$
$$= \cos^2 \omega \frac{T}{n} \quad (10.67)$$

According to Rule 10 of Chapter 3 the state vector then immediately collapses to $|\xi >$ [Eq. (10.65)] again. The probability that the second measurement also yields +1 is therefore also given by Eq. (10.67). And so on. So that

$$p = (\cos^2 \omega T/n)^n \quad (10.68)$$

For $n \to \infty$

$$p \to 1 \quad (10.69)$$

which implies that when such repeated observations are infinitely closely packed they simply prevent the spin from precessing. A similar argument leads to the conclusion that "continuously watching" (or, more precisely, repeatedly observing at infinitely short time intervals) an unstable particle to check whether it is still there, results in preventing it from decaying [27]. For obvious reasons this is sometimes called the *watch-dog effect*. The alternative names *Zeno effect* or *quantum Zeno paradox* refer to the Greek philosopher Zeno who, by analyzing the motion of an arrow in shorter and shorter time intervals, claimed he could prove that it never reached its target.

In real life watching the arrow does not stop it. Indeed it does not even slow its motion. Zeno's paradoxical claim is nowadays quite generally attributed to the fact that he did not know about convergent geometrical series. Is the present case similar? In the foregoing calculation have we been misled, when we referred to the wave packet being reduced *upon measurement*, by a sort of mixing up of physics with psychology? Prima facie such qualms are not quite groundless. Let us therefore show that an, at least partial, quantum Zeno effect is indeed something *purely physical*. This can conveniently be done by making use of a simplified, explicitly soluble, measurement model [28] built up along the lines of the one described here in Section 10.2. Let us therefore assume that during a very short time interval extending from time $t_{1-} = t_1 - \varepsilon$ to time $t_{1+} = t_1 + \varepsilon$ the precessing spin interacts with an instrument constituted as described in this section and that the interaction Hamiltonian H' [Eq. (10.9)] is then so large that the interaction of the spin with the constant magnetic field B can be neglected compared to it. Moreover, let the spin be, at time $t_0 = 0$, in the quantum state

$$|\phi_0> = \sum_m c_m |\phi_m>, \qquad (m = +1, -1) \qquad (10.70)$$

In order to simplify the formulas let us introduce a slight change in the notation: the measured spin component is to be called S_z (instead of S_x) and, of course, the magnetic field that induces the precession will be assumed not to lie along the Oz axis (for definiteness it may, for example, be assumed to lie along the Ox axis, but this specification plays no role in what follows).

Calling U the time-evolution operator describing the evolution of the spin between $t = t_0$ and $t = t_{1-}$ and $|\Psi(t)>$ the state vector of the composite system, we have

$$|\Psi(0)> = g(\xi)|\phi(0)> \qquad (10.71)$$
$$|\Psi(t_{1-})> = g(\xi)U|\phi(0)>$$
$$= \sum_m g(\xi)|\phi_m> <\phi_m|U|\phi_0> \qquad (10.72)$$

(since, during this time interval, U does not operate on ξ), and

$$|\Psi(t_{1+})> = \sum_m g(\xi - a_m)|\phi_m> <\phi_m|U|\phi_0> \qquad (10.73)$$

From time t_{1+} on, the spin again processes freely, until, at a time t_{2-}, it encounters, we assume, another instrument quite similar to the first one but with state vector $h(\eta)$, with which it interacts from time $t_{2-} = t_2 - \varepsilon$ to time $t_{2+} = t_2 + \varepsilon$. We now have to consider the state vector $|\Phi(t)>$ of the composite system made up of the spin and the two instruments. From Eq. (10.73) it follows

$$|\Phi(t_{1+})> = h(\eta)\sum_m g(\xi - a_m)|\phi_m> <\phi_m|U|\phi_0> \qquad (10.74)$$

whence

$$|\Phi(t_{2-})> = h(\eta)\sum_m g(\xi - a_m)|\hat{U}|\phi_m> <\phi_m|U|\phi_0> \qquad (10.75)$$

where $\hat{U}$ is the time-evolution operator describing the evolution of the spin from t_{1+} to t_{2-}. This equation can also be written

$$|\Phi(t_{2-})> = \sum_{m,k} h(\eta)\, g(\xi - a_m)|\phi_k>$$
$$< \phi_k|\hat{U}|\phi_m><\phi_m|U|\phi_0> \qquad (10.76)$$

whence

$$|\Phi(t_{2+})> = \sum_{m,k} h(\eta - a_k)\, g(\xi - a_m)|\phi_k>$$
$$< \phi_k|\hat{U}|\phi_m|><\phi_m|U|\phi_0> \qquad (10.77)$$

The probability that, after t_{2+}, ξ is found with value x, η with value y and S_z—upon some ultimate, here unanalyzed, measurement—with value $j\,(j = \pm 1)$ therefore is

$$p(x,y,j) = |\iint \delta(\eta - y)\delta(\xi - x) < \phi_j|\Phi(t_{2+})> d\eta d\xi|^2$$
$$= |\sum_{m} h(y - a_j)\, g(x - a_m)$$
$$< \phi_j|\hat{U}|\phi_m><\phi_m|U|\phi_0>|^2 \qquad (10.78)$$

The probability that ξ be found in interval $[\alpha]$, and η in interval $[\beta]$ is obtained by summing this $p(x,y,j)$ over j and within intervals $[\alpha]$ and $[\beta]$. It is

$$p(\alpha,\beta) = < \phi_0|U^{\dagger}(\sum_{m,n}|\phi_m> K_{mn}^{\alpha\beta} < \phi_n|)\, U|\phi_0> \qquad (10.79)$$

with

$$K_{mn}^{\alpha\beta} = <\phi_m|\hat{U}^{\dagger}\sum_{j}(|\phi_j> f_j^{\beta} < \phi_j|)|\hat{U}|\phi_n> M_{m,n}^{\alpha} \qquad (10.80)$$

$$M_{m,n}^{\alpha} = \int_{[\alpha]} g^*(x - a_m)\, g(x - a_n)\, dx \qquad (10.81)$$

$$f_j^{\beta} = \int_{[\beta]} |h(y - a_j)|^2\, dy \qquad (10.82)$$

From these formulas the probabilities $p_{\xi}(\alpha)$ that ξ be found in interval $[\alpha]$ irrespective of η and $p_{\eta}(\beta)$ that η be found in interval $[\beta]$ irrespective of ξ are immediately obtained by summing over all possible values of the disregarded variable. Note that extending this way α to the whole $\xi'\xi$ axis amounts to replacing the two-by-two matrix M, whose elements are $M_{m,n}^{\alpha}$ by the two-by-two matrix $(M_{m,n})$ whose elements are

$$M_{1,1} = M_{2,2} = 1$$
$$M_{1,2} = M_{2,1} = \int g^*(x - a_1) \ g(x - a_2) \, dx \qquad (10.83)$$

For reasons that will soon appear, we considered up to this point the general case, where $g(x - a_1)$ and $g(x - a_2)$ are not fully orthogonal. When they are, or, more precisely, when the functions $g(\xi)$ and $h(\eta)$ are sufficiently peaked and the intervals between consecutive a_k's sufficiently large that the overlaps of functions $g(x - a_m)$ with different values of m and those of functions $h(y - a_j)$ with different values of j are vanishingly small, we say that the instruments are *ideal* (or "perfect"). In such cases, of course, the foregoing formulas simplify. Let us, for example, set $c_1 = +1$ and $c_2 = 0$ in formula (10.70); that is, let us assume that at time $t_0 = 0$ S_z is +1 so that $|\phi_0 > = |+ >$, and let us, moreover, take for both intervals $[\alpha]$ and $[\beta]$ the positive semi-axes Ox, Oy. We then get

$$M_{+,+}^+ = 1, \ M_{+,-}^+ = M_{-,+}^+ = M_{-,-}^+ = 0 \qquad (10.84a)$$
$$f_j^+ = \delta_{j,+} \qquad (10.84b)$$

so that

$$p(+,+) = | < +|\hat{U}|+ > |^2| < +|U|+ > |^2 \qquad (10.85)$$
$$p_n(+) = | < +|\hat{U}|+ > |^2| < +|U|+ > |^2$$
$$+| < +|\hat{U}|- > |^2| < -|U|+ > |^2 \qquad (10.86)$$

These formulas are interesting because they are at variance with what a simple (too simple!) argument might well suggest. A proponent of this argument would begin with the remark that, in the model described in Section 10.2, the Hamiltonian representing the interaction between the spin and the instrument was constructed in such a way that the interaction in question should *not* change the spin state it informs us about (see the comments there). From this he would "infer" that, generally speaking, this interaction has only a registering, passive effect, and simply *does not affect* the spin, full stop. However, formula

(10.86) shows that this inference is not correct for if it were, if really the first instrument had no effect on the spin whatsoever, then the probability $p_n(+)$ would be equal to what it is when no instrument is present, namely

$$p'_n(+) = | < +|\hat{U} U|+ > |^2 \qquad (10.87)$$

which it is not. In fact, although formula (10.86) is quite elementary what it shows is, in some sense, even more at variance with "commonsense" views than the Zeno effect is. In the latter the outcomes of all the successive (here the two) measurements are taken into account, whereas formula (10.86) applies to the case where only the *existence* of the intermediate one is. As for the Zeno effect proper, it is easily recovered by generalizing formula (10.85) to a large number $n(n \rightarrow \infty)$ of closely packed successive measurements (see also Home and Whittaker [29]). In fact, with $t_1 - t_0 = t_2 - t_1 = T/2$, the right-hand sides of formulas (10.67) (with $n = 2$) and (10.85) coincide, as is easily checked.

All these calculations are made with the sole help of the Schrödinger equation, applied to the spin-plus-instruments composite system. One point they make clear is that there is nothing inherently paradoxical in the quantum Zeno effect and related phenomena, since they are ascribable just to the normal, Schrödinger-like time evolution of composite systems, without there being any need to refer to such notions as that of "increase of information" or even "wave-function collapse."

10.10 Imperfect Measurements

The foregoing analysis makes it rather clear that the quantum Zeno effect is just as "physical" as any other, and is ascribable to the interactions necessarily taking place between the measured system and the instruments of observation. But then, the very notion that lies at the basis of the effect, namely the idea of an infinite number of measurements taking place in a finite time, cannot but raise some questions. To be specific [28], suppose again that the observable to be thus repeatedly measured is the z component S_z of the spin of a spin 1/2 particle and let the measurements in question be made by means of Stern–Gerlach instruments. Since the particle has a nonzero velocity, if the number n of instruments is increased while the time interval T within which all the measurements take place remains the same, at some stage the Stern–Gerlach devices must obviously be brought very close to one another. But if several such devices are very close, their magnetic fields must obviously be very strong for the two beams issuing from any one of them to be well separated before the particle enters the next one. At the limit $n \to \infty$, T finite, the strength of these fields should be infinite, which of course is an unphysical idealization.

Such considerations point to the existence and even, in some circumstances, to the unavoidable use of measurements that can well be called "imperfect," in that two instrument states corresponding to neighboring microsystem states may not be completely orthogonal to each other. There exists quite an elaborate theory of such measurements. It is called the "theory of operations and effects" and is reported on in the next section. Here let us merely note that, as used in the foregoing calculations, the measurement model described in Section 10.2 easily accommodates such imperfect measurements, which it simply describes by means of wave functions $g(\xi)$ that are not peaked strongly enough for the "displaced" functions $g(\xi - a_m)$ with different a_m's to be practically orthogonal [28]. As a simple example of the use of this method let us consider the extreme case in which the first measurement is so imperfect that it is practically no measurement at all. This means that the functions $g(\xi - a_m)$ very nearly coincide. Assuming the second measuring instrument to be "perfect" [so that Eq. (10.84b) holds] we get

$$M^+_{+,+} \approx M^+_{-,+} \approx M^+_{+,-} \approx M^+_{-,-} \approx 1 \qquad (10.88)$$

and with the help of formula (3.11) (expansion of the unit operator) it is easily checked that, with Eqs. (10.84b) and (10.88), Eq. (10.85) yields

$$p(+,+) \approx | < +|\hat{U}\, U|+ > |^2 \qquad (10.89)$$

which, when compared to Eq. (10.87), can be interpreted as meaning that in this case the behavior of the measured system is influenced neither by the fact that the coordinate of the first instrument moves toward the $+Ox$ rather than the $-Ox$ direction, nor indeed even by the fact that the measured system did interact with the instrument in question.

On the other hand, this is of course a limiting case. In general, even for quite "imperfect" instruments, the "displaced" functions $g(\xi - a_m)$ will be far from coinciding, so that the actual value of $p(+, +)$ will lie between those given by the right-hand sides of Eqs. (10.85) and (10.89).

To sum up: The content of this section shows that a complete Zeno effect is not to be expected and should be viewed as essentially unphysical, but, as mentioned above, a *partial* Zeno effect is, on the contrary, something quite likely to be real.

10.11 The Formalism of Operations and Effects

The formalism laid out in Section 6.7 describes the effects on a quantum system of a measurement performed on this system by means of an instrument considered as being *external* to the formalism itself, in the sense that it is not attributed a Hilbert space. The only Hilbert space taken into consideration, both before and after the measurement act, is that of the quantum system on which the measurement is performed: whence the necessity of explicitly bringing collapse into the picture. Moreover, the measurement in question is an ideal—or "perfect"—one, in the sense that if before the measurement the quantum system happens to be in an eigenstate of the quantity which is to be measured, then, after the measurement process is over, the quantum system still is in this eigenstate, and the instrument contains an unambiguous record of what this state is. As we just saw, in most cases such a description is very much of an idealization, and real, physical measurement processes are likely to be more or less "imperfect."

For some purposes it may be useful to dispose of a formalism equivalent to the one of Section 6.7—in the sense that applying it does not necessitate bringing a quantum description of the instrument explicitly into the picture—even in the case of *imperfect* measurements. Of course, the use and significance of such a formalism will be subject to the same general restriction as that of the formalism of Section 6.7, namely, the instrument must be complex enough—or, in other terms, sufficiently tightly bound to its environment—that correlations between it and the quantum system defined by quantities such as M [Eq. (10.37)] should not be measurable in practice. But still, it is an interesting question to know whether or not such a generalization of the formalism of Section 6.7 exists. The answer is yes. It is called the *formalism of operations and effects* [30,31] In fact, as we are just about to see, introducing a quantum description of the instrument happens to be necessary for *building up* the formalism in question. But, to repeat, such a reference to the quantum nature of the instrument disappears from the "end result" once built up, the formalism can be applied without any reference made to the Hilbert space of the instrument.

As we saw at various places and, in particular, in Section 8.3, strictly speaking a state operator—or density matrix—can only be associated with an ensemble. There are circumstances, however, in which we can, without inconvenience, simplify our language and speak of "the system"—instead of "the ensemble of systems"—associated with a given state operator. This is here the case and, to construct the formalism, we shall therefore (following closely the presentation of Ref. [31]) freely refer to individual systems rather than ensembles. Let us then consider a measurement process in which a system S, initially described by the state operator ρ, interacts with a measuring instrument A, initially described by the state operator ρ_A. After the interaction the joint state operator of the $S + A$ system is

$$\rho_{tot} = U\rho \otimes \rho_A U^\dagger$$

where U is the joint evolution operator.

Let then an apparatus observable G, with eigenvalues g_m, be observed. Let P_m be the corresponding projector. According to formula (6.15) the probability of obtaining result g_m is

$$p_m = \text{Tr}[\rho_{tot}1 \otimes P_m] = \text{Tr}[\rho \otimes \rho_A U^\dagger 1 \otimes P_m U] \qquad (10.90)$$

which can be written as

$$p_m = \text{Tr}_S[\rho F_m] \equiv \text{Tr}_S[F_m \rho] \qquad (10.91)$$

with

$$F_m = \text{Tr}_A[\rho_A U^\dagger 1 \otimes P_m U] \qquad (10.92)$$

where the symbol Tr_A (Tr_S) means partial tracing within the Hilbert space of $A(S)$ only, so that F_m is an operator in the Hilbert space of S.

Formula (10.91) is similar to formula (6.15). The essential difference is that F_m is, in general, not a projector ($F_m^2 \neq F_m$). Operators such as F_m are called *Effects*.

Let us now assume that our observation of the instrument A has yielded the particular value G_m. The state operator ρ_m of the composite $S + A$ system immediately after the measurement is then given by formula (6.32):

$$\rho_{m,tot} = (1 \otimes P_m)\rho_{tot}(1 \otimes P_m)/p_m \qquad (10.93)$$

At this stage, let us take into account the fact that we are only interested in describing S, which means that what we want to know is the state operator yielding the probabilities p_α' of the various possible outcomes of future measurements made on S. These probabilities are of the form:

$$\text{Tr}[\rho_{m,tot}P_\alpha' \otimes 1]$$

where the P_α' are the corresponding projectors; and it is easily verified that this can be written

$$\text{Tr}_S[\rho_m P_\alpha'] \qquad (10.93a)$$

with

$$\rho_m = \text{Tr}_A[\rho_{m,tot}]$$

Hence, in virtue of Eq. (10.93):

$$\rho_m = \text{Tr}_A[(1 \otimes P_m)\rho_{tot}(1 \otimes P_m)/p_m \qquad (10.94)$$

To within normalization, the numerator can be considered as a linear mapping of the system state operator ρ before measurement into the (unrenormalized) state operator ρ_m after measurement. For this reason Eq. (10.94) is conventionally written

$$\rho_m = F\rho/p_m \qquad (10.94a)$$

the map

$$F\rho = \text{Tr}_A[1 \otimes P_m \; \rho_{tot} \; 1 \otimes P_m] \qquad (10.95)$$

being called an *Operation*. Equation (10.94a)—with definition (10.95)—thus generalizes Eq. (6.32) to the imperfect measurement case, same as Eq. (10.91) generalizes Eq. (6.22). Note that from Eq. (10.95) it follows that

$$Tr_S[F\rho] = Tr[1 \otimes P_m \, \rho_{tot} \, 1 \otimes P_m] = Tr[\rho_{tot} 1 \otimes P_m]$$

and comparison with Eq. (10.90) shows that

$$p_m = Tr_S[F\rho] \tag{10.96}$$

Since, as we saw, imperfect measurements are particularly to be expected in the cases in which the measurements are repeated, it is interesting to see how this formalism applies in an instance of repeated measurements such as the one investigated in the foregoing section. For this purpose let us first write down the probability $p_\xi(\alpha)$ that the pointer of the first instrument is found to lie in the interval $[\alpha]$. From standard quantum mechanics this probability is easily found to be:

$$\int_{[\alpha]} d\xi \sum_m |g(\xi - a_m)|^2| < \phi_m|U|\phi_0 > |^2$$

and the point is that it can also be written as

$$p_\xi(\alpha) = Tr\,[F_{[\xi,\alpha]}\rho_1]$$

where $\rho_1 = U|\phi_0 >< \phi_0|U^\dagger$ is the spin state operator at time t_{1-} and

$$F_{[\xi,\alpha]} = \sum_m |\phi_m > \int_{[\alpha]} |g(\xi - a_m)|^2 < \phi_m|d\xi \tag{10.97}$$

is an Effect. If the first pointer is seen to lie in interval $[\alpha]$, then the state of the composite system at t_{1+} is, from formulas (3.20) and (10.73):

$$p_\xi(\alpha)^{-1/2} \int_{[\alpha]} d\xi' \delta(\xi - \xi') \sum_m g(\xi' - a_m)|\phi_m >< \phi_m|U|\phi_0 >$$
$$= p_\xi(\alpha)^{-1/2} \sum_m g(\xi - a_m)|\phi_m >< \phi_m|U|\phi_0 > \quad \text{for } \xi \in [\alpha]$$

and zero elsewhere, so that the corresponding state operator is

$$p_\xi(\alpha)^{-1} \sum_{m,n} g(\xi - a_m)g^*(\xi' - a_n)|\phi_m>$$
$$<\phi_m|\rho_1|\phi_n><\phi_n| \qquad (10.98)$$

for ξ and $\xi' \in [\alpha]$ and zero elsewhere, where, again

$$\rho_1 = U|\phi_0><\phi_0|U^\dagger$$

is the spin state operator at time t_{1-}. The state operator of the spin at time t_{1+} is obtained from expression (10.98) by partial tracing over the instrument variable ξ, which means setting $\xi' = \xi$ and summing over ξ. It is therefore:

$$\rho_\alpha = \mathbf{F}_\alpha\rho_1/p_\xi(\alpha)$$

with

$$\mathbf{F}_\alpha\rho_1 = \sum_{m,n} \int_{[\alpha]} d\xi g(\xi - a_m)g^*(\xi - a_n)|\phi_m>$$
$$<\phi_m|\rho_1|\phi_n><\phi_n| \qquad (10.99)$$

$\mathbf{F}_\alpha$ is an *Operation*.

The second measurement is identical to the first, and so the conditional probability that the second pointer is found to lie in interval $[\beta]$ if the first one was found to lie in interval $[\alpha]$ is

$$p(\alpha|\beta) = \mathrm{Tr}_S[F'_{[n,\beta]}\hat{U}\rho_\alpha\hat{U}^\dagger].$$

where $F'_{[n,\beta]}$ is an operator similar in form to $F_{[\xi\alpha]}$ but with $h(n)$ and β playing the roles of $g(\xi)$ and α, respectively. Hence the joint probability for the two measurements is

$$p(\alpha, \beta) = p(\alpha|\beta)p_\xi(\alpha) = \mathrm{Tr}_S[F'_{[n,\beta]}\hat{U}\,\mathbf{F}_\alpha\rho_1\hat{U}^\dagger] \qquad (10.100)$$

and it is easily checked that this formula yields the same value for $p(\alpha, \beta)$ as formula (10.79).

As this comparison makes it apparent, the formalism of Operations and Effects yields the same predictions as the standard rules of quantum mechanics. It cannot therefore be expected that it should, all by itself, remove any of the conceptual difficulties enumerated in Section 10.1. It is true that, in a way, it conceals them, but this is only because it is built up on the model of the ideal measurement theory described in Section 6.7, and because, like this theory, it is valid only in the case in which the possibility of future measurements bearing on correlations between the measured system and quantum mechanically described parts of the instrument is not even considered.

It is also true, however, that the formalism of Operations and Effects shows, in a way, more flexibility than the standard rules of quantum mechanics, and the question then arises whether this flexibility could be made use of in order to remove the aforementioned difficulties. The flexibility in question consists of the fact that to one Effect it is possible to associate several Operations, out of which only one corresponds to the quantum mechanical predictions concerning future measurement outcomes. For example, instead of the Operation (10.99) the Operation

$$\mathbf{G}_\alpha \, \rho_1 = \sum_{m,n} \left(\int_{[\alpha]} d\xi |g(\xi - a_m)|^2 \right)^{1/2}$$

$$\left(\int_{[\alpha]} d\xi |g(\xi - a_n)|^2 \right)^{1/2} |\phi_m\rangle\langle\phi_m|\rho_1|\phi_n\rangle\langle\phi_n| \quad (10.101)$$

could be considered. This Operation—considered in Ref. [32]—reproduces the Effect (10.97) for a single measurement. It therefore yields the same probability, namely $p_\xi(\alpha)$, as $\mathbf{F}_\alpha$ concerning the first measurement. But of course it yields an outcome differing from the right-hand side of Eq. (10.100) for the joint probability $p(\alpha, \beta)$ of the outcomes of two successive measurements. The calculations reported in the foregoing section—which yielded an unambiguous result—therefore show that using $\mathbf{G}_\alpha$ to describe processes governed by the interaction Hamiltonian (10.9) would imply a violation of the predictive rules of quantum mechanics. This answers the question considered in this paragraph. If the predictive rules in question are not to be altered, the flexibility of the formalism of Operations and Effects cannot be made use of—at least not in the way considered in this example—in order to remove the conceptual difficulties listed in Section 10.1.

Problems in Relativistic Measurement Theory

As was seen in Chapter 3, standard textbook quantum theory implies *wave-packet reduction* or *collapse* (Rules 10 and 10a). The peculiar problems that any tentative relativistic measurement theory has to cope with are related to the fact that such a collapse cannot be purely local. All the probabilities that are associated to a ket in virtue of formulas such as (3.3) suddenly change, including those relative to possible measurements performed at places distant from the one where the considered measurement is made [if, e.g., $\psi(x)$ is an extended one-particle wave function and if I observe the particle to be at point x', the probabilities for it to be found anywhere else suddenly vanish].

This has far-reaching consequences that were nicely summed up by Fleming [33]. For example, following I. Bloch [34] let us consider two spatially separated space-time regions R_1 and R_2, in each of which a measurement is made on a one-particle system. This system must be described by some wave function $\psi(x,t) = < x|\psi(t) >$ so let us ask what this wave function is at a space-time point P lying outside R_1 and R_2, on a spacelike line that passes through both regions. From the geometry, there exist reference frames in which any space-time point in R_1 is earlier than any space-time point in R_2. In such a reference frame F', space-time point P, with coordinates x'_0, t'_0, lies *after* the measurement in R_1 has been performed. According to Rule (10) or (10a) (collapse), in F' the wave function $\psi'(x'_0, t'_0)$ must therefore depend on the outcome registered in R_1 and cannot depend on the outcome of the not-yet-performed measurement in R_2. But there also exist reference frames in which R_2 is totally earlier than R_1. Let x''_0, t''_0 be the space-time coordinates of P in one such frame F''. In F'' the wave function $\psi''(x''_0, t''_0)$ must depend on the outcome of the measurement in R_2 and be independent of that of the measurement performed in R_1. Hence $\psi'(x'_0, t'_0)$ and $\psi''(x''_0, t''_0)$ are functions that depend on totally different parameters and clearly cannot be related by any kinematic transformation. Since the coordinates (x'_0, t'_0) and (x''_0, t''_0) are Lorentz transforms of each other (remember they are the coordinates of one and the same point, in F' and F'' respectively) it would seem that the internal consistency of the formalism requires the wave functions ψ' and ψ'' (which describe the same system in, respectively, F' and F'') to be Lorentz transforms of each other. That this is impossible in view of the just-made remark must therefore be felt as being quite a serious difficulty.

In order to remove the difficulty in question, the idea could be considered that, when a measurement involves a space-time interaction region R, the thereby-induced collapse, instead of being instantaneous, develops with light velocity, that is, occurs along the forward light cone envelope of R. Since such a light cone envelope has the same description in any reference frame, this hypothesis *does* restore Lorentz invariance. On the other hand if, for example, the system at hand is an electron and the measurement is that of a position, the hypothesis in question implies that, in one given reference frame F', immediately after the measurement performed at a place x'_0 has yielded outcome "yes," the electron charge is concentrated at x'_0 and yet, all the same, the probability is still nonzero that the electron is found elsewhere. If we imagine observers at rest in F' and scattered at large distances from x'_0, there is a nonvanishing chance that one of them will observe the electron. If, afterward, this observer sets out to meet the one who was at x'_0 and if the two compare notes on their observations, they will come to the conclusion that the law of conservation of electric charge is violated in the process.

Does the idea (put forward by Hellwig and Kraus [35]) that collapse occurs along the *past* rather than the forward light cone of R remove this difficulty? Aharonov and Albert have convincingly shown [36] that it does not. One of their arguments is based on considering an electron trapped in a double-well potential and lying in a quantum superposition with equal amplitudes of two wave functions, each of which vanishes in one well. If a position measurement is performed in one of these wells at some time t_1 (in the reference frame F' where the wells are at rest) and if its outcome is positive, then, according to this idea, there must be times before t_1 at which the probability was only $1/2$ that the electron would have been found in that well if a test for its presence had been performed, while at such a time the probability of finding it in the other well was already zero. What is more, on a *two* spin $1/2$ particles system lying in a spin zero state and each element of which is trapped in one of these wells they could show the following: although the total spin is a nonlocal observable, still it *can* be checked, at any time before the time t_1 at which a measurement on one of the particles is made and without thereby disturbing the system, that this total spin still has value zero This is inconsistent with the implication of the Hellwig–Kraus idea that in the well where the measurement was not made the collapse already had occurred at some such time (see Fig 10 1, where t_0 labels such a time)

Aharonov and Albert pointed out that nowithstanding such difficulties no real violation of relativity theory—when conceived of as an operational theory—is to be feared since such "Lorentz covariance violating processes" can neither entail superluminal energy transfers nor allow for superluminal signaling In view of all this, they first suggested that in the relativistic domain the whole state vector formalism should be given up, and that the calculation of probabilities of well-specified measurement outcomes should be carried on by more abstract means Later [37] they stressed the fact that in spite of appearances to the contrary the measurement process *is* covariant Its covariance may be made manifest by describing states, not by state vectors any more but—in the manner once advocated by Tomonaga and Schwinger—by functionals on sets of spacelike surfaces For the case of the two spin 1/2 particles in a singlet state the corresponding calculations were explicitly made by Ghose [38] For the general case Fleming [33] has developed a theory in which special such spacelike surfaces, namely spacelike hyperplanes, have a privileged role, and which admits of the same general conclusion

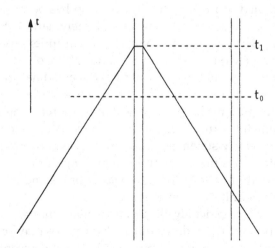

Figure 10.1. A double-well potential

Taking these developments into consideration, it must be granted that when all is said and done the measurement process is to be seen as being formally covariant However, this does not mean that no lesson is to be drawn from the difficulties reviewed

To make out more clearly what such lessons may consist of, the relativistic case must be compared with the nonrelativistic one on a very specific point, which is as follows In nonrelativistic quantum mechanics let E be a pure case in the sense defined in Section 6 2 In other words, let E be an ensemble of systems S described, at some time t, by a state vector Moreover, let this state vector, $|\psi>$, be one of the eigenkets $|\psi_{k,r}>$ with eigenvalue a_k of some observable A Formula (3 3) then informs us that if we measured A immediately after time t we would with certainty get result a_k, which means of course that we *actually* know in advance what the outcome of such a measurement would be As pointed out in Section 4 3, under these conditions we may consistently assert that at time t and on each element S of E considered separately A *has* the value in question, namely a_k At least, such an assertion conflicts neither with the (nonrelativistic) quantum rules in general nor with any particular piece of information we may possess It is true that nothing forces us to utter such a "realistic" assertion and that, for making predictions concerning outcomes of future measurements of any kind the assertion in question is redundant once $|\psi>$ is known Nevertheless, assertions of this kind, in which no reference is implied to the way the information has been gained or will be used, are in many instances extremely natural Statements such as "The energy of a hydrogen atom in its ground state (or in any well-specified state) is so and so" are of the considered type In view of all this, it would have seemed natural if, also in *relativistic* quantum physics, "realistic" assertions could have been validated in the same way, that is, by referring to state vectors (remember that in classical relativity theory such realistically interpretable statements exist, at least as concerns events) But the difficulties reviewed at the beginning of this section show that, in spite of the formal covariance of the measurement process, this definitely is not the case

More generally, the extent to which the difficulties in question force us to water down our realistic views may be appreciated by looking at Fleming's approach [33] This author considers two spacelike hyperplanes Π' and Π'' say, both containing the point P of

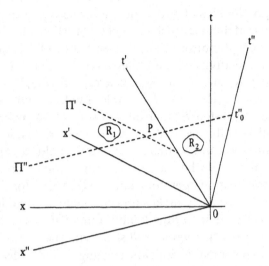

Figure 10.2. Two hyperplanes, Π' and Π'', both containing point P, and such that for Π' R_1 lies entirely in the past and R_2 entirely in the future, and vice versa for Π''

Fig. 10.2 and such that for one of them, Π', R_1 lies entirely in the past (i.e., all space-time points of R_1 have a negative time coordinate in the reference frame F' in which all the points of Π' are simultaneous) and R_2 lies entirely in the future; and vice versa concerning Π'' He grants that—as pointed out above—under these conditions the wave function $\psi'(x_0', t_0')$ describing the considered system in the reference frame F' (considered above) where Π' corresponds to constant t', is altogether different from the wave function $\psi''(x_0'', t_0'')$ describing the same system in a reference frame F'' in which Π'' corresponds to constant t'' But, he points out, these wave functions yield the probability amplitudes for the position of a particle at a definite time in each reference frame and such probability amplitudes can be empirically corroborated only by instruments which, in effect, search for the particle at a definite time. Now, in two relatively moving reference frames such as F' and F'' such searches entail two distinct, and in fact incompatible experimental arrangements. *In keeping with Bohr's insistence on the need to take into account the whole experimental setup,* Fleming argues that therefore there is a priori no reason to expect a simple kinematic relationship to hold between $\psi'(x_0', t_0')$ and $\psi''(x_0'', t_0'')$

We already have been interested in the limitations on the possibilities of a realistic interpretation of quantum mechanics. It is at the stage just reached that further such limitations that are specially introduced (over and above those already known) by the relativistic requirements come into full light. As mentioned before, in the nonrelativistic case, if $|\psi(\mathbf{x}, t) >$ happens to be an eigenvector of an observable A with eigenvalues a_k, knowledge that a system S (or an ensemble E of systems) is actually describable by this $|\psi >$ allows us to consider as valid the assertion that, at time t and on this S (or E), A has value a_k, *full stop*. If $|\psi(\mathbf{x},t) >$ happens to be the eigenvector common to a complete set of commuting operators, knowledge that S (or E) is describable by $|\psi >$ similarly makes it permissible to consider as valid the assertion that, at time t and on this S (or E), all these observables *have* definite values *full stop* (with the corollary that $|\psi >$ itself then somehow appears as being "real," at least in some sense). Here and above, "full stop" is of course intended to signify that there is no need to specify under what circumstances depending on the observer the assertion in question is to be considered as meaningful and true, and this is undoubtedly a condition that any assertion bearing on objects "as they really are" is expected to fulfill. But, as we just saw, it is met by *no* assertion similar to the ones considered here as soon as the relativistic requirements are duly taken into account. Instead, there are then the Fleming *Bohr-like* conditions, about which more will be said in further chapters but that clearly introduce—via reference to mutually exclusive experimental arrangements—some dependence on circumstances ultimately depending on the state (at least, on the "state of motion") of the observer.

The conclusion must be that, in spite of the formal covariance of the process, relativistic quantum mechanical systems cannot, as Aharonov and Albert put it [36] "be described in terms of some subset of their observable physical variables even at times when the system is not directly being observed." Such a conclusion does not alter in any way the predictive content of the theory. But it undoubtedly has a bearing on its permissible interpretations. To quote the same authors once more, "Even if one can predict with certainty that any measurement of the total charge of the system will yield value e, still neither this charge nor any other physical property may consistently be attributed to the state" of the system as such.

References

1. B. d'Espagnat, *Conceptual Foundations of Quantum Mechanics*, Addison-Wesley, Reading, Mass., 2nd ed., 1976.
2. B. d'Espagnat, *Reality and the Physicist*, Cambridge U. Press, 1989
3. J.M. Jauch, *Helv. Phys. Acta* **37**, 293 (1964).
4. F. London and E. Bauer, *La théorie de l'observation en mécanique quantique*, Hermann, Paris, 1939.
5. E.P. Wigner, *Symmetries and Reflections*, Indiana U. Press, Bloomington, 1967
6. K. Gottfried, *Quantum Mechanics*, W.A. Benjamin, Reading, Mass., ch. 20, 1966.
7. J.S. Bell, "Against 'Measurement'" in *Sixty two Years of Uncertainty*, A.I. Miller (Ed.), Plenum, New York, 1990.
8. E.P. Wigner, *Z. Phys.* **133**, 101 (1952).
9. H. Araki and M. Yanase, *Phys. Rev.* **120**, 622 (1961).
10. A.J. Leggett, *Contemp. Phys.* **25**, 583 (1984).
11. A.J. Leggett, *The Current Status of Quantum Mechanics at the Macroscopic Level in Proceedings of the 2nd International Symposium Foundations of Quantum Mechanics in the Light of New Technology*, M. Namiki (Ed.), Physical Society of Japan, Tokyo, 1987
12. H. Primas, *Chemistry, Quantum Mechanics and Reductionism*, Springer-Verlag, Berlin, 1981.
13. W. Heisenberg, *Physics and Philosophy*, Harper and Brothers, New York, 1958.
14. B. d'Espagnat, *Nuovo Cim., Supplemento*, 4828 (1966).
15. H. Zeh, *Foundations of Physics* **1**, 67 (1970).
16. W.H. Zurek, *Phys. Rev. D* **26**, 1862 (1982).
17. B. d'Espagnat, *Found. Phys.* **20**, 1147 (1990).
18. E. Joos and H.D. Zeh, *Z. Phys.* **B59**, 223 (1985).
19. B. d'Espagnat, *In Search of Reality*, Springer-Verlag, New York, 1983.
20. W.H. Zurek, *Phys. Today*, October 1991, p.36
21. K. Hepp, *Helv. Phys. Acta* **45**, 237 (1972).
22. J.S. Bell, *Helv. Phys. Acta* **48**, 93 (1975).
23. W.G. Unruh, "Quantum Measurement," in *New Techniques and Ideas in Quantum Measurement Theory, Annals of the New York Academy of Sciences*, **480**, 242 (1986).
24. B. d'Espagnat, *Found. Phys.* **22**, 1495 (1992).
25. A. Peres, "When Is a Quantum Measurement?" in *New Techniques and Ideas in Quantum Measurement Theory, Annals of the New York Academy of Sciences*, **480**, 242 (1986).
26. A. Peres, *Am. J. Phys.* **48**, 931 (1980).
27. B. Misra and D. Sudarshan, *J. Math Phys. (N.Y.)*, **18**, 756 (1977).
28. B. d'Espagnat, *Found. Phys.* **16**, 351 (1986).
29. D. Home and M.A.B. Whittaker, *J. Phys. A*, **25**, 657 (1992).
30. E. Davies, *Quantum Theory of Open Systems*, Academic, London, 1976; G. Ludwig, *Foundations of Quantum Mechanics*, Springer-Verlag, Berlin,

1983; K. Kraus, *States, Effects and Operations: Fundamental Notions of Quantum Theory*, Springer-Verlag, Berlin, 1983.

31. S.L. Braunstein and C.M. Caves, "Quantum Rules: an Effect can have more than one Operation," *Found. Phys. Lett.*, April 1988.

32. A. Barchielli, L. Lanz, and G.M. Prosperi, *Nuovo Cim.* **72B,** 79 (1982).

33. G.N. Fleming, in *The Mini-Course and Workshop on Fundamental Physics*, held at the Colegio Universitario de Humacao, Universidad de Puerto-Rico, 1985.

34. I. Bloch, *Phys. Rev.* **156,** 1377 (1967).

35. K.E. Hellwig and K. Kraus, *Phys. Rev. D* **1,** 566 (1970).

36. Y. Aharonov and D.Z. Albert, *Phys. Rev. D* **21,** 3316 (1980); *Phys. Rev. D* **24,** 359 (1982).

37 Y. Aharonov and D.Z. Albert, *Phys. Rev. D* **29,** 228 (1984).

38. P. Ghose in *Bell's Theorem and the Foundations of Physics, Proceedings of the International Conference in Memory of John Bell Held in Cesena, 7–10 October 1991*, World Scientific, Singapore, 1993.

Other Reading

P. Busch, *Macroscopic Quantum Systems and the Objectification Problem*, Symposium on the Foundations of Modern Physics 1990, World Scientific, Singapore, 1991.

Variations on a Bohrian Theme

11 In this chapter an attempt is made at analyzing the guiding ideas of Bohr's epistemological approach to the conceptual problems of conventional (no-hidden-variables) quantum mechanics. Some recent theories constructed more or less along these lines are then reviewed.

11.1 Bohr-like Definition of Dynamical Properties

Many twentieth-century philosophers, the Vienna Circle positivists foremost, rightly stressed that the verbs 'to have' and 'to be' are all too often used to draw up apparently meaningful sentences that closer inspection shows are meaningless. This is especially true in physics, where many current, commonsense concepts prove to have but a limited domain of validity. The phrase "Electrons are yellow" is a clearcut, often used example. Concerning it the physicist's objection is not that "it is merely conjectural", that "we do not really know whether electrons are yellow or green or red   " The criticism is more basic. It is that, while we have a clear notion of what the property "being yellow" means when macro-objects are at stake, we cannot even define it in connection with electrons, so that to say they *have* this property just makes no sense.

This being so, we must be careful. All the more, in fact, as we have no "God-given" knowledge about what actually determines the domain of validity of a concept. In general its specification will, of course, involve that of the set of objects to which the concept applies. But it may possibly also involve that of some contingent external circumstances. In particular, this may well be true concerning the concept "value of a dynamical physical quantity." And indeed already in Section 4.3, the question was examined under what conditions the verb 'to have' may be used in sentences such as "on system S, quantity A has value a." One— still rather broad—necessary condition for this is of course (as pointed out there) that a possibility must exist of measuring A on systems similar to S. In fact this condition should even be considered as an element of the very definition of property A. It is usually expressed by asserting that the definition of properties must be operational, which rules out the "Electrons are yellow" kind of sentences.

In practice, during the classical age of physics this "operationality" condition was, roughly speaking, considered as being also a sufficient one. If a quantity could thus be operationally defined relatively to a given type of systems—as is for example the case concerning the quantity momentum relatively to mass-points—then it could safely be considered that on any such system the property in question had some definite value; that, in the example, any mass-point had, at any time, some definite, known or unknown, momentum. With the advent of quantum mechanics physicists had to become more cautious concerning this. It is a fact that the momentum of an electron can be measured but the quantum physicists do not infer from this that an electron has, in any circumstances, a definite, known or unknown, momentum.

This is why, already in Section 4.3, the necessity of adding some further condition was considered. As may be remembered, the condition considered relies on our actually knowing with certainty what the outcome of a measurement would be if it were performed. For example, when electrons are shot from an accelerator, the accelerating device is often so contrived that any experimentalist working with it can be informed of the result he would get if he measured their momentum. If he considers the above condition as sufficient, this piece of knowledge allows him to assert that the momenta of these electrons all *have* this value. Here, one important point that should be noticed is the above use of the conditionals "if it were," "if he measured," and so on. The use of this conditional tense implies in fact considerations of a counterfactual nature, since, with the experimental preparation procedure at hand, the experimentalist is allowed to speak of the electron momentum even in cases in which he neither performs the measurement nor, in fact, intends to do so. For this reason we shall refer to this defining procedure as the counterfactual one.

This counterfactual way of defining the meaning of sentences such as "The value of the momentum of this electron is so and so" is the one that is implicitly taken up in practically all the textbook descriptions of standard quantum mechanics.

But is it the one Niels Bohr had in mind? In spite of some obscurities in the writings of this author there are serious reasons to doubt it was; and because of this it seems necessary to make a clear distinction between standard textbook quantum mechanics on the one hand and the conceptions of Bohr and some present-day theorists (whose views we shall shortly review) on the other hand. In fact, one of the points Bohr stressed most was the necessity he felt we are in of referring to the complete experimental setup whenever we speak of an atomic phenomenon. As the use of the word "complete" implies and as the contexts of this often reiterated statement of Bohr clearly show, the experimental setup thereby referred to is not just the preparation device (the accelerator in our example). Indeed, he claimed [1] the word "phenomenon" should be made use of "only for reporting on observational data obtained in strictly well-defined conditions, the description of which necessitates that of the *full* experimental arrangement," and he correlatively *rejected* as inappropriate the frequently occurring phrase "to create through a measurement the physical attributes of a system." Since, in the electron-accelerator example, the precise value of the electron momentum may be considered as due to an inbuilt automatic measurement of it, performed by the accelerator itself, it is clear that, although, to our knowledge, Bohr did not explicitly consider the example in question, his rejection of the phrase quoted must, for consistency, have induced him to reject in this case the idea of the escaping electrons *having* a definite momentum as a physical attribute.

It is true that this issue is not absolutely clear. In his reply to Einstein, Podolsky, and Rosen (EPR) [2], Bohr considered a system of two particles the sum of the momenta of which is known as well as the difference of their position coordinates, and wrote: "in this arrangement, it is   clear that a subsequent single measurement either of the position or of the momentum of one of the particles will automatically determine the position or momentum, respectively, of the other particle." Since it is quite certain that Bohr did not consider these quantities to be (both) predetermined (together) prior to measurement, this sentence cannot be interpreted as meaning that the measurement in question will just *inform* us of the value the corresponding physical quantity *has and previously had* on the other particle. Hence at first sight it seems to mean that the measurement on the first particle has indeed *created*, on the other particle, the physical attribute consisting in it having a definite position (or momentum, as the case may be)—in obvious contradiction with Bohr's own rejection (noted above) of the idea of a measurement creating a physical attribute. This is an example of the difficulty that sometimes occurs in bringing together in a consistent whole the contents of Bohr's qualitative assertions. Indeed, in the present case we apparently have no other choice than to decide for ourselves which one of these two apparently contradictory statements we consider as expressing *the* real Bohr doctrine. When proceeding to do so, it seems the reasonable choice is to prefer the general statements of this author, those he formulated for the purpose of explaining what he considered as the general guiding ideas of the new physics, to one he made, in a way, incidentally, perhaps just in order to make himself quickly and easily understood at a place in his reasoning where the question of the creation of attributes by measurements was not immediately the critical point at issue. With this choice in mind, we shall henceforth interpret Bohr's last quoted sentence as merely meaning that, under the stated conditions, if a measurement of the physical quantity at issue were performed on the second particle such and such well-defined value would be found. This is a counterfactual statement, but we shall not view it as defining a physical attribute of the particle in question. In fact, we may say that, because of the strict correlations prevailing in the EPR example, there are two equivalent ways of measuring the momentum, say, of one of the two involved particles: either by directly measuring it or, indirectly, by measuring the momentum of the other one: and that we are allowed to speak of the value this momentum has, only if the experimental arrangement is such as to make possible either one or the other of these two equivalent operations.

By interpreting Bohr's assertions in this manner we feel we are faithful to the overall spirit of his approach, as expressed by his general, epistemological statements, and we then see how crucial is, for him, the role of the instruments of observation in creating the experimental conditions under which, and only under which, an assertion of the type "such and such a dynamical quantity *has* such and such a value" is meaningful. We can then fully appreciate the correctness of one of L. Rosenfeld's statements when, in explaining Bohr's views, he wrote that in them the existence of the quantum of action "imposes a limit on the analysis of the interaction between the system and the experimental equipment that determines the circumstances under which we observe it. It is therefore the undivisible whole composed of the system *and the instruments of observation* that defines the 'phenomenon' On the atomic scale the phenomenon must thus be conceived in a lump.   " (emphasis added) [3].

An appropriate comment to this is that if so, if such a concept as that of a dynamical physical property can only have a meaning in connection with the actual setup of the instruments of observation, then, obviously, any "conditionality" or "counterfactuality" is strictly out of the question for defining these properties. The fact that only those properties of an individual system are meaningful that are associated with some actual setup rules out any possibility of defining a dynamical property of this system by referring to what the outcome would be of a measurement performed with some setup that is not actually in place.

As we see, the conditions under which Bohr could consider it as meaningful to speak of the value a dynamical quantity *has* on a microsystem seem to be even more stringent than those considered above in connection with standard textbook quantum mechanics. I observed [4] that in this respect Bohr's views are in close relationship with ideas put forward by authors such as Carnap and Hempel [5] in connection with the problem of the definition of properties. The problem these philosophers were interested in was how to give operational definitions of the "dispositional terms," that is, terms such as 'magnetic,' 'soluble' and so on, that describe a property some objects may have, of "behaving" in such and such a way when submitted to such and such conditions. At first sight this seems easy, nay, even trivial. The definition of the phrase "this object is soluble in water" that consists in stating: "If this object is immersed in water then it dissolves" seems to fit appropriately. But, according to basic formal logic the sentential connective *if a then b* is equivalent to "either *not-a* or *b*." In other words, its truth value is *no* only if *a* has truth value *yes* and *b* has truth value *no*. It is *yes* in all three other cases. In particular, it is *yes* whenever *a* has truth value *no*, which, according to the considered definition, would imply that any object that is not immersed in water is soluble in water. Clearly this is not what we mean.

To prevent such nonsense most people would accept without qualms the idea of replacing the formal logic interpretation of the connective *if then* by some more appropriate one, that is, by one involving somehow the notion of necessity: the notion that, even if the object is not actually immersed, its physical nature is such that if it were immersed then (in due time) it would dissolve. They will say that this is what we really mean when we say the object is soluble; and of course they will agree that this notion of necessity is, in this context, in obvious relationship with that of counterfactuality. Accordingly the upholders of such a commonsense answer would sympathize with the advocates of what was, above, called the counterfactual procedure for defining dynamical properties of microsystems.

But the misgivings of Carnap, Hempel, and their followers on this point was that their general philosophy was based to a very large extent on the use of conventional formal logic, that such notions as those of necessity and counterfactuality formally belong to a much more complex and not so generally accepted theory called "modal logic" and that, for this and other reasons, they considered it as proper to make as little use as possible of the notions in question. As a result, they thought it advisable to reject the commonsense way (described above) of defining dispositional terms and suggested replacing it by what they called the "partial definition" method. Basically this method consists in exclusively considering actual states of affairs. Magnetism, for example, gets manifested, as already noted in Chapter 1, in several types of experiments, calling for different experimental equipments. Within the partial definition method the phrase "Object *X* is magnetic" is exclusively defined in the corresponding contexts. It is a questionable (and questioned) point whether or not the method fits its purpose as well as its inventors hoped it would: at least, it must be granted that its applicability to some dispositional words—such as "soluble" for example!—is not immediately obvious. But still the method works well in many cases and logicians took it seriously, mainly because of the positive point that it can be formalized using but the highly precise and uncontroversial language of ordinary (i.e., nonmodal)

symbolic logic. Its existence at least shows that such problems
as concern definitions of dispositional properties can be consistently dealt with without any reference—explicit or implicit—
being made to counterfactuals. There is no reason not to try
and extend this partial definitions procedure to the definition
of the values of dynamical properties of microsystems. And, in
fact, the Bohr approach described amounts to just this [4], since
the basic role it ascribes to the instruments of observation obviously parallels the one of the operational conditions under
which, in the Carnap–Hempel scheme, the dispositional properties are manifested. For future reference it is convenient to give
such a definition procedure of dynamical properties a name. Let
us call it the *Bohr postulate* ("postulate" because, as we saw in the
foregoing pages, for defining the values of dynamical properties, general procedures other than this one are available, which
shows we do not here have to do with a logically necessary rule).
Its concise formulation is as follows.

The Bohr postulate. A quantum system has no dynamical properties of its own whatsoever (it is therefore meaningless to think
of it as having some unknown ones). When associated with a
given experimental setup it can be said to have the dynamical
property this setup is appropriate for measuring. The, properly
so-called, measurement event (the actual interaction with the
instrument) then reveals the value this dynamical property has
on the system.

• *Remark 1.* The name "Bohr *postulate*" may at first sight look
surprising. Admittedly, Bohr himself did not consider the necessity of taking the instruments of observation into account for
defining the properties of a system to merely be a *postulate*. He
saw it as *deriving* from the physical theory itself, and in particular from the wholeness of the quantum of action. On this
point, however, he was mistaken. One counterexample is sufficient to establish this, and such a counterexample exists: it is the
pilot-wave theory of Louis de Broglie and David Bohm. As we
shall see in Chapter 13 this theory quantitatively reproduces all
the experimentally verifiable predictions of elementary quantum mechanics. And although this hidden-variables theory is
"nonlocal" and "contextualistic" (see Chapter 13), in it the representative point in configuration space (the 3N dimensional

space of the coordinates of the N constituting particles) has, at any time, both a well-defined position and a well-defined velocity, without it being necessary to bring the notion of instruments into the picture. It is true that, for reasons explained in Chapter 13, very serious doubts exist that this theory is more than an ingenious model, with a low degree of convincing power. This, however, is not the point at issue here. As a counterexample showing that Bohr's standpoint, described above, is merely a postulate (though perhaps a very good one!) it serves its role in a decisive way.

• *Remark 2:* Clearly, the Bohr postulate rules out any possibility of ascribing by thought to a quantum system variables "possessed by it" independently of whether they will be measured or not (in particular it disposes at one stroke of the hidden-variables hypothesis while it is consonant with weak completeness, as defined in Section 4.2: Remark 3). An important consequence is that, according to it, a measurement performed on a quantum system is something very different from what, in ordinary life, we consider a measurement consists of, since it prevents us from thinking the measurement just records a property the system had beforehand, independently of the external conditions to which we subject it for measuring. Correlatively the postulate implies that, contrary to what takes place in "ordinary life," it is meaningless to make a clear distinction between the (contingent) objective state of this or that quantum system (the set of all the dynamical properties it happens to have at the considered time) and the conditions that define the type of predictions that can be made concerning it (with the avowedly anthropocentric undertone of the word "prediction"). This last consequence played a crucial role in Bohr's response [2] to EPR so that it is convenient to give it a name also. Let us call it *the Bohr corollary*. The Bohr corollary can be expressed in terms borrowed practically literally from the text of the response in question.

The Bohr corollary. The conditions which define the possible types of predictions regarding the future behavior of a quantum system are an inherent element in the description of any phenomenon to which the expression 'physical reality' can meaningfully be attached.

Comment. Since the advent of the Kochen and Specker theorem and of Bell's "other theorem" (not Bell 2 or Bell 3, but the theorem referred to in Section 13.4), both proving contextuality, there are cogent reasons for considering that what was just called the Bohr corollary, while remaining a conjecture, still rather adequately reflects some necessary (although not immediately manifest!) consequences of the formal structure of the set of predictive rules that make up quantum mechanics. But these theorems date back only from the 1960s. In Bohr's times the reasons just referred to did not exist. Although, in his answer to EPR, Bohr put forward the stated corollary as if it were an established truth, in fact it then was definitely not one. Admittedly it could be derived from the "Bohr postulate" stated above but this, as already noted, was only a postulate. Hence it is a question which should be most wondered at: Bohr's extraordinary intuition, which led him to an essentially correct guess (what nowadays we call contextuality, see Section 13.4) or his surprising self-assurance in setting forth a postulate as if it were quite an unquestionable truth.

• *Remark 3.* In its form, the foregoing analysis of Bohr's conceptions differs from more conventional ones in that it does not set the notion of *complementarity* on the forefront. But of course this is not meant to undervalue the role this notion had in Bohr's thought. As we know (Chapter 1), complementarity is the idea that whatever we call quantum reality cannot be totally described by specifying the values of just one set of simultaneously measurable dynamical quantities and that in fact at least a duality of mutually incompatible such sets is necessary for this purpose. Of course, if by "the values" we mean the results that measurements of these quantities would yield, complementarity is a straightforward consequence of the quantum mechanical formalism and is therefore true within any interpretation. However, in conventional (von Neumann inspired) quantum mechanics it merely appears as a—remarkable but somewhat useless—consequence of the formalism. By contrast, in Bohr's thought it plays quite a central role and, as we claim here, this is due to Bohr's strict rejection of the counterfactual definition procedure, which forced him to define the, for him, most basic elements of the theory—the values of the dynamical quantities—directly by reference to the experimental setup.

11.2 Bohr and EPR

As we noted, what we called here the Bohr corollary was given a basic role in Bohr's response to EPR. Analyzing this role is interesting. It should provide us with a better understanding of the nature and content of the Bohr postulate and its corollary and, at the same time, of the reasons why this response of Bohr's could not convince persons such as Einstein who (implicitly) rejected the postulate.

It is convenient to perform this analysis not on the example used in both EPR and Bohr's original 1935 papers but on the Bohm standard example with the two spins introduced in Section 9.1. Let us therefore consider this example once more, using the notations of this section. Along with their criterion, EPR, as already explained, postulated what we called Einsteinian separability (the real factual situation in space-time region R_V cannot be changed by anything done in a space-time region R_U spatially separated from R_V) or, at least, an equivalent idea of no interaction at a distance. According to Bohr the question is: Did EPR appropriately (i.e., unambiguously) formulate this item? Or, to put it squarely· Is Einsteinian separability or something of the same kind a really tenable idea when we have to do with quantum systems? In effect, what Bohr claimed is that it is not. There is no question, he granted, of a mechanical disturbance of system V in R_V by the measurement made on system U in R_U, but the measurement made in R_U, all the same, modifies the conditions that define the possible types of prediction regarding the future behavior of V In the Bohm example, as Andås and Gjøtterud [6] pointed out, the mentioned condition is the conservation of the total spin: for (a) as long as this conservation rule holds good it does make it possible to predict the outcome of a future measurement of any spin component of V by first permorming a measurement of the same spin component of U, and (b) this conservation is "modified" (in fact destroyed) by the measurement that is actually performed of one well-specified component of the spin of U. If we do accept Bohr's corollary we must therefore grant that this measurement—made on U—in fact modifies an inherent element of whatever may be called the "reality" of V Hence the EPR criterion cannot apply.

On the other hand, it is clear on this Bohm example that the validity of the just reached conclusion crucially depends on our accepting the Bohr corollary, that is, presumably, also the Bohr postulate, which provides a logical basis for the corollary. To see this it suffices to consider a naively realistic model—let us call it "model M"—in which this conclusion is not warranted. In model M, the two spins are just two material arrows U and V that fly off from the source with opposite orientations and both keep their respective orientation until each one encounters some measuring instrument. Let us assume, moreover, that these instruments only measure the *sign* of the spin component along some definite direction, and at the same time reorient the arrow they interact with along this very direction. Finally, let us also assume that we have to do with an isotropically distributed ensemble of such spin-pairs.

In model M, just as in Bohm's standard quantum example, a strict (negative) correlation is predicted between the outcomes of the thus defined "spin measurements" performed along the same direction on the U and V components of each pair. This can be expressed in a language similar to the one Bohr uses, by pointing out that, in model M, the fact the two arrows composing each pair are oppositely oriented constitutes a condition that makes it possible to infer predictions concerning any spin component of V from a measurement of the same spin component of U Hence we must say that such a condition is present in model M just as in Bohm's example. Moreover, in model M just as in Bohm's example the condition in question is modified (the possibility of making the mentioned inference is removed) by any actual measurement performed on U, since this destroys the initially present correlation between the orientations of the two spins within a pair. Consequently, the Bohr corollary would force us to grant that, also in model M, the measurement performed on U modifies an inherent element of whatever may be called the reality of the (distant) system V

However it is clear that, concerning model M, this conclusion is just nonsense. The measurement performed on U only disturbs U. Actual contraptions materializing model M can easily be produced and, when looking at them, nobody in his right mind would say such a measurement also changes V In other words, nobody would agree that the modification this measurement entails of the "conditions that make it possible to derive predictions concerning any spin component of V from the outcome of a measurement performed on U" changes the objective reality of V, everybody would say these conditions relate only to our *knowledge* of system V Otherwise said, everybody, faced symbolic logic. Its existence at least shows that such problems as concern definitions of dispositional properties can be consistently dealt with without any reference—explicit or implicit—being made to counterfactuals. There is no reason not to try and extend this partial definitions procedure to the definition of the values of dynamical properties of microsystems. And, in fact, the Bohr approach described amounts to just this [4], since the basic role it ascribes to the instruments of observation obviously parallels the one of the operational conditions under which, in the Carnap–Hempel scheme, the dispositional properties are manifested. For future reference it is convenient to give such a definition procedure of dynamical properties a name. Let us call it the *Bohr postulate* ("postulate" because, as we saw in the foregoing pages, for defining the values of dynamical properties, general procedures other than this one are available, which shows we do not here have to do with a logically necessary rule). Its concise formulation is as follows.

The Bohr postulate. A quantum system has no dynamical properties of its own whatsoever (it is therefore meaningless to think of it as having some unknown ones). When associated with a given experimental setup it can be said to have the dynamical property this setup is appropriate for measuring. The, properly so-called, measurement event (the actual interaction with the instrument) then reveals the value this dynamical property has on the system.

• *Remark 1:* The name "Bohr *postulate*" may at first sight look surprising. Admittedly, Bohr himself did not consider the necessity of taking the instruments of observation into account for defining the properties of a system to merely be a *postulate*. He saw it as *deriving* from the physical theory itself, and in particular from the wholeness of the quantum of action. On this point, however, he was mistaken. One counterexample is sufficient to establish this, and such a counterexample exists: it is the pilot-wave theory of Louis de Broglie and David Bohm. As we shall see in Chapter 13 this theory quantitatively reproduces all the experimentally verifiable predictions of elementary quantum mechanics. And although this hidden-variables theory is "nonlocal" and "contextualistic" (see Chapter 13), in it the representative point in configuration space (the 3N dimensional with a material realization of model M, would reject the idea of applying to it the Bohr corollary, as well as, of course, for consistency, the Bohr postulate

It goes without saying that model M is a "hidden-variables" model that (contrary to the pilot-wave model), is too crude to faithfully reproduce all the quantum mechanical observational predictions, and even all those merely concerning Bohm's standard example In particular it does not reproduce those bearing on correlations of spin measurements outcomes along directions that are neither parallel nor orthogonal to each other, and the arguments developed in Chapter 9 conclusively show that this defect of the model cannot be cured by any refinement that would preserve Einsteinian separability and the EPR criterion, even at the price of introducing hidden variables But the impossibility of doing so is a special case of Bell's theorem Nonseparability cannot be proved by merely referring, in a qualitative, Bohr-like way, to the existence of the quantum of action except if, in addition, the validity of Bohr's postulate is assumed or some kind of completeness is postulated So that there seems to be a kind of a vicious circle in Bohr's attempt at falsifying EPR on the sole basis of the fact that the quantum rules of prediction work Consequently it is impossible to follow the above-quoted authors [6] when they claim that Bell's theorem is superfluous, Einsteinian separability having being disproved already by Bohr

Concerning, more generally, the Bohr–EPR controversy, it emerges from the foregoing analysis that what Bohr succeeded in doing was essentially to show his *own* approach was consistent That is, he could show his assumptions implied no contradiction and, if made, did entail *a form* of completeness What Bohr did *not* succeed in doing (although most physicists thought he did) was to show that his approach was the *only* consistent one compatible with quantum mechanical data, and that, in particular, the EPR one was not For his approach assumed, right at the start, ideas that, admittedly, did entail his conclusions (concerning the reality of objects being influenced by conditions fixed at places possibly quite distant from them in other words, a kind of nonseparability) But, being simply assumed, these ideas were of course not compulsory And to repeat, the mere existence of the pilot-wave theory (see Chapter 13) suffices to show that, in principle at least, all the experimentally verifiable predictions of elementary quantum mechanics can be recovered on the basis of general ideas that are seriously at variance with Bohr's starting points.[1]

Objects and the Truth Criterion Within Bohr's Approach

Within Bohr's approach, and particularly within the just described interpretation of his views concerning the dynamical properties of microsystems, a crucial epistemological role is imparted to the instruments of observation. By assumption their setting can be described "as it is, full stop." Which means that the values of some at least of their contingent dynamical properties—position and orientation coordinates, etc.—are meaningful concepts per se, the definition of which are not conditioned by the actual settings of some "superinstruments" with the help of which we should be supposed to observe them. In Bohr's words they "must be described in purely classical terms." This implies that in the Bohr-like views the problems of the "quantum-classical cut" we encountered in Section 10.1 (when discussing the break of the "von Neumann chain") take a special turn. Concerning them it is essential to note that at the place [1] where Bohr made a conceptual analysis of some *gedanken* experiments involving movable diaphragms, what he suggested in this connection is that instruments and their parts are to be considered either as classical or as quantum systems not according to their physical properties but just according to our *point of view.* They are quantum systems if we carry on observations on them by means of other instruments. They are classical systems if we use them as instruments of observation.

This, of course, raises some delicate problems, concerning both the notion of object and that of instrument. What "part" of a phenomenon can properly be called the (quantum) object? This question has been made the theme of some debate. Here

---

1. The assertion to which I gave the name "Bohr's corollary" above was precisely stated and made use of by Bohr only when he felt he had to respond to EPR. Since this response of Bohr's constitutes one of the main texts in which the latter's conceptions are expressed, some physicists tend to consider it as containing all the essentials of Bohr's approach. If this view is taken, then it can be conjectured that, perhaps, Bohr postulated *only* what is in the assertion in question, *not* what I called his "postulate." But it is questionable whether a logically consistent scheme can be constructed along these lines and anyhow, examination of *other* texts by Bohr strongly suggests, as shown in Section 11.1, that Bohr did make the assumption called here his postulate.

let it simply be noted how J. Roldàn [7] contributed to somewhat clarify this point: he noted first that, of course, within any
well-specified phenomenon we feel it necessary to distinguish
between the instruments of observation and "all the rest." But his
idea is that we must be careful not to rashly identify this "rest"
with the object, as we might well be tempted to do. Following
him, let us just call it "the inside" of the phenomenon. His point
is that, as we just noted, the cut between the instrument and
the "inside" depends on what phenomenon is considered: in a
two-slit experiment with a movable diaphragm (and of the type
Bohr considered), it leaves the diaphragm on the side of the "inside", on the contrary, in the case in which the same diaphragm
is fixed, it leaves it on the instrument side. Hence it is only in
the latter case that the "inside" coincides with the diffracted particle and can therefore properly be called "the object." To get
at, within Bohr's views, a suitable general definition of "the object" we must therefore make one more move. Roldàn suggests
defining the object as the feature of the "inside" that remains the
same when two complementary phenomena are considered. In
the example under study the phenomena with fixed and movable diaphragms are, in a way, complementary in Bohr's sense
and their "insides" do have a common part, namely the particle. We see therefore (what was not obvious at the start!) that
even within Bohr's approach the notion of a quantum object can
consistently be kept.

As for the ticklish problems concerning the nature of the instruments, in Bohr's approach they are obviously tightly intermingled with basic epistemological problems and will therefore
not be analyzed at the present stage. In (the remaining part of)
this section and the following one the existence of classically
describable instruments is just taken for granted, exactly as in
Bohr's writings, and we focus our attention on other points. The
one to be examined here is the very considerable conceptual
differences existing between the "Bohr-like" and the "counterfactual" definitions of dynamical properties, and the correlative
differences in the types of theoretical approaches that each one of
these two standpoints makes possible. The importance of the differences in question can be brought into full light by observing
that while both in classical physics and in conventional textbooks quantum mechanics the truth criterion stated in Section
1.6 is considered as valid (and essential), in Bohr's approach it

makes, at the quantum level, no sense. This is simply due to the fact that it refers, not only to the actually performed measurements but also to those that could be done, which is inconsistent with Bohr's implicit ruling out of counterfactual considerations in the very definition of what is meant by a dynamical property. From what we saw it appears that, when the Bohr postulate is used for defining values of dynamical properties, attributing to the quantum system such and such a definite dynamical property (whose value the measurement will reveal) is essentially a mental act that is both harmless and useful when some suitable experimental setup is in place but becomes inconsistent as soon as some other experimental setup, incompatible with the one in question, is assumed being in place instead.

In contrast, in conventional textbook quantum mechanics our (Section 1.6) truth criterion is normally considered as valid. Indeed most of the quantum theoretical developments of the last sixty years or so were made using the mathematical tools, and therefore also the general assumptions, originally put forward by Paul Dirac and John von Neumann, including the idea (underlying them) that quantum systems may, in definite circumstances, have properties of their own, independently of what the instruments are by means of which these systems *will* eventually be observed. According to our analysis this implies that the theories thus developed are based in part on the counterfactual defining procedure and that consequently, in them, the truth criterion is normally considered as being valid. As a matter of fact, in the analysis of the role of the environment made in Section 10.6, it was by (implicitly) applying this truth criterion that we reached the conclusion there expressed. As it may be remembered, this conclusion was that the statement "Macroscopic systems always have macroscopically well-defined positions in space" cannot be true if the completeness assumption is made and the verb "to have" is given its popular, that is, ontological, sense. Clearly the first of the two proposals sketched in Section 10.6: Remark 1 in order to remove, or at least alleviate, this difficulty is tantamount to a watering down of the said truth criterion. In Chapter 15 this suggestion will be developed.

On the other hand, some recent theoretical proposals can be viewed as constituting a partial "comeback" to Bohr's general conceptions and in particular to the Bohr-like idea of defining dynamical properties of quantum systems by reference to the instruments of observation. To a limited extent this is already true of at least the first version of Griffiths's "consistent histories" theory, and it is clearly the case concerning Omnès's "consistent logics" theory, both of which theories are briefly analyzed below. In view of the above it is then to be expected that in these theories our Section 1.6 truth criterion may not apply as it stands and may have to be replaced by some other one. In Sections 11.4 and 15.4 we shall reexamine this question.

11.4 Some Remarks on Griffiths's and Omnès's Theories

It is impossible to do justice in a few pages to these two new and interesting proposals and the reader must be referred to the original articles [8, 13, 9, 15] for a comprehensive picture. Here, we merely point out a few distinctive features of these approaches and try and form an opinion on what kind of "reality" they can be assumed to describe.

Griffiths's Approach

Earlier (see in particular Sections 1.6 and 6.7 as well as Remark 3 in Chapter 3) we saw that in standard textbook quantum theory the operation conventionally called "measurement" is not really a measurement in the familiar, classical sense: in general it informs us of the value the measured dynamical quantity has *after* and not *before* the operation in question.

Not so for Griffiths. This author wants a measurement to "reveal properties that already existed." This requirement led him to put forward a theoretical formalism conceptually quite different from conventional quantum mechanics although it uses the same mathematical tools and aims at reproducing its (experimentally confirmed) results. By way of an introduction to it, let us imagine for a moment a classical but indeterministic world. In this imaginary world, for example, if a ball is made to fall upon a pin stuck horizontally into the wall there is an *intrinsic* probability that it will be pushed off to the right and an intrinsic probability that it will be pushed off to the left, where "intrinsic" means that which one of the two events does actually take place is not determined by any small parameters with unknown values. Of course the ball may be thought to undergo several such happenings successively. This implies it will have a definite history, which will in general differ from the history of

another ball initially placed in the same state and falling upon
the same lattice of pins. Each one of these histories has a definite probability, that a theory of the effect should enable us to
compute. And—needless to say—at any time one definite ball
always really *has* one only of the positions allowed by the structure of the contraption. A snapshot taken at that time, that is,
a measurement of that position, would just reveal it without in
any way creating it. In a sense Griffiths's theory is an attempt at
building up a quantum mechanical formalism that would conceptually resemble as much as possible such a nondeterministic
classical theory.

To this end Griffiths focused his attention on the above considered notions of 'histories' and 'probabilities of histories.' For
him, histories are just successions of events, where by "event"
we must understand the actual state of a system at a given time,
characterized by the values some observables have on it. As
for the probability law for histories, he essentially took up the
well-known Wigner formula (6.38), which he rediscovered and
reinterpreted. In conventional quantum mechanics this formula
yields, as we saw, the probability that if successive measurements are made of observables $A, B,$ at specified times $t_1, t_2,$
, the outcomes of these measurements are $a_j, b_k,$ Within his
tentatively more "realistic" approach Griffiths had to assume
that the formula in question yields the probability that: "the
dynamical quantity A *has* value a_j at time t_1 *and* the dynamical
quantity B *has* value b_k at time t_2 *and* etc."

Of course, not all conceivable such probabilities are open
to such a "realistic" interpretation. Those that are are said to
refer to "consistent histories." Necessary conditions for a history
to be consistent were written down by Griffiths. They can be
derived by different means, some more formal than others. The
just mentioned identity of Griffiths's and Wigner's probability
formulas can be made use of to obtain them—at least in the
simpler cases—in a direct way.

To this end (d'Espagnat [10]), let us consider, within the
realm of the conventional view, an experiment X in which two
observables, $A,$ with eigenvalues spectrum $\{a_j\}$, and $B,$ with
eigenvalue spectrum $\{b_k\}$, are measured successively at times
t_1 and t_2 respectively, and in which only the second result is
noted; and let us compare it with an experiment Y in which
only the second measurement is performed. It is clear that in

general the probability of getting outcome b_k at time t_2 is different in the two experiments, for in experiment X the state of the system is changed because of its "material" interaction with the A-measuring instrument. There may, however, be cases in which the two probabilities are equal. Only in such cases can we say that *if*, in experiment Y, A *had* been measured (without the outcome being registered, that is, if Y had been X) nothing would have been changed as regards the outcome of the B measurement. Only in such cases can we therefore be sure that if, concerning experiment Y, we say· "At t_1 the unobserved quantity A *had* a definite value which we do not know" we thereby utter a statement that, even though it has no meaning in the eyes of a strict operationalist, at least is a harmless one, not inconsistent with what we already know. Hence a condition of this type must be fulfilled for a realistic interpretation—such as Griffiths's—of the considered sequence of events to be possible (at least concerning the event $A = a_j$ in the history of experiment Y).

With the help of Wigner's formula (6.38) this condition reads:

$$\sum_j \text{Tr}[P_k^B \, P_j^A \, \rho \, P_j^A] = \text{Tr}[P_k^B \, \rho] \qquad (11.1)$$

since obviously the right-hand side of this relation is the probability of outcome b_k in experiment Y and the left-hand side is the probability of outcome b_k in experiment X, irrespective of what the, unregistered, outcome of the A measurement was (account has been taken of the fact that the various possible outcomes of the A measurement are mutually exclusive so that a sum over all the corresponding relevant probabilities yields the overall probability of outcome b_k). Of course, in (11.1) P_j^A is the (Heisenberg) projector onto the eigenspace corresponding to eigenvalue a_j of A and similarly as concerns P_k^B Using the fact that

$$\sum_j P_j^A = 1 \qquad (11.2)$$

the right-hand side of Eq. (11.1) can be written as

$$\sum_{j,j'} \text{Tr}[P_k^B \, P_j^A \, \rho \, P_{j'}^A]$$
$$= \sum_j \text{Tr}[P_k^B \, P_j^A \, \rho \, P_j^A]$$
$$+ \sum_{j,j'}' \text{Tr}[P_k^B \, P_j^A \, \rho \, P_{j'}^A] \qquad (11.3)$$

where Σ' means a summation over *different* j and j' values only. This implies that the so-called *consistency condition* (11.1) can be rewritten as

$$\sum\nolimits'_{j,j'} \text{Tr}[P_k^B \, P_j^A \, \rho \, P_{j'}^A] = 0 \qquad (11.4)$$

Note that, since, for any operator A, $\text{Tr}(A) = [\text{Tr}\,(A^\dagger)]^*$ and, for any pair A, B of operators, $\text{Tr}(A\,B) = \text{Tr}(B\,A)$, we have:

$$\begin{aligned} \text{Tr}[P_k^B \, P_j^A \, \rho \, P_{j'}^A] &= \text{Tr}[(P_{j'}^A)^\dagger \, \rho^\dagger \, (P_j^A)^\dagger \, (P_k^B)^\dagger]^* \\ &= \text{Tr}[(P_k^B)^\dagger \, (P_{j'}^A)^\dagger \, \rho^\dagger \, (P_j^A)^\dagger]^* \qquad (11.5) \end{aligned}$$

which can be written suppressing the † signs since the projectors and the state operator are Hermitean operators; so that the consistency condition (11.4) can also be written

$$\text{Re} \, \sum\nolimits'_{j,j'} \text{Tr}[P_k^B \, P_j^A \, \rho \, P_{j'}^A] = 0 \qquad (11.6)$$

This is one form of the consistency condition for this simple case. When the consistency conditions are fulfilled, the probabilities under study are meaningful in the following sense. Let us consider a quantum system on which an initial measurement is performed at a time t_0 (thus defining the state operator ρ) and a final measurement is performed at time t_f, and, in between these two times, let us consider a sequence of time values $t_1, t_2,$ to each one of which we associate the idea of a dynamical quantity $A_1, A_2,$ These various observables not being necessarily compatible with one another. If the consistency conditions are fulfilled we can, without creating inconsistencies between the observable predictions computed in this or that way, think of the dynamical quantities in question as actually taking up, at the corresponding times, some definite values (the simple case analyzed above corresponds to $f = 2$), so that the system may be thought of as, at least in a restricted sense, having a history; although, of course, this history is not determined since there is only a *probability*, expressed by Wigner's formula (6.38), that it is realized.

Note that if we set $t_f = t_2$, choose t_1 to be a time just immediately before t_2, and decide to consider the *same* observable, C, with eigenvalue spectrum $\{c_k\}$, at times t_1 and t_2 (i.e., if in formula (6.38), we identify both A and B to observable C), then by making use of this same formula it is easily shown that (in accordance with expectations!) the conditional probability

$$\text{Prob}[A = c_k \quad \text{if} \quad B = c_k] \tag{11.7}$$

is equal to 1. In conventional quantum mechanics this simply means that if C is measured twice within a vanishingly short time interval the outcomes of the two measurements must be the same. Within Griffiths's conceptions (and if the consistency condition is met) it obviously means something else, since at t_1 no measurement is assumed to be made: in fact it means that at time t_1 observable C already had the value c_k that the measurement at time t_2 reveals. This result is easily generalized to any consistent history theory with more than two times, thus confirming that the theory under study does indeed meet the requirement set forth at the beginning of this section.

But of course it is known from past experience that to build up an inconsistency-free "realistic" interpretation of quantum mechanics is not an easy enterprise, so that it should not come as a surprise that Griffiths's theory meets, at some points, with difficulties. A serious one, acknowledged by Griffiths himself, is most simply described using once more the paradigmatic Bohm example of two flying-apart spin 1/2 particles U and V, created at time 0 in a spin zero state. In this case, the probability that at a time $t_2 > 0$ a measurement of the component S_b^V of the spin of V along direction b yields outcome m_b $(= \pm 1/2)$ does not depend on whether or not, at a time $t_1 < t_2$, a measurement on U (again: with no outcome registration) is made. Hence it can be expected (according to the argument given) that the history $\{S_a^U = m_a$ at $t_1, S_b^V = m_b$ at $t_2\}$ is consistent, and indeed it is. The same holds for more complex histories such as $\{S_a^U = m_a$ at $t_{1/2}, S_a^U = m_a$ at $t_1, S_b^V = m_b$ at $t_{3/2}, S_b^V = m_b$ at $t_2\}$, with $0 < t_{1/2} < t_1 < t_{3/2} < t_2$, and it also holds, because of the strict correlation induced by the initial state, for histories such as, for example,

$$H_1 \equiv \{S_a^U = m_a \text{ at } t_{1/2}, \quad S_a^V = -m_a \text{ at } t_{1/2},$$
$$S_a^U = m_a \text{ at } t_1, \quad S_b^V = m_b \text{ at } t_2\} \tag{11.8a}$$

and

$$H_2 \equiv \{S_b^U = -m_b \text{ at } t_{1/2}, \quad S_b^V = m_b \text{ at } t_{1/2},$$
$$S_a^U = m_a \text{ at } t_1, \quad S_b^V = m_b \text{ at } t_2\} \tag{11.8b}$$

(here we may think of measurements of S_a^U and S_b^V being actually done at times t_1 and t_2 respectively; the consistency of the histories H_1 and H_2 can then be shown by a straightforward generalization of the above described argument). Moreover, in the present case the conditional probabilities that at time $t_{1/2}$ the spin components of U and V along directions **a** and **b** have values $m_a, m_b, -m_a$, or $-m_b$ given that, at $t_1 S_a^U$ has value m_a and, at t_2, S_b^V has value m_b, can be calculated along the same lines as above. In accordance with straightforward expectations we thus get, using again the notation $(X|Y)$ for 'the probability that X *if* Y'

$$(S_a^U = m_a \text{ at } t_{1/2}|x \& y) = 1 \tag{11.9a}$$
$$(S_b^V = m_b \text{ at } t_{1/2}|x \& y) = 1 \tag{11.9b}$$
$$(S_b^U = -m_b \text{ at } t_{1/2}|x \& y) = 1 \tag{11.10a}$$
$$(S_a^V = -m_a \text{ at } t_{1/2}|x \& y) = 1 \tag{11.10b}$$

where x and y are shorthand abbreviations for events $(S_a^U = m_a$ at $t_1)$ and $(S_b^V = m_b$ at $t_2)$ respectively, and "&" means "and." The validity of relationships (11.10a) and (11.10b) is of course due to the fact that ρ describes a spin-zero state, which implies strict correlations between the spin components of U and V in any given direction.

The inescapable paradox in this theory, is that, under the conditions stated, x and y are both true, so that Eqs. (11.9) and (11.10) imply that at time $t_{1/2}$ both S_a^U and S_b^U have definite values. Whenever **a** is different from **b** this, however, is impossible since, in Griffiths's theory just as in conventional quantum mechanics, the fact that a set of physical quantities all simultaneously have definite values is associated with a projector in Hilbert space, whereas no projector can be associated with the conjunction of the two facts considered above.

While Griffiths acknowledged this difficulty, he nevertheless maintained that Eqs. (11.9a) and (11.10a) are both true, and claimed that notwithstanding this (and still assuming x and y to be true) the proposition

$$[(S_a^U = m_a \text{ at } t_{1/2}) \& (S_b^U = -m_b \text{ at } t_{1/2})] \tag{11.10c}$$

is meaningless. It must frankly be pointed out that this is logically impossible [10,11]. First, observe that as soon as a proposition D *entails* a proposition B it logically follows (the logicians call this *modus ponens*) that if a system S is considered on which D happens to be true, then B is necessarily true on S. Second, note that if two propositions B and B′ are true (hence meaningful) it also follows, in any known logic, that the proposition (B & B′), which is the conjunction of both, is also meaningful and true. Under these conditions, if it is the case that D implies B and that D also implies B′, and if a system S is considered on which D is true, then, on S, the proposition (B & B′) is *necessarily* meaningful and true. In fact the opposite claim, which amounts to Griffiths's with the appropriate symbol identifications, could only be valid in some as yet unspecified logic, of which it is not even known how it could be self-consistent.

Note that the foregoing objection can also be developed [10] on the simpler example of just one free spin 1/2 particle whose spin components S_a and S_b along two different directions **a** and **b** are measured at two different times t_1 and t_2 respectively, with outcomes m_a and m_b. Denoting x and y the propositions ($S_a = m_a$ at t_1) and ($S_b = m_b$ at t_2) respectively, the same argumentation as above shows that both equations

$$(S_a = m_a | x \& y) = 1 \qquad\qquad (11.11a)$$

and

$$(S_b = m_b | x \& y) = 1 \qquad\qquad (11.11b)$$

hold true, where the propositions in the left-hand sides of the brackets are relative to some time t between t_1 and t_2. And of course this statement falls under the criticism expressed above.

In two articles [12, 13] both of them more recent than the ones referred to above, Griffiths has developed further and somewhat modified his theory, partly with the purpose of answering the preceding objection. Concerning this, and more generally concerning the basic ideas underlying his conceptions, the more explicit of these two is [13]. What emerges from it is that Griffiths's approach can be said to be inconsistency-free only at the price of basic alterations of our normal way of thinking.

Basically these alterations are

1. We have to accept that a history in which Schrödinger's cat ends up being in a quantum superposition of live and dead is a history that may very well be physically true.

2. The very notion of truth is relative. For this reason Griffiths calls it Truth. But the capital does not mean it is *more* absolute than good old truth. It is in fact just the reverse. Truth (even of macroscopic past events) is fully relative to what we "prefer to discuss." It largely coincides with what Omnès calls "reliability within one given consistent logic" (see below), but with an important difference concerning the status of the macroscopic events.

3. Griffiths's theory is one about closed systems. Given a consistent history, we are allowed to ask what we would see if we measured this or that (or what probability there is that the outcome be this or that) only if this question does not violate the consistency conditions relative to this history. In contrast, in "ordinary," rule-of-the-game quantum mechanics this is a question that can always be asked.

The relativity of Truth, as defined in point 2, may be said to remove the objection described above on the basis of Eqs. (11.9, 11.10). But it is a high price to pay. More generally, the worrying character of points 1–3 can hardly be denied. It may make one doubtful that what positive element Griffiths's approach may bring in is worth the price of having to accept them. The main motivation Griffiths puts forward in favor of acceptance is that he claims locality can then be saved and he prefers saving locality to saving Aristotelian logic and the absolute nature of truth. Of course, Griffiths is right in pointing out that his theory makes a radical break with the hypotheses under which the Bell theorems are derived and that locality can be saved this way. But it is not so clear that realism can be saved together with locality, for it is a basic requirement of traditional realism that there are facts that are true quite independently of the conventions we decide to make as to which consistent family of histories we prefer to discuss. Giving up this requirement amounts to putting forward another theory, in which physics does not describe reality in the sense realists impart to this word. The independently developed Theory of Empirical Reality, to be described in Chapter 15, consists in part in openly and avowedly taking this step.

Omnès's Theory

Particularly in its first version [9], Omnès's theory was very much inspired by the one reviewed above and its formal structure is still much the same as Griffiths's, which spares us the task of describing it in detail. One significant difference between these two authors is that, while initially Griffiths obviously aimed at building up a "realistic" theory (a theory compatible with *physical realism*), Omnès's approach was, from the beginning, a more formal one. In this spirit he, first of all, renamed "consistent representations of logic" (or, for short, "consistent logics") what are essentially Griffiths's 'consistent histories.' Contrary to Birkhoff's and others' "quantum logics," these "consistent logics" are Boolean and, by using mathematical measures defined in them in the same way as Griffiths used probabilities (including conditional ones), Omnès was able to define implications (symbol ⇒) in such a way that, apart from very special cases that do not occur in the examples used here, any relation bearing on conditional probabilities $(X|Y)$ and of the form

$$(X|Y) = 1$$

valid in Griffiths's theory is also valid in Omnès's theory, where it is written:

$$Y \Rightarrow X.$$

As a consequence it is clear that the objection described above in connection with Eqs.(11.9a) and (11.10a) reappears here, in the form that while

$$x \,\&\, y \Rightarrow (S_a^U = m_a \text{ at } t_{1/2}) \tag{11.12}$$
$$x \,\&\, y \Rightarrow (S_b^U = -m_b \text{ at } t_{1/2}) \tag{11.13}$$

yet, because of completeness:

$$x \,\&\, y \not\Rightarrow [(S_a^U = m_a \text{ at } t_{1/2}) \,\&\, (S_b^U = -m_b \text{ at } t_{1/2})] \tag{11.14}$$

since the assertion within square brackets corresponds to no Hilbert-space projector so that the two propositions $(S_a^U = m_a$ at $t_{1/2})$ and $(S_b^U = -m_b$ at $t_{1/2})$ cannot both be true together. This shows (d'Espagnat [11]) that either the theory is inconsistent or, at best, it lacks something: in fact it lacks a "criterion for the truth of a property" that would have the effect that the propositions in question, while partaking a bit of truth (because they never lead to self-contradiction) are not true in the proper sense. A qualitative way of formulating such a criterion—different from, though compatible with, the general truth criterion put forward in Chapter 1—is the following [11]:

Restrictive Truth Criterion 1 (RTC 1). A statement bearing on a property attached to a physical system can be said to be "true" only if it is valid per se, that is, independently of how we choose to associate the idea it states with other ideas.

In the example considered above, while proposition ($S_a^U = m_a$ at $t_{1/2}$) holds good in one consistent logic L_1, corresponding to History H_1 (formula 11.9a), still, it cannot be added to propositions constituting other consistent logics (such as the logic, L'_2, corresponding to a history H'_2 obtained from H_2 by changing $t_{1/2}$ to some $t_{3/4} > t_{1/2}$) In other words, if associated to L'_2 it would make L'_2 inconsistent. Under these conditions RTC1 indeed implies that the proposition in question cannot be called true.

However, as already noted, propositions that, as this one, are elements of a consistent logic partake of the notion of truth to some appreciable extent. It has been proposed (d'Espagnat [11]) that it should essentially be for referring to such propositions that the qualificative "reliable"—originally given a less specific sense by Omnès [9][2]—should be used: And to, consequently, speak of propositions "reliable" (or "trustworthy" [14]) *within such and such a consistent logic."*

On the other hand, at this stage a question obviously arises: Are there true propositions, and, if so, what are they?

In Omnès's conception, a first element of an answer is that, according to him, there are *facts*, that is, classically describable phenomena. This assertion he grounded on the theory of environment-induced classical properties of macrosystems, a theory to be critically reviewed in Chapter 12. However, for

---

2. The word "reliable" already appeared in the first version [9] of Omnès theory along with some indication as to the conditions under which it had to be used in lieu of the word "true." However, as they were stated in this version these conditions did not set, on the validity of the concept of truth, bounds sufficiently definite to guarantee self-consistency of the theory.

the sake of the argument, let us here accept the notion of classical facts as a working hypothesis. With the notions of *facts* and
trustworthy propositions at our disposal, the guiding idea of RTC1
can be expressed in a somewhat more precise manner by stating
that, to be true, a proposition must be trustworthy within *any*
consistent logic containing all the true facts. In fact, to get at a
strict and fully consistent formulation additional care must be
exerted here. This was done in Omnès's updated theory [15].
For future reference, let us call the thus elaborated final version
of the criterion "Restrictive Truth Criterion 2." To express it (full
details are to be found in [15]) Omnès first defined a special
class of logics, those containing all the actual facts plus, possibly, other propositions. He called them "sensible logics" and
assumed them to be consistent. His criterion is then expressed
as follows:

Restrictive Truth Criterion 2 (RTC 2).
(α) Actual facts are taken to be true.
(β) For another property a to be true it must be the case that a
satisfies the two following conditions:
(i) One can add a to any sensible logic while preserving consistency.
(ii) In all the thus augmented logics a is logically equivalent to a
factual phenomenon.[3]

It is clear that RTC 2 is quite stringent—so much so indeed
that the question arises whether, in addition to facts, there are any
other true propositions. In particular, what about the outcome
of a (quantum) measurement? A distinction must of course be
made between the *position of the pointer*, which is a fact, and any
property that may be correlatively attributed to the measured
system itself. In the theory, it is consistent, Omnès claimed, to
assert that at the (exact) time when the measurement actually
begins, the proposition that the measured quantity *has* the value
the measurement will reveal is true. Also true, of course, are the
propositions describing present classical properties, as well as
those relative to such past classical properties as can be recon-
structed in a deterministic way from present classical records.

_____

3. Logical equivalence of two propositions a and b is simply defined as "$a \Rightarrow b$
and $b \Rightarrow a$."

But it seems [15] that this list is exhaustive. All other propositions bearing on quantum systems and which are considered as being true in conventional quantum theory are merely "reliable" in Omnès's theory.

This latter fact has quite far-reaching implications as regards the possible meaning of the sentences appearing in the theory. Although, in the latter, such words as 'property' and 'state' are freely used and applied to microscopic quantum systems, they are not, there, understood in the usual sense. When, in ordinary language and also in conventional textbook quantum mechanics, we speak of the property P of a microscopic physical system S, such as an atom or a proton, we usually (rightly or wrongly) imply that there is a variety of circumstances under which the proposition "system S has property P"can meaningfully be said to be true; and we consider it as obvious that if P *is* true on S at some time t the thus described state of affairs does not depend on whether or not S is, at time t, just beginning to interact with an instrument of observation suitable for measuring P In other words we feel that P would remain true even if, without changing anything else, we would remove the instrument. Obviously this is not the case in Omnès's theory any more than it is the case within Bohr's interpretation of quantum physics, since in both approaches the presence of the instrument is a necessary condition for the very possibility of P being true.

A consequence of this is that, as used in Omnès's theory, the word 'state' is somewhat of a misnomer since in normal language when we say a system is in such and such a state we mean it has some properties, and the assertion that it has them is then considered as true. Omnès's theory parts with such an acceptation of the word in quite a radical way. The 'states' it describes by kets or state operators obey the familiar rules (tensor products, Schrödinger time evolution, and so on) but as mere mathematical devices that in general cannot be thought of as corresponding to any property being true on the system under study. The one-spin example considered above, where a free spin 1/2 particle is considered to be at time t_1 in the state $|u_1 >$ corresponding to $S_a = +1/2$ and whose S_b spin component is measured at a later time t_2, may serve to illustrate this. There, the familiar quantum law of time evolution results in the particle remaining in state $|u_1 >$ until it interacts with something. Nevertheless, we are not allowed, as we saw, to consider that during this time (i.e., from

t_1 to t_2) the statement "S_a is equal to $1/2$" is a *true* one. The 'state' (in the mathematical sense) of the system is therefore in no way a description of the "state the system is in" in the usual sense we give to this expression: so that—as, it must be said, in *other* thoughtful interpretations of quantum mechanics—this 'state' cannot be interpreted here as "something real."

In Section 11.1 the relationship between Bohr's standpoint and the partial definitions procedure was pointed out. It is clear that a similar relationship also exists between this procedure and Omnès's views since, there also, a property of a microsystem can only be true if an instrument is at hand for measuring it. It is also clear that, correlatively, both Bohr and Omnès had to somehow, at least implicitly, discard anything resembling the counterfactual definition procedure that is so natural to us in daily life. This leads them to standpoints that may well be viewed as paradoxical, such as the one about a property of a system ceasing to be true if and when we remove the instrument with which the system was just about to interact. To follow them along these lines implies renouncing counterfactuals in a really quite drastic way.

However, the fact remains that these paradoxes are not *logical* ones. They are paradoxes only with respect to our normal ways of thinking, which do involve belief in free choice concerning future, unchangeability of the past, and so on. Hence the fact that the theory under study *could* be consistently built up may be viewed as an additional corroboration of something that has become increasingly apparent in modern times: the fact that rational thought is not at all bounded by limits due to apparently "obvious" notions, as we intuitively tend to think. In this also, of course, it is in line with Bohr's teaching.

But there is also another, more precise, lesson to be drawn from this theory. It again has to do with what was earlier called the "Bohr-like" definition of dynamical properties. In Bohr's writings this definition procedure is expressed only in a qualitative way, by means of assertions of a great generality, and when trying to make it more precise we soon encounter a conceptual difficulty. This is that in the simple case in which the operator of the measured quantity A commutes with the Hamiltonian of the system this Bohr-like definition seems to imply that A already had a definite value (the one that the measurement will reveal) some finite time before the actual measurement took place. If this were true, genuine logical paradoxes could well emerge. One of the interests of Omnès's work is that, by changing Bohr's insight into a precise theory, it makes it possible to assert that the implication at issue is in fact not a valid one. Note that we can clearly see the reason why it is not (d'Espagnat [14]) by taking up again (in reverse order) the one-spin example and considering in detail what can be said concerning the second measurement. Any measurement takes place during some finite time interval so let this one take place between times t_2 and $t_2 + \Delta t_2$. Omnès's theory tells us that at time t_2 the proposition $P \cdot (S_b = m_b)$ [which we then may note $P(t_2)$] is true. But is it true at a time $t_2 - \varepsilon$? To answer this question we must inquire whether or not some consistent logic exists to which Proposition P *cannot* be added without destroying its consistency. In this case the answer is that such logics indeed exist. An example is the consistent logic:

$$L \equiv \{(S_a = m_a \text{ at } t_1),$$
$$(S_a = m_a \text{ at } t_2 - \varepsilon/2),$$
$$(S_b = m_b \text{ at } t_2)\} \qquad (11.15)$$

(indeed, it is easily verified that if we actually inserted an S_b measuring instrument I at time $t_2 - \varepsilon$ the mere interaction of I and the system would modify the probabilities of the subsequent events). Therefore Proposition P $(t_2 - \varepsilon)$ is not true. It is merely "reliable in some logics." Of course, one of the most striking features of Omnès's updated theory [15]—namely the aforementioned drastic limitation of the types of propositions that can be "true"—can be considered as being a generalization of this.

To sum up, we see, first of all that a restrictive criterion for truth such as (for the sake of preciseness) RTC2, is an essential element in the (present-day version of) Omnès's theory, since it makes it contradiction-free. In view of the fact that this criterion severely limits the extension of the set of true propositions, it is clear that the theory in question cannot, in its present stage, be interpreted as yielding a possibility of reconciling quantum mechanics with strong objectivity. In other words it cannot be classified among the ontologically interpretable *quantum* theories (see Chapter 13). On the other hand, the formal preciseness of its content makes it an inspiring and efficient tool for further investigations in the very field we are interested in in this book, namely the borderline between physics and epistemology. In fact, we shall below make substantial use of this theory at the place (Chapter 15) where we put forward and discuss the notion—central in our analysis—of empirical reality.

References

1. N. Bohr, *Atomic Physics and Human Knowledge*, Science Editions, New York, 1961.
2. N. Bohr, *Phys. Rev.* **48**, 696 (1935).
3. L. Rosenfeld, in *Louis de Broglie, physicien et penseur*, Albin Michel, Paris, 1953.
4. B. d'Espagnat, *In Search of Reality*, Springer-Verlag, New York, 1983; *Reality and the Physicist*, Cambridge U. Press, 1989.
5. C.G. Hempel, *Methods of Concept Formation Science*, International Encyclopedia of United Science, University of Chicago Press, 1953.
6. H.E. Andås and O.K. Gjøtterud, *Found. Phys. Letters* **6**(1), 55 (1993).
7 J. Roldán, *Langage, mécanique quantique et réalité, un essai sur la pensée de Niels Bohr*, Université de Paris, Panthéon-Sorbonne, 1991.
8. R. Griffiths, *J. Stat. Phys.* **36**, 219 (1984).
9. R. Omnès, *J. Stat. Phys.* **53**, 893, 933, 957 (1988).
10. B. d'Espagnat, *Phys. Lett. A* **124**, 204 (1987).
11. B. d'Espagnat, *J. Stat. Phys.* **56**, 747 (1989).
12. R. Griffiths, *Phys. Rev. Letters* **70**, 2201 (1993).
13. R. Griffiths, *Found. Phys.* (in press).
14. B. d'Espagnat, *Found. Phys.* **20**, 1147 (1990).
15. R. Omnès, *Rev. Mod. Phys.* **64**, 339 (1992).

Quantum Mechanics as a Universal Theory, Classical Appearances in a Quantum World

12 In this chapter we review and discuss the essential aspects of three theories—or rather clusters of theories—that, however distinct, all lay special emphasis on more or less the same points. These points essentially are that if (as assumed by their authors) quantum theory is to be applicable to the world as a whole, this implies that it should be expressible with no reference to classical instruments lying outside the analyzed system and correlatively that it should somehow account for the emergence of classical appearances within this world. The theories we are interested in are the Everett relative state theory, the theory of environment-induced classical appearances (Zeh, Joos, Zurek, Unruh, Caldeira and Leggett)—briefly referred to as the "environment" theory—and the Gell-Mann and Hartle cosmological theory.

12.1 The Relative States Theory

Detailed accounts of Everett's "relative state" interpretation of quantum mechanics can be found in many articles and textbooks [1, 2], so that here we can keep to its most essential aspects. Let it be emphasized right at the start that there is no single generally accepted version of this theory· some, such as Bryce De Witt, call it a "many-worlds" theory. Others, such as Everett himself (at least at most places in his articles) rather describe it as a theory of *one* (branching) universe. Here, for brevity sake and because we see it as more promising, we only consider the latter (tentative) interpretation.

One of Everett's main ideas seems to have been to include the observers and their consciousnesses as purely physical systems in the theory. For this purpose he assumed that the universe as a whole, including the observers, is completely described by a wave function referred to as the "wave function of the universe." This wave function, or state vector, obeys the first law of evolution of quantum mechanics (Schrödinger's equation or its generalizations). In other words, no wave-packet reduction ever occurs.

In view of all our preceding discussions, particularly those of Chapter 10, such a statement seems, at first, very strange. Even in a simplified universe consisting merely of a system S and an instrument A, the evolution would, in general, produce a state in which the pointer of A would be spread over several of its possible macroscopically distinct positions. Similarly, if A involved an observer, this observer would in general be expected to evolve into a superposition of distinct states of consciousness. The originality of the theory under discussion is that it serenely accepts these apparently absurd conclusions. Yes, it asserts, measurement-like interactions occur all the time between the various components of the universe. Yes, as a result this universe of ours is continuously splitting into branches. Yes, we, as parts of this universe, are ourselves continuously splitting.

To those of us who would object that we do not feel ourselves split, the proponents of the theory answer by proving a proposition: To the extent that we can be regarded as mere automata, *the laws of quantum mechanics do not allow us to register any trace of this split* (hence, presumably, to "feel" it). Since this is obviously one key point, let us first show that the proposition does indeed follow from the principle of the theory.

Let us again consider our simplified world, in which there exists only one system S, with state vector $|\psi>$, and one instrument A, whose possible state vectors are $|m,r>$ To the extent that observers are simply automata, A can also, of course, be an observer. Let us assume moreover that if, initially,

$$|\psi> = \sum_m c_m |\psi_m>$$

the evolution in time of systems S and A is described, as in Chapter 10, by the transition

$$(\sum_m c_m |\psi_m>) \otimes |0,r> \to \sum_m c_m |\psi_m> \otimes |m,s_{m,r}> \quad (12.1)$$

It must be stressed that, contrary to the case in Chapter 10, the state vectors in (12.1) describe, not ensembles but just *one* system $S + A$, namely the simplified universe under consideration. It must also be stressed that, correlatively, in a simple process like the one described in (12.1) there is as yet no place for a statistical interpretation.

Following Everett, let us rewrite the right-hand side of (12.1) as

$$\sum_m |\psi_m> \otimes |\phi_m>$$
(12.2)

where

$$|\phi_m> = c_m |m, s_{m,r}>$$
(12.3)

is what this author calls a "relative state" of the instrument. Obviously, (12.2) associates any $|\phi_m>$ with a given value of m to the $|\psi_m>$ that has the same value of m. This is the reason why $|\phi_m>$ is said to be relative to $|\psi_m>$ and conversely. When all the c_m are zero except just one, c_1 say, (12.2) has, of course, the simple and straightforward interpretation that S is in the state $|\psi_1>$ and that, correlatively, A is in the state $|\phi_1>$. When several of the c_m are different from zero, on the other hand, the association between $|\psi_m>$ and $|\phi_m>$ cannot just be described in terms of the familiar notion of a *correlation* since we lack here the statistical element. However (12.2) does formally imply an association between state vectors of S and state vectors of A, and this is sufficient to allow us to proceed.

The next step is to consider iterated measurement processes that can be described by relation

$$(\sum_m c_m |\psi_m>) \otimes |a_0> \otimes |b_0> \rightarrow (\sum_m c_m |\psi_m> \otimes |a_m>) \otimes |b_0> \rightarrow$$
$$\sum_m c_m |\psi_m> \otimes |a_m> \otimes |b_m>$$
(12.3a)

This category of processes encompasses the measurements by a second instrument B of the "pointer position" of instrument A after A has performed a measurement on S. It also includes processes of repeated measurements, when B measures again, on S, the observable that A has already measured. B and A can then, of course, be considered as two distinct parts of the same instrument so that (12 3a) also describes a mere repetition of a measurement already made once Then the states $c_m |a_m > \otimes |b_m >$ appearing in (12 3a) are just the relative states of the instrument In spite of the fact that no reduction of the wave packet ever takes place, it is of course apparent from (12 3a) that the formal association mentioned also holds between the outcomes of the two separate instruments In other words, with a given $|\psi_m >$ (12 3a) associates unambiguously relative states of A and B that are labeled by the same index m When this is the case we say that *by definition* "the results of the two measurements agree with one another " The split is there, but no trace of it is registered

More complex questions can be asked For instance, let us assume that a complex system S is split into two parts, U and V, as in the examples of Chapter 8, that U and V fly far apart, that a quantity L^U is measured on U by an instrument A that registers the result on a memory tape, that a quantity L^V—strongly correlated with L^U—is similarly measured on V by an instrument B, and that, in the end, A measures the value registered in the memory of B and compares it with the value it has already registered in its memory upon measurement of L^U Because of the one-to-one correspondence we assumed between L^U and L^V (the aforementioned "strong correlation"), the two results obtained by A should agree with one another To verify that this is the case let us write the system $U + V$ as (with obvious notations)

$$\sum_m c_m |u_m > \otimes |v_m > \tag{12 4}$$

Let $|b_i >$ describe the possible states of the instrument B, and let $|a_{j,k} >$ describe those of instrument A In $|b_i >$ the value $i = n$ refers to the state into which B would evolve as a consequence of its interaction with V if all the c_m were zero except just one, labeled n In $|a_{j,k} >$ the index j has the same significance (only, now, with respect to A and to U) and the index k refers in a similar way to the measurement made by A on B it takes the value $k = p$ if B is observed by A to be in state $|b_p >$ Then, if the initial state is

$$|a_{0,0} > \otimes |b_0 > \otimes \sum_m c_m |u_m > \otimes |v_m > \tag{12 5}$$

the states just after the first, second, and third measurement processes are, respectively, described by

$$\sum_m c_m |a_{m,0}> \otimes |b_0> \otimes |u_m> \otimes |v_m>$$ *(12.6a)*

$$\sum_m c_m |a_{m,0}> \otimes |b_m> \otimes |u_m> \otimes |v_m>$$ *(12.6b)*

$$\sum_m c_m |a_{m,m}> \otimes |b_m> \otimes |u_m> \phi |v_m>$$ *(12.6c)*

this being a mere consequence of the Schrödinger-like time evolution of the complete state vector.

The fact that the *same* index m appears twice in the first factor of Eq. (12.6c) shows that indeed the two results obtained by A do agree, as they should. Moreover, it has been shown [2, 3] that, the usual statistical interpretation of the coefficients c_m *emerges from the formalism itself*, provided that the—very natural looking—assumption is made that state vectors with zero norm correspond to nonexisting branches (for a proof see, e.g., [4]). Cogent arguments have been put forward to the effect of showing that this assumption is not so "innocent" as it seems, but we shall not enter here into such debates.

The distinctive feature of the theory is of course that in equations such as (12.4) all the "branches" $|u_1> \otimes |v_1>$, $|u_2> \otimes |v_2>$, , $|u_n> \otimes |v_n>$ simultaneously exist. This statement can, however, be understood in two very different ways. First we can understand it in the material sense that neither the number of instruments nor the number of systems is constant during the process: in each measurement-like process each one is multiplied by N, the number of branches in the process. This solution is the one favored by De Witt. Another possibility, the only one to be considered here as we said, is to assert that the number of systems and the number of instruments is always conserved. This view is apparently the one Everett had in mind and is also the one that most of us would accept without qualms as long as S and A are two quantum systems. But when it comes to the question of whether this view should finally be extended to systems $S + A$, the A part of which incorporates conscious observers a difficulty appears. It is due to the unquestionable fact that we have impressions that are definite (even if and when they are delusory). If we want to avoid collapse it seems we have only two possibilities. Let us here summarily describe the one Everett seems to have favored (although he did not make the point quite explicit) The second one is described in the next section

The solution Everett seems to have favored is that each part of the wave function of $S + A$ somehow corresponds to one possible state of awareness of the observer's memory This means that when the wave function is of the form (12 1) with several nonvanishing c_m, Everett must assume A has a definite property, namely this very "state of awareness," whose remarkable property is that it can be put in correspondence with one component only of the total Hilbert space vector This is very much at variance with the general rules of standard quantum mechanics, which imply (see Section 4 3 and Rule 9 of Chapter 3) that for an observable to have a definite value a necessary condition is that the system it belongs to is in an eigenstate of that observable It is clear therefore that Everett's approach amounts to considering "consciousness"—or "awareness"—as a property the nature of which is qualitatively different from that of ordinary physical properties as conceived of in conventional quantum mechanics (where we do not think of, say, the position of a particle as having a definite value when the wave function is extended) In fact it is difficult, for this reason, not to think of such a 'state of awareness' as of a "supplementary"—or "hidden"—variable of some sort, although one of a special kind (of course, the term "hidden" is especially inappropriate in this context, and must be understood only in its "technical" sense) This implies in particular that, contrary, as it seems, to Everett's expectations, this author's approach does not reduce mind to ordinary quantum mechanical processes On the other hand it must be granted that, if we accept this, then, as shown above, the theory does account for the so-called intersubjective agreement consisting of the fact that we normally agree on such things as whether or not we see a teapot on a table (see Chapter 1) or the graduation interval in which we observe an instrument pointer to lie A description of a *hidden variables* theory treating consciousnesses as supplementary variables is given in Section 16 6

There are many problems of a semitechnical kind connected with Everett's theory The two main ones bear on the possibility of deriving (without implicit postulates) the probability rule (Rule 6 in Section 3 1) from the basic views of the theory and the problem, already met with in Section 10 1, of what determines the Hilbert-space basis according to which our (common) awareness, divides into branches. These two interesting but quite difficult problems have been studied extensively by many authors. It cannot be said that they have been solved to general and complete satisfaction, but still, with all the work that has been made on them it now seems they are solvable. For more information the reader is referred to the detailed and exhaustive work of Y. Ben Dov [5].

12.2 An Explicitly Dualistic Model

As stressed above, in the Everett theory it seems impossible to completely reduce mind to ordinary quantum processes. This being so, there is some rationale for approaching this theory from a point of view (d'Espagnat [4: ch. 23])[1] that openly considers mind as something different from the physical reality quantum mechanics is supposed to describe.

In such an approach there would, of course, be no point in ascribing a wave function to consciousness or making the latter an element of some kind of a wave function. Let us therefore describe the evolution of a composite system $S + A$, where A is endowed with a consciousness C, by the system of relations

$$\sum_m c_m |\psi_m> |a_0> \rightarrow \sum_m c_m |\psi_m> |a_m> \qquad (12.7)$$

$$C_0 \rightarrow C_m \qquad (12.7a)$$

the symbols C_0 and C_m standing respectively for the initial and final states of A's consciousness C (for simplicity the degeneracy index in the kets describing A have been suppressed).

When an observer performs measurements on several—say a large number of—identical systems $S^{(1)}, S^{(2)}, \quad S^{(M)}$ that initially are all in the same state, Eq. (12.7), according to the standard Everett scheme, generalizes to:

$$(\sum_m c_m |s_m^{(1)}>)(\sum_n c_n |s_n^{(2)}>) \quad (\sum_r c_r |s_r^{(M)}>)|a_{0,0,...0}> \rightarrow$$
$$\sum_{m,n,...,r} c_m c_n \quad c_r |s_m^{(1)}> |s_n^{(2)}> \quad |s_r^{(M)}> |a_{m,n,...,r}> \qquad (12.8)$$

---

1. Recently, this dualistic model was also considered, from a somewhat different angle, by Albert [6].

But in the present model the state of consciousness of A is in correspondence with only one of the kets $|a_{m,n,...r}>$ Now it has been shown—this is part of the proof alluded to below Eq. (12.6c)—that, when the number, M, of the systems S goes to infinity, the relative number of the indexes $m, n, \quad , r$ in the ket $|a_{m,n,...r}>$ that have a given value p tends to $|c_p|^2$ This result here implies that the observer's consciousness C contains $M|c_p|^2$ "marks," indicating that he has had the impression of interacting with a system in state $|s_p>$ Hence this model exactly reproduces the predictions of the conventional theory.

On the other hand, the model also has some features of its own, one of which, at least, is somewhat surprising. It concerns the phenomena of correlation at distance in which two observers participate and one, A, gets informed of the impression the other, B, has had, in the way described above, as concerns instruments, by means of Eqs. (12.5–12.6c). For the sake of maximum clarity let us assume that the two measurements performed by A and B take place in spacelike separated regions and that A afterward gets informed, by ordinary means, of the impression B has had. Then, in view of the facts that all the branches of the overall wave function are present (no collapse) and that, when B performs his measurement, he has no way of being informed of the outcome of A's measurement and conversely, to assume that A and B get impressions originating from the same branch would be quite arbitrary and groundless. In the model, therefore, it is natural to consider that in general the outcomes in question originate from different branches and are therefore different and uncorrelated. In other words, in a reference frame in which the time order of the measurements is still the one considered when writing Eqs. (12.5–12.6c), expression (12.6a) is now replaced by the system:

$$\sum_m c_m |a_{m,0}> \otimes |b_0> \otimes |u_m> \otimes |v_m>, \qquad C_i^{(A)}, \quad C_0^{(B)}$$

where $C_i^{(A)}$ means that the state of consciousness of A is the one that corresponds to the branch $m = i$. Similarly, expression (12.6b) is replaced by the system:

$$\sum_m c_m |a_{m,0}> \otimes |b_m> \otimes |u_m> \otimes |v_m>, \qquad C_i^{(A)}, \quad C_j^{(B)}$$

where j has no reason to be equal to i.

But does this imply that the model predicts—in the considered example—a violation of the intersubjective agreement between A and B will be observed, in violation of the quantum mechanical predictive rules? Not in the least, for A can never *know* that B does not have the same impression he has. The reason is that any transfer of information from B to A—including even an answer by B to a question asked by A—unavoidably proceeds through physical means. It necessarily takes the form of a measurement made by A on B (more precisely on the neuronal system and communication organs of B) the description of which is entirely contained in the wave function. Under these conditions, as we already checked by means of Eq. (12.6c), A necessarily gets a response (answer) that agrees with his own perception. But in the present conceptual framework there is no reason that this apparent agreement should correspond to something real.

In Section 1.4 the intersubjective agreement problem was touched upon in general, philosophical terms and it was noted that quantum mechanics offers a formal solution to it that is different from the usual realist answer that "we all see one teapot on the table because there really *is* one teapot on the table." However this solution referred to collapse, and indirectly therefore to the idea that there are classical instruments lying outside the realm of what is describable by quantum mechanics. Here we have an example of a solution that involves neither collapses nor classical instruments. It describes the whole of the physical world in purely quantum mechanical terms. But, for all that, it does not restore the commonsense realistic explanation of intersubjective agreement. It is interesting to observe that, quite the contrary, it makes this agreement an appearance.

• *Remark 1.* Albert and Loewer [7] (see also Albert [6]) have put forward a modification of this model based on the assumption that every sentient physical system there is is associated not with a single mind but with an infinity of minds. Further assumptions in this theory have the (interesting and surprising) effect of restoring locality. As explained by these authors, this is not a violation of Bell's theorems because the latter are based on the assumption that there are matters of fact about the outcomes of a pair of measurements: these outcomes are considered as physical facts so that any strict correlation existing between them is also a physical fact But, as is apparent from the above, this is not the case here Even in the model described, in which *A* and *B* each have but one mind, there is no factual correlation between what each of them observes The correlation is only between *B*'s state of consciousness and the belief *A* develops about what *B* observed But this belief he grounds on information he gets by "normal" physical means involving no violation of local causality

12.3 Emergence of Classical Appearances Through Interaction with Environment

In the foregoing short summary of Everett's approach a rather obvious but important point was left implicit It is clear that, to provide a foundation for states of awareness, some kind of (machinelike or brainlike) processing and storing of data, necessarily associated with complex systems, is necessary Everett did acknowledge this (he did use the concept of "memories") On the other hand it must be granted that he did not actually explain how data and facts stable and definite enough to be associable with definite states of awareness, emerge from the "undefinable cloud" of the universal state vector The now to be reviewed theory, which deals with just this, may be said to fill in this gap But needless to say it is primarily interesting for its own sake

Its guiding idea is to focus on the nonaccessibility of the environment variables However, in quantum measurement theory the nonaccessibility of the exact initial state of the instrument—which includes the environment—does not, by itself solve the measurement riddle (see Section 10 4) Similarly, while in classical statistical mechanics the random behaviour of the systems is often attributed to an uncontrollable influence exerted *on* them *by* the environment, the quantum measurement riddle cannot be removed in such a way To this end, the idea that works (see Section 10.6) is indeed the reverse of the latter: taking the action *of* the system *on* the environment into account.

Now, measurement being considered as performed by means of macroscopic instruments with "classical" properties, the measurement problem is obviously tightly linked with the more general one of explaining the existence of the classical properties of macroscopic objects (even including, though as a limit case, the structure of large biomolecules), and what holds true concerning the first problem may be expected to extend also to the second: in other words no reference to the uncontrollable influence of the outside world is likely to account for the classical properties of macroscopic systems. In fact the theory we are about to consider, due to Zeh, Joos, Zurek, and others [8–11], takes just the opposite standpoint. Its guiding idea is as follows. The macroscopic system under study—call it S—modifies the state of the environment in much the same way as, in a measurement process, the measured system modifies the state of the instrument (plus, possibly, that of the environment, of course). As a consequence, the state vector of S gets entangled with that of the environment so that the state operator ρ_S describing S alone (obtained by partial tracing over the environment variables) is not a projection operator. Just as in measurement theory (see Sections 10.1 and 10.6) it is such that an ensemble of such systems is an *improper mixture*. If, now, we decide to forget about the variables (such as M in Section 10.6) the measurements of which might conceivably reveal the entanglement, we can, at least for operational purposes, identify the improper mixture with a proper one and say that the ensemble in question is a mixture of subensembles and that the elements of these subensembles have definite values of the observables the eigenvectors of which diagonalize ρ_S. Under appropriate circumstances these observables will hopefully coincide with the familiar, classical ones.

This approach is new in two ways. First it attributes the fact that some systems exhibit features we call classical not to actual properties that these systems would "inherently possess" (as did classical physics) but just to the fact that the systems in question unavoidably interact with their environment in a *non*negligible way. Second, according to this approach what is relevant in the interaction mechanism is not the influence of the environment on the system (which, in limiting cases, could even be thought of as vanishing, same as what takes place in ideal measurement processes, see Section 10.2 where the influence of the instrument on the system also vanishes). As stressed by Joos and Zeh [9] and already noted, it is, in a way, the opposite: the influence of the system on the environment.

The way these general ideas apply concerning one of the most important observables, namely position, is as follows (Joos [12]). Consider a "free," small but still macroscopic, object S, such as a typical interstellar dust grain. This grain is not totally isolated since it, at least, encounters and scatters photons from the 3°K cosmic background radiation, and presumably other particles as well. Consider one of these scattered particles and let $|\chi >$ be its state vector before collision. After the collision its state vector $|\chi_x >$ will depend on the location x of the dust grain center of mass. If this location were infinitely precise such a transition could be described as

$$|x > \otimes |\chi > \rightarrow |x > \otimes |\chi_x > \qquad (12.9)$$

where the mass of the dust grain is considered as infinite, so that the collision is recoil-free ($|x >$ is unchanged). The arrow describes the time evolution of the composite system as governed by the time-dependent Schrödinger equation, much as in Eq. (10.2). When the dust grain is not in an eigenstate of its center of mass position operator but is, instead, described by a state vector $|\phi >$ such that $< x|\phi >$ is not a delta-function (which is the general case, of course), the linearity of the Schrödinger equation entails, just as in the case of a measurement [Eq. (10.4)], that

$$|\phi > \otimes |\chi > \equiv \int dx\, |x > < x|\phi > \otimes |\chi >$$
$$\rightarrow \int dx < x|\phi > |x > \otimes |\chi_x >$$
$$\equiv \int dx\, \phi(x)|x > \otimes |\chi_x > \qquad (12.10)$$

where Eqs. (3.3d) and (3.17) have been made use of. The state operator ρ of the composite system is then $|\Psi > < \Psi|$, where $|\Psi >$ is the right-hand side of Eq. (12.10). Hence

$$\rho = \iint dxdx'\ \phi(x)\phi^*(x')\ |x > < x'| \otimes |\chi_x > < \chi_{x'}| \quad (12.11)$$

Since we are only interested in the behavior of the dust grain, we can work just with its state operator ρ_S, which as we know (Section 7.2) is the partial trace of ρ over the scattered particle Hilbert space. Hence, from (7.11):

$$\rho_S = \iint dxdx'\ \phi(x)\phi^*(x')\ |x > < x'| < \chi_{x'}|\chi_x > \quad (12.12)$$

Equation (6 23) then yields the corresponding density matrix, here also called ρ_S

$$\rho_S(x'', x''') = \iint dx dx' \quad \phi(x)\phi^*(x')$$
$$\delta(x - x'')\delta(x' - x''') < \chi_{x'}|\chi_x >$$
$$= \phi(x'')\phi^*(x''') < \chi_{x'''}|\chi_{x''} > \qquad (12\ 13)$$

This shows that, because the dust grain encounters and scatters photons and/or other particles, its center of mass density matrix, which would be

$$\rho_S = \phi(x)\phi^*(x') \qquad (12\ 14)$$

if the grain were isolated, changes to

$$\rho_S = \phi(x)\phi^*(x') < \chi_{x'}|\chi_x > \qquad (12\ 15)$$

The multiplicative factor $< \chi_{x'}|\chi_x >$, being a scalar product of two states scattered from scattering centers at different positions x and x' is of course small as soon as $|x' - x|$ is large It can be shown that it is approximately zero as soon as $|x' - x| \gg \lambda$, the incoming particle wavelength But even slow scattered particles can appreciably modify ρ_S if sufficiently many are scattered per unit time In the case of randomly distributed scattered particles (e g , photons) Joos and Zeh have calculated from Eq (12 15) that if there are a great many such scattering processes the overall effect is an exponential damping of the type

$$\rho_S(x, x', t) = \rho_S(x, x', 0) \exp[-\Lambda t(x - x')^2] \qquad (12\ 16)$$

with a so-called *localization rate* Λ given by

$$\Lambda = k^2 \sigma_{eff} N v / 8\pi^2 V \qquad (12\ 17)$$

where k is the wave number of the incident particles, σ_{eff} is a typical cross section and Nv/V is the incoming flux For a dust particle with radius 10^{-5}cm cosmic background radiation alone gives $\Lambda = 10^{-6}$cm^{2}sec^{-1}, thermal radiation at room temperature gives $\Lambda = 10^{12}$cm^{2}sec^{-1} and scattering of air molecules gives $\Lambda = 10^{19}$cm^{2}sec^{-1} in the best available laboratory vacuum (numbers taken from Joos [12])

Clearly when Λt is large the exponential factor in (12.16) has the effect of squeezing ρ_S along the first diagonal in the **x, x'** coordinates. What is the *physical* meaning of such a change in a density matrix? This is a general question, the interest of which is not restricted to the particular theory here under study. But this theory offers us an opportunity to look into the matter.

For this purpose, consider first the very idealized, simple case of a (unit normalized) wave function $\psi(x)$ that is a constant within some interval and zero elsewhere. Specifically let

$$\psi(x) = \begin{cases} (Na)^{-1/2} & \text{for } 0 < x < Na \\ 0 & \text{elsewhere.} \end{cases} \qquad (12.18)$$

N being a positive integer.

The corresponding density matrix

$$\rho = \psi(x)\psi^*(x') \qquad (12.19)$$

is then equal to Na within the square $(0 < x < Na, 0 < x' < Na)$ and zero elsewhere.

In order to simulate the change from Eq. (12.14) to Eq. (12.15) (i.e., the "squeezing" of ρ along the first diagonal) let us replace this ρ by a matrix ρ' such that

$$\rho'(x,x') \propto \rho(x,x')$$

when the point P with coordinates x, x' lies within any one of the small squares defined by $(n-1)a < x < na$ and $(n-1)a < x' < na$, n being a positive integer $n = 1, \quad ,N$ (these squares are obviously aligned along the first diagonal $x' = x$), and

$$\rho'(x,x') = 0$$

whenever P lies elsewhere. The thus defined density matrix can obviously be written:

$$\rho' = \sum_n N^{-1}\psi_n(x)\psi_n^*(x') \qquad (12.20)$$

with

$$\psi_n(x) = \begin{cases} a^{-1/2} & \text{if } (n-1)a < x < na \\ 0 & \text{elsewhere} \end{cases} \qquad (12\ 21)$$

When we want to calculate the probability that the described particle be found in such and such an interval we can of course use ρ' instead of ρ since we must then set $x' = x$ On the other hand, Eq (12 20) can also be written

$$\rho' = \sum_n p_n \psi_n(x)\psi_n^*(x') \qquad (12\ 22)$$

with

$$p_n = N^{-1}$$

and can therefore be interpreted as describing a mixture with probabilities p_n of states ψ_n each of which is localized within an interval $[(n-1)a,\ na]$ much narrower than the one $[0,\ Na]$ within which the original ψ is

The foregoing is an indication that a decrease of the nondiagonal terms of a position density matrix [such as the one that corresponds to replacing Eq (12 14) by Eq (12 15)] may mean replacing an ensemble of poorly localized particles by one in which the particles are individually more sharply localized However, it is just an indication, and we shall discover in a moment that things are not quite as simple as this Actually, what necessarily follows from the fact that a density matrix is squeezed along the first diagonal is not quite this but a fact we normally consider as a consequence of such a localization, namely the fact that interference terms get suppressed

That they indeed are, can be shown as follows Let some particles propagate along Ox and let

$$\Pi_\alpha(z) = \begin{cases} 1 & \text{if } \tfrac{1}{2}[-a - (-1)^{\alpha+1}\Delta z] < z < \tfrac{1}{2}[a - (-1)^{\alpha+1}\Delta z] \\ 0 & \text{elsewhere} \end{cases} \qquad (12\ 23)$$

with $\alpha = 1$ or 2 In a Young-type two-slit experiment Π_1 and Π_2 thus describe the apertures of two slits of width a opened at a distance Δz from one another in a plane diaphragm with abcissa $x = 0$. If, before it reaches the diaphragm, the ensemble of the considered particles is described by the density matrix

$$\rho(z, z') \qquad\qquad (12.24)$$

immediately beyond it, it must be described by

$$\rho'(z, z') = N[\Pi_1(z)$$
$$+ \Pi_2(z)]\rho(z, z')[\Pi_1(z') + \Pi_2(z')] \qquad (12.25)$$

where N is a normalization factor (note that here as in Chapter 2, only the z variable is taken quantum mechanically into account, motion along the x axis being implicitly treated as classical). The density matrix $\rho''(z, z')$ at an abcissa x_1 beyond the screen is obtained from $\rho'(z, z')$ by making use of formula (6.17) (more precisely· of its "translation" in density-matrix terms). Because of the fact that the operations described are linear, this $\rho''(z, z')$ is, just as ρ', a sum of four terms corresponding to $(\alpha, \alpha') = (1, 1), (1, 2), (2, 1)$, and $(2, 2)$ respectively, and, of course, it yields interference fringes only if the terms $\alpha' \neq \alpha$ do not vanish, which implies that the same must be true also in Eq. (12.25). On the other hand, in the right-hand side of Eq. (12.25) a term such as

$$\Pi_1(z)\rho(z, z')\Pi_2(z') \qquad\qquad (12.26)$$

can obviously be appreciable only if ρ is itself appreciable for at least *some* values of z and z' obeying $|z - z'| > \Delta z - a$. This shows that the squeezing of the density matrix along the first diagonal does indeed suppress interference fringes that otherwise would have been there.

This brings us back to the localization question. Admittedly, an obvious way to make sure that a Young-type two-slit experiment will show no fringe is never to shoot on the diaphragm anything but beams whose wave functions are localized in z better than Δz. However, there is no proof that this sufficient condition is a necessary one as well. And in fact it is not. There are mixtures, some elements of which are z-localized much more poorly than this and whose density matrix is nevertheless sufficiently squeezed along the first diagonal as to obey the no-fringes condition. An interesting example of this state of affairs is the case in which the density matrix describing our ensemble of dust grains is a Gaussian of the form

$$\rho_S = K \exp[-(Ay^2 + Cz^2)] \qquad (12.27)$$

with

$$y = x - x', \qquad z = x + x' \qquad (12.28)$$

where the conditions

$$A \approx \Lambda t \qquad (12.29)$$
$$C \ll A \qquad (12.30)$$

must be met for ρ_S to be a special instance of the ρ_S shown in Eq. (12.16).

ρ_S is obviously such that no Young diaphragm with $\Delta z - a \gg A^{-1/2}$ can produce appreciable fringes. However if, following Joos and Zeh [9], we diagonalize ρ_S, we get

$$\rho_S = \sum_n p_n \phi_n(x) \phi_n^*(x') \qquad (12.31)$$

where

$$p_n = \frac{2C^{1/2}}{A^{1/2} + C^{1/2}} \left(\frac{A^{1/2} - C^{1/2}}{A^{1/2} + C^{1/2}} \right)^2 \qquad (12.32)$$

and where the ϕ_n are the harmonic oscillator eigenfunctions

$$\phi_n(x) = N H_n[2(AC)^{1/2}x] \exp[-2(AC)^{1/2}x^2] \qquad (12.33)$$

(the H_n are the Hermitean polynomials). Clearly, when $C \ll A$ the domains in which these ϕ_n have appreciable values are much larger than $A^{-1/2}$ and if $\Delta z - a < 2^{-1/2}(AC)^{-1/4}$ each separate subensemble corresponding to one definite ϕ_n would, if alone, generate fringes. Qualitatively, considering the expressions these ϕ_n have, we may, for these reasons and following again Joos and Zeh, say that they are inappropriate to represent localized particles. Note, moreover, that of course they are orthogonal to each other, but that, contrary to the ψ_n (Eq. (12.21) of the elementary example above, their supports are not disjoint.

At first sight, this result is disappointing It may look as if the theory were not capable of yielding localization, or, at least, not a sufficient one, and not of the expected kind The oddity here, however, is not as great as it may seem A point not explicitly made—to my knowledge—in the corresponding literature is, at this stage, quite important[2] It is that, in fact, the theory allows for a stricter localization than the one the ϕ_n in Eq (12 33) suggests This is because of a general fact pointed out in Chapter 6 a "proper" density matrix (i e , one that is not a projection operator) corresponds not to just one but to several—indeed an infinity of—different proper mixtures, in the sense that it constitutes an adequate description of each one of them Otherwise said when it is not requested that the ϕ_α be mutually orthogonal, any statistical operator ρ can be expressed in the form (6 1) in an infinity of ways A meaningful question then is In the case of our squeezed ρ_S, is there a choice of ϕ_α's and associated weights p_α such that this ρ_S should be expressible in terms of them, the ϕ_α being sufficiently localized—that is, localized to within the "reasonable" length scale $A^{-1/2}$?

The answer is that at least one such choice exists It is obtained by turning α into a continuous index-variable X and setting (up to appropriate normalizing factors)

$$\phi(x,X) = \exp[-2A(x - X)^2]$$
$$p(X) = \exp(-DX^2) \qquad (12\ 34)$$

with $D \ll A$ for expression (6 1) then reads

$$\rho_S(x,x') = \int dX\ \exp[-2A(x - X)^2]$$
$$\exp(-DX^2)\ \exp[-2A(x' - X)^2] \qquad (12\ 35)$$

that is (again within some renormalizing factor)

$$\rho_S(x,x') \cong \exp\{-[A(x - x')^2 + \frac{AD}{4A + D}(x + x')^2]\}$$

Expression (12 27) is then recovered by setting

$$C = D(4 + D/A)^{-1}$$

and with $D \ll A$ the "squeezing" condition $C \ll A$ is met Of course the $\phi(x, X)$, where X serves as a continuous "index,"

---

2 In fact it *was* made! The reference is E Joos, Phys Rev **D 36**, 3285 (1987); (*added note*)

are not orthogonal to each other. But there exists no compelling reason that would make, here, orthogonality necessary.

It is important to try and determine exactly what the preceding analysis has established and what it leaves as mere possibilities of speculation of a more or less "conceptual" type. What it has strictly established is that if an *ensemble* of dust grains (to keep to our specific example) interacts with its environment in the unavoidable way we saw it has to, and if no measurement involving this environment is assumed made [the partial tracing leading to Eq. (12.12)], then the ensemble in question is not distinguishable from a proper mixture of ensembles E_X of dust grains such that each E_X is a pure case, described by the corresponding wave function $\phi(x, X)$ [Eq. (12.34)] and therefore approximately localized within a domain of extension $A^{-1/2}$ around point X (which, of course, accounts also for our previous result that no interference fringes can be observed in a Young-like two-slit experiment if the distance between the slits is somewhat larger than $A^{-1/2}$).

Does the analysis show more? And would we like it to show more? To the last question we may be tempted to answer yes, since after all our dust grain just serves here as a typical representative of the class of macroscopic systems, and macroscopic systems do exist individually. Hence, if we could have shown that, under the conditions stated, an individual dust grain must appear as being localized, this would have been even nicer. But this, clearly, the theory does not achieve, just simply because it is a theory about ensembles. So that, about individual macroscopic systems it says nothing: neither that they are localized *nor that they are not!* It is true that in most problems the fact that quantum theory is basically one about ensembles is not felt as leading to a significant limitation of its power. Usually we can just simply forget in the end about this "detail." Here however the situation is different. As remarked previously, there are many proper mixtures that are described by the density matrix ρ_S associated with our ensemble of dust grains, and the pure subensembles composing most of them are not, or are insufficiently, localized. Arbitrarily selecting, as we did, the one (or ones) that does (do) have this property, saying it is sufficient that *one* should have it for accounting for (observed) localization, hence constitutes an intellectual further step, that we have to take here and which does not quite have its equivalent in the usual applications of quantum theory. An intelligent demon who would know the Schrödinger equation and such things but would not have the concept of locality ingrained in his mind, if he did the calculations reported above, would obviously not derive this concept from them. When we investigate the question of determining what, within our human knowlege, comes from the "outside world" and what from "us," this is a point that we shall have to keep in mind.

In the calculations reported above no account has been taken of the well-known "natural" growth in time of the wave packet describing the center of mass wave function of a free system. This effect tends of course to counterbalance the one studied here and which points toward a finer localization. Joos and Zeh have investigated this point and shown that in general the environment effect dominates over the internal dynamics of the system. More precisely they considered the *coherence length*, which measures the distance beyond which the grain should show no interference effect and which from Eq. (12.16) is

$$b = (8\Lambda t)^{-1/2} \tag{12.36}$$

in the simplified model, and showed that its limiting value for large times, with internal dynamic and recoil taken into account, is equal, not to zero but to the thermal de Broglie wavelength of the dust grain (the wavelength $\lambda = h/mv$ where m is the grain mass and v its average thermal velocity), a very small quantity in general (10^{-14}cm for a dust grain with radius 10^{-5}cm at room temperature).

The same theory similarly accounts for the fact that large molecules can be attributed a well-defined spatial structure and that some molecules have chiral states as their ground state.

For obvious reasons (lack of space in particular) it is of course impossible to report here on the extensive and also most significant work on these matters by Caldeira and Leggett [10], Zurek [11], Primas [13], Unruh and Zurek [14], and a few others. Just because they are important they cannot be summarized in a few pages. All of them go to prove that the interaction of the macroscopic systems with their environment contributes a great deal to create the appearance of a classical world. This is a considerable achievement for—as some philosophers would ask—why should an appearance be called a "mere appearance" if it can be shown to be universal and the same for everybody? On the other hand, let us be careful not to interpret these results in a careless and unwarranted way. They cannot and should not be given an ontological interpretation. The truth of this statement follows from a number of points that have been mentioned. We shall have occasions to comment on it in later chapters.

12.4   Remarks Concerning the Gell-Mann and Hartle Quantum Cosmological Theory

The fact that quantum mechanics works so remarkably well as a tool for making reliable predictions in all sorts of domains is of course felt by many physicists as a powerful incitement to the quest of a quantum theory of the universe as a whole. It is therefore not surprising that a number of attempts in this direction are made. Everett's theory was the first one, but, as explained in Section 12.1, it cannot be considered as having been entirely successful since it does not adequately account for the fact that our impressions are definite and more generally for the appearance of a classical world. Griffiths's and Omnès's theories go along the same lines, and, as we saw, they encountered difficulties of a somewhat similar nature, which they could overcome only at the price of introducing new forms of logic. More recently Gell-Mann and Hartle proposed a scheme [15] that has a number of points in common with Griffiths's and Omnès's models but in which the need for a new logic does not appear, or at least, is not mentioned explicitly. It may well be expected that the scheme in question is not the last of the series and that other theories with the same objective will be developed. For this reason, we shall not enter here into the detailed formalism of the last-mentioned proposal, but shall merely mention what, in its approach, is most relevant concerning the relationship of the human mind and the universe.

In so doing, the similarity between the formalism of this theory and that of Griffiths will be helpful. Gell-Mann and Hartle make use of a consistency condition that is essentially equivalent to the one proposed by Griffiths. That is, they use a condition, which they call the "decoherence condition," essentially generalizing condition (11.4) to a large number, n, of be shown to be universal and the same for everybody? On the other hand, let us be careful not to interpret these results in a careless and unwarranted way. They cannot and should not be given an ontological interpretation. The truth of this statement follows from a number of points that have been mentioned. We shall have occasions to comment on it in later chapters.

12.4 Remarks Concerning the Gell-Mann and Hartle Quantum Cosmological Theory

The fact that quantum mechanics works so remarkably well as a tool for making reliable predictions in all sorts of domains is of course felt by many physicists as a powerful incitement to the quest of a quantum theory of the universe as a whole. It is therefore not surprising that a number of attempts in this direction are made. Everett's theory was the first one, but, as explained in Section 12.1, it cannot be considered as having been entirely successful since it does not adequately account for the fact that our impressions are definite and more generally for the appearance of a classical world. Griffiths's and Omnès's theories go along the same lines, and, as we saw, they encountered difficulties of a somewhat similar nature, which they could overcome only at the price of introducing new forms of logic. More recently Gell-Mann and Hartle proposed a scheme [15] that has a number of points in common with Griffiths's and Omnès's models but in which the need for a new logic does not appear, or at least, is not mentioned explicitly. It may well be expected that the scheme in question is not the last of the series and that other theories with the same objective will be developed. For this reason, we shall not enter here into the detailed formalism of the last-mentioned proposal, but shall merely mention what, in its approach, is most relevant concerning the relationship of the human mind and the universe.

In so doing, the similarity between the formalism of this theory and that of Griffiths will be helpful. Gell-Mann and Hartle make use of a consistency condition that is essentially equivalent to the one proposed by Griffiths. That is, they use a condition, which they call the "decoherence condition," essentially generalizing condition (11.4) to a large number, n, of time-ordered "events" [Eq. (11.4) corresponds to $n = 2$]. To a sequence of projectors P_j^A, P_k^B, P_l^C, they, by definition, give the name "history." But to such a history we cannot, in general, assign a probability just as, in a two-slit experiment, we cannot, in general, assign probabilities to the two alternative histories in which the electron passed through one or the other slit, since—barring hidden variables—such an assignment is incompatible, as we know, with the existence of interference terms.

At several places in Chapter 10 we saw that, in the opinion of a great number of the theorists who worked on quantum measurement, a sufficient condition for a quantum measurement theory to reach its ends is that it should yield conditions under which cross-terms of this type are negligible or, at least, small. Apparently, this is also the authors' standpoint. In the two-slit example, such a condition is that the slits should be large enough, which is nothing else than a condition of coarse graining concerning the localization of the place where the electron traverses the diaphragm. In the projector formalism such a coarse graining implies that one "history" is a *coarse graining* of another if some of the P_α representing the first "history" are sums of the P_α of the second history, and it can be shown that the vanishing of the cross-terms corresponds, in this formalism, to the decoherence condition being satisfied (though decoherence can obtain also in more general circumstances). The corresponding histories are then, in Griffiths's language, "consistent" ones, and can be assigned probabilities, the values of which are yielded by Wigner's formula, (6.38).

With this schematic summary of the main ideas of the Gell-Mann and Hartle theory we are in a position to try and analyze its epistemological background in the light of the content of the foregoing chapters and Sections 12.1 and 12.3.

There are several points to which we must turn our attention. One of them bears on the nature of the coarse graining which, as the authors point out, is a necessary condition for decoherence and therefore for the very possibility of considering consistent histories. As they write, a completely fine-grained theory is specified by giving the values of a complete set of operators at all time. And there are at least three common types of coarse graining: (1) specifying observables, not at all times but only at some times; (2) specifying at any one time not a complete set of observables but only some of them; (3) specifying for these observables not precise values but only ranges of values.

A question of interest then is: "Are these three types of coarse-graining objective or subjective?" Since their definitions—as just stated—are obviously subjective, the question rather is: "Can they, or at least some of them, *also* be conceived of, at least in some circumstances, as objective?"

To illustrate the nature and bearing of this question, consider for example, the histories with observables specified at but one time t. Eq. (11.4) (with P_k^B removed) is then automatically satisfied so that such histories automatically decohere, and, in the terminology of the authors, this is due to coarse graining of type (1): specifying observables not at all times but only at some times. This coarse graining is quite obviously subjective. Nothing prevented us from considering also other times. The case of the two-slit example makes this point clear. Admittedly, if the *only* observable we decide to take into account is the position of the particle *when it passes through the diaphragm*, then formula

$$p_j = \text{Tr}[\rho P_j^A]; \qquad (j = 1, 2)$$

yields the probability p_j that if counters were set immediately behind each slit, counter j would be seen to click. And if, once and for all, we have decided we shall *not* be interested in anything else, nothing can prevent us from saying that, in an ensemble of N electrons, there actually are Np_j of them that do objectively pass through slit j. But this, of course, does not imply that in the real experiment, where these counters are not in place and in which we observe the impacts on the screen, the electron passes either through slit 1 or through slit 2. Indeed we know quite well that such an interpretation cannot be valid. In other words the here considered coarse graining is not objective in any sense. It depends on our individual choice, so that it is neither "strongly" nor even "weakly" objective, in the senses defined in Chapter 1. Consequently, the fact that the two histories decohere is neither strongly nor weakly objective either. It is purely subjective. Similar remarks could be made concerning the other types of coarse graining, labeled (2) and (3) above. Specifying some observables only, or some ranges of values only, is obviously a subjective process, and the question whether this, in some cases, corresponds to objective facts remains open.

In the article reviewed here the authors did not make any quite clear distinction between objective and subjective coarse graining and decoherence (not even to mention the subtler one between weak and strong objectivity). But they introduced an idea that parallels this distinction when they wrote: "We have the impression that the Universe exhibits a[ ] set of decohering histories, *independent of us*, defining a sort of classical domain" [emphasis added]. And it is quite clear that there are conditions under which they do interpret coarse graining and decoherence as being objective. This is apparent in the examples of decoherence they consider, namely the localization of the moon or of a typical dust grain, where they attribute decoherence to the interaction of the considered macroscopic object with its environment and refer to the work of Joos and Zeh reviewed above. By elaborating on such arguments they argue that "quasiclassical, maximal sets" of decohering histories can be defined such that the notion of roughly classical orbits makes sense for the corresponding systems.

But is this weak or strong objectivity? Their reference to the Joos and Zeh "environmental" theory makes it appropriate, in this connection, to quote some careful statements made by Joos and Zeh in their 1985 paper [9], analysed above. "The interference terms still exist" these authors wrote. "The interaction with the environment cannot describe a non-unitarity in the 'total' system. .. The collapse could then be based on an assumption of how a nonlocality is experienced subjectively by a local observer." This—conjectural—reference to the locality of observers is typical of what we called "weak" objectivity, and, in the cosmological theory of Gell-Mann and Hartle it would, in fact, be difficult to discover—on this question of objectivity—anything going beyond what Joos and Zeh could establish in their paper; so that we must consider that the said theory is weakly objective only. And indeed this might well be the standpoint implicitly adopted by its authors for, at the only place in their paper where they mention the nonlocality problem, their argument for setting it aside is as follows: "The problem with the local realism that Einstein would have liked is not *locality* but *realism*."

Now, if *this* is really the authors' standpoint—in other words, if they do identify science with just a description of "empirical reality," as defined in Chapter 1 (and commented on in Chapter 15)—a question arises as to whether or not this standpoint is fully consistent with their stated purpose of building up a theory of the whole universe *including thought*.

Let this point be considered more precisely. The authors take the view that both singly and collectively we are examples of complex, adaptive systems of the type they call Information Gathering and Utilizing Systems (IGUSes). These systems use probabilities for histories; that is, they use decohering sets of alternative histories. To this end, they perform further coarse graining on an already quasi-classical domain, that is, they exploit a particular quasi-classical domain or a set of such domains. This leads the authors to the assertion that: "IGUSes, including human beings, occupy no special place and play no preferred role in the laws of physics. They merely utilize the probabilities presented by quantum mechanics in the context of a quasi-classical domain."

The meaning of this assertion would be entirely clear within a realist standpoint; and when reading it we almost automatically take up such a standpoint. Thought is then conceived of as emerging from a reality which is logically prior to it. In particular, the emergence of the quasi-classical maximal sets of decohering histories is then viewed as logically as well as chronologically prior to that of IGUSes, and therefore of consciousness. This, however, is definitely *not* in accordance with the way Joos and Zeh understand the emergence of the classical properties since, as we saw, they ultimately conceive of it as following from a property of locality that *consciousness* is assumed to have. Since the basic arguments Gell-Mann and Hartle produced in favor of their notion of histories that decohere objectively (in contrast with those that decohere just because of individual choices of coarse graining) are of the Joos and Zeh type, Gell-Mann and Hartle were, to repeat, quite consistent in discarding realism (in the usual sense of metaphysical realism) as we saw they explicitly did. But then the meaning of their statement that "IGUSes occupy no special place and play no preferred role in the laws of physics" becomes obscure or, at least, highly nontrivial. We can no longer understand it in the manner that seemed the obvious one at first sight, and it is not at all clear whether there is any other meaning that could consistently be imparted to it. This remark is not meant to imply there is none—only that the question remains entirely open. The authors' quoted article contains no hint as to a possible answer.

• *Remark:*   Of course, we could try and remove ourselves the difficulty in question by adding to the theory some elements not explicitly stated by its authors, or by slightly altering, if necessary, some of its statements. It is clear that, to this end, we should strive to change the theory into a strongly objective one; and, as we saw, some of its authors' assertions and claims seem to indicate that they implicitly have in mind some idea akin to this. But, on the other hand, in a strongly objective theory it seems difficult to dispense with the notion of definite objective *states* (at any given time, in a given reference frame) of at least *some* individual physical systems. Such states cannot be identified in a general way with the Schrödinger-picture quantum states of standard quantum mechanics since, in a general measurement process, the quantum state of the $S + A$ ("measured" system *plus* apparatus) system is, after the S, A interaction has taken place, a superposition of distinct macrostates. Moreover, as we saw in Section 10.8, the Heisenberg picture is of no real help on this issue. Hence it seems that, in trying to attain our end, we could not avoid introducing a concept of state foreign to standard quantum physics. The name "dynamical state" has been put forward by some physicists for qualifying a concept of state not essentially different from the one that seems here appropriate. On "dynamical states" more will be reported in Chapter 13. Here, let it merely be stressed that, as will there become apparent, this concept raises nontrivial questions, so that it should not be kept implicit. Since it seems no notion of this sort is mentioned or alluded to in the quoted paper of Gell-Mann and Hartle, the conclusion that the theory they propose is weakly objective only seems an unavoidable one.

References

1. H. Everett III, *Rev. Mod. Phys.* **29**, 454 (1957).
2. B.S. De Witt and R.D. Graham, *The Many-Worlds Interpretation of Quantum Mechanics*, Princeton U. Press, Princeton, N.J., 1973.
3. B.L. De Witt, in *Batelle Rencontres I, 1967; Lectures in Mathematics and Physics*, M. De Witt and J.A. Wheeler (Eds.), W.A. Benjamin, New York, 1968.

4. B. d'Espagnat, *Conceptual Foundations of Quantum Mechanics*, 2nd ed., Addison-Wesley, Reading, Mass., 1976.

5. Y. Ben Dov, Versions de la mécanique quantique sans réduction de la fonction d'onde: la théorie d'Everett et l'onde-pilote [main text in English], Thèse de doctorat, Université Paris VII, 1987

6. D.Z. Albert, *Quantum Mechanics and Experience*, Harvard U. Press, Cambridge, Mass., 1992.

7 D.Z. Albert and B. Loewer, *Synthese*, **12**, (1988).

8. H.D. Zeh, *Found. Phys.* **1**, 69 (1970).

9. E. Joos and H.D. Zeh, *Z. Phys.* **B59**, 223 (1985).

10. A.O. Caldeira and A.J. Leggett, *Phys. Rev.* **A31**, 1057 (1985).

11. W.H. Zurek in *Frontiers of Nonequilibrium Statistical Physics*, G.T. Moore and M.O. Scully (Eds.), Plenum, New York, 1986.

12. E. Joos, "Quantum Theory and the Appearance of a Classical World," in *New Techniques and Ideas in Quantum Measurement Theory, Annals of the New York Academy of Sciences* **480**, 242 (1986).

13. H. Primas, *Chemistry, Quantum Mechanics and Reductionism*, Springer-Verlag, Berlin, 1981.

14. W.G. Unruh and W.H. Zurek, *Phys. Rev.* **D40**, 1071 (1989).

15. M. Gell-Mann and J.B. Hartle, "Quantum Mechanics in the Light of Quantum Cosmology," in *Proceedings of the Santa Fe Institute Workshop on Complexity, Entropy and the Physics of Information, May 1989*.

Other Reading

P. Mittelstaedt, *The Objectification in the Measuring Process and the Many Worlds* Interpretation in *Symposium on the Foundation of Modern Physics 1990*, World Scientific, Singapore (1991).

Ontological Approaches (Hidden Variables and All That)

13 "I saw the impossible done." This is how John Bell [1] described the surprise he had when, in 1952, he read the, then just published, Bohm's articles [2].

Why "the impossible"? Because, just as any of us, Bell had been taught that indeed—as Feynman put it even later—"nobody can explain more than we have explained. We have no idea whatsoever of a more basic mechanism from which the foregoing results [the interference fringes] could be deduced." If we are to believe Feynman (and Banesh Hoffmann and many others who expressed the same idea in both learned and popular books), Bohm's theory cannot exist.

Still, it exists, and is even older than Bohm's papers. In fact its essential idea had been advanced by Louis de Broglie as early as 1927 in his pilot-wave picture [3]. Since it offers explanations for things that are declared unexplainable in high circles it is worth being taken into consideration, even by physicists who (as is the case with the present author) do not believe it gives us the final clue as to how "reality really is." In this chapter its main aspects and the corresponding difficulties are considered. From this starting point we then proceed to examine other theories with the same, explicit, ontological claims, as well as more general problems concerning locality within a realistic outlook.

13.1 | Pilot-Wave Theory (De Broglie, Bohm)

A concise, clear account of this theory has been given by John Bell [4, 5]. Let us work with a simplified model in which the whole world is simply a large number N of particles with Hamiltonian

$$H = \sum_n \frac{\mathbf{p}_n^2}{2M_n} + \sum_{m>n} V_{mn}(\mathbf{r}_m - \mathbf{r}_n) \qquad (13.1)$$

The world wave function $\psi(r,t)$, where r stands for all the r's, evolves according to

$$\frac{\partial}{\partial t}\psi(t) = -iH\ \psi \qquad (13.2)$$

A purely mathematical consequence of this is that

$$\frac{\partial}{\partial t}\rho(r,t) + \sum_n \frac{\partial}{\partial \mathbf{r}_n} j_n(r,t) = 0 \qquad (13.3)$$

where

$$\rho(r,t) = |\psi(r,t)|^2 \qquad (13.4)$$

$$j_n(r,t) = M_n^{-1}\ Im(\psi^*(r,t)\frac{\partial}{\partial \mathbf{r}_n}\psi(r,t)) \qquad (13.5)$$

Up to this point, this is just conventional quantum theory. The "pilot-wave" idea is that the thus defined world wave function drives the particles. The latter have at any time well-defined positions and well-defined velocities. We label their positions by means of the c numbers $x_1, x_2,\ \ x_N$ in one-to-one correspondence with the r's and we assume that their velocities are

$$\frac{d}{dt}x_n = \frac{j_n(x,t)}{\rho(x,t)} \qquad (13.6)$$

This, then is a deterministic system in which everything is fixed by the initial values of the particle configuration x and the wave ψ, the latter (or rather both its real and "imaginary" components) being considered as "just as real and objective— Bell tells us—as, say, the fields of classical Maxwell theory."

Let us now imagine a statistical, Gibbsian, ensemble of worlds, in which at time zero the configuration x is distributed according to the configuration density

$$\rho(x,0) \qquad\qquad (13.7)$$

It is a consequence of Eqs. (13.3) and (13.6) that then, at any time t, the configuration x is distributed according to the density

$$\rho(x,t) \qquad\qquad (13.8)$$

so that a hypothetical outside observer who would look at our Gibbsian ensemble of worlds would, at any time, see exactly what the axioms of conventional quantum mechanics predict he would observe if he were to perform simultaneous ideal measurements of the positions of all the particles on all the worlds composing the ensemble.

So far, so good. However, the following question arises: What is the interest of either theory giving distributions over a hypothetical ensemble of worlds when we have only one world? Bohm and Bell answer along substantially the same lines, which presumably are also those Gibbs would have followed; and Bell makes them quite clear by using an analogy with an alpha particle track. A track—he points out— is on the one hand a single event, but it is at the same time an ensemble of single scatterings. When tracks are considered as single events, each one of them is one single experimental result. To test the theoretically predicted probabilities we then need in principle a whole ensemble of such tracks. But at the same time, if we are interested in, say, the probabilities of such and such scattering angles, then, if one track is long enough it can, for this testing purpose, be regarded as a collection of *many* independent single experimental results (single scatterings). Of course, among the tracks there could occasionally be freaks, tracks with all scatterings up, or all down, etc., but the typical one will show scattering angles whose distribution will statistically be in accordance with the theoretical prediction. In Bell's analogy a track corresponds to a world. In principle, in order to test the probabilistic predictions of *either* conventional quantum mechanics or the pilot-wave theory we would need a whole Gibbsian ensemble of worlds. But if we postulate that our actually existing world is typical, in the same way as a randomly chosen track is assumed to be typical, then the predictions in question are also valid concerning the statistical distribution *within* our world.

Consequently, at least for the simplified world considered here, the assumptions (13.6) and (13.7) guarantee, at any rate if this world is "typical" in the foregoing sense, that the deterministic behavior of its constituent particles will be such that their distribution will be the quantum-mechanical one (13.8) at all times. In particular this theory therefore quantitatively "explains" the existence of interference fringes in the two-slit experiment, their disappearance if one slit is blocked, and so on.

• *Remark 1.* Although the descriptions of the pilot-wave theory by Louis de Broglie, Bohm and Bell are essentially the same with regard to basic ideas, still there are some differences of emphasis between the three concerning the relative importance of the concepts they introduce. Between Bohm's and Bell's approaches one such difference concerns the status of the wave function. It is true that they both view it as a real entity and describe its real and imaginary parts as "fields." But in Bohm's pedagogical descriptions of his theory the fact is often stressed that these fields are derivable from a "quantum-mechanical potential" which—in the theory—must be added to the classical potentials and whose role thereby at first seems just to be to modify the interactions of the particles. Within Bell's approach, on the contrary, the fields in question are, as we saw, given a very primeval status, quite on equal footing with the particles. This difference has some outward similarity with the familiar one, known to exist between the Newtonian conception of a field of forces and the Maxwellian conception of what a field is. But for reasons to appear below, this analogy is superficial and in fact we here merely have to do with a slight difference of points of view. But still, as such it is not totally irrelevant. Concerning, for example, the locality problem, if we take up Bell's standpoint, then the fact that the wave function of a system of N particles, far from being composed of one-point functions, is a N-points function that generally does not factor, immediately leads to the conclusion that reality is nonlocal. If we take up Bohm's standpoint, this is not so immediately obvious (although it is also true there, of course, due to the fact that the quantum potential often does not decrease when the distance increases).

The reason why the aforementioned difference is merely formal is worth mentioning. It consists in the fact—very much emphasized by Bohm—that in the latter's theory the quantum potential depends on the state of the whole system in a way that cannot be defined simply as a preassigned interaction between the particles. In other words, contrary to what is the case in, say, Newton's gravitational theory, knowledge of (a) the dynamical properties and relative situations in space and so on of the particles making up a system and (b) the general laws of forces (Hamiltonian, etc.), even supposing we had it, would not be sufficient for providing us with a real knowledge of the system. For this, we should know the "quantum state," that is, the wave function, of the whole system, and this knowledge is not derivable from the aforementioned quantities. For example, as Bohm and Hiley pointed out [7], in the hydrogen atom the quantum potential in the s state depends on the distance r only whereas in a superposition of s and p states it also depends on the relative angular coordinates θ and ϕ. Obviously there is no preassigned function of r that would simultaneously represent the interaction between the particles in both s and p states. This shows quite clearly that, in Bohm's approach just as in Bell's, the (real and imaginary parts of the) wave function must in fact be considered as just as "real" and primeval as the particles. To the theory a feature of wholeness is thereby imparted that Bohm and Hiley consider as being distinct from the one of nonlocality. The latter would also be there if the potentials and fields were mere emanations of the particle, provided only that one of these—here the quantum potential—were distance-independent. Clearly the wholeness character goes over and beyond this.

• *Remark 2:* The most conspicuous feature of this theory is, of course, that it is deterministic. But there is another one, which is, basically, even more important: it is that no reference to the observers, or to "measurements" or to anything of the kind, appears in the "axioms." This of course is a considerable difference with conventional textbook quantum theory. In our language, we say that whereas the latter is a weakly objective—or intersubjective—theory, the pilot-wave theory is a strongly objective one.

13.2 | Difficulties

As is well known, there are many difficulties. One of the best advertised is of course the fact that while the formulas in this theory are more complicated than in the conventional one, and the theory itself less flexible and adaptable to new problems, still, even in the domain in which it *is* applicable nobody could, up to now, ever derive from it any testable prediction not already derived from the conventional theory and well verified by experiment. But difficulties of an even more basic nature exist.

One of the main ones has to do with relativity. As it is described above, the theory is quite obviously nonrelativistic: formula (13.6) sets no limit to the velocity a particle can take up. At first sight this does not look as a big subject of worry. After all, conventional quantum mechanics is also nonrelativistic, and this defect could to a great extent be cured by generalizing to quantum field theory. Here however the difficulty is more serious, due to the very fact, noted above as a "virtue," that the theory is strongly objective. As we saw in Chapter 9, no strongly objective local theory can reproduce all the quantum theoretical predictions. Hence any strongly objective theory that *does* reproduce these predictions must necessarily be nonlocal, and must therefore meet with some trouble concerning relativity. The reason why this does not apply to the relativistic generalizations of quantum mechanics is just that they are *not* strongly objective and that, since nonseparability does not allow for superluminal transfer of signals, weakly objective theories obeying the no-faster-than-light rule remain possible.

These kinds of troubles of course reflect in all the numerous attempts that have been made at somehow reconciling relativity with hidden-variables theories, at the price of some alterations of both. One example of such attempts can be found in the paper that Bell dedicated to Bohm [8]. The model he develops there has the remarkable property that it exactly reproduces all the, in principle, testable predictions of relativity theory such as shrinking of rods, slowing of clocks, and so on. But still, for its very formulation it requires the introduction of a *real* universal time and a *real* Euclidean space. We, of course, can never know whether or not we are at rest in this space, since the phenomena look for us the same in any Galilean frame, as relativity implies. But still, since this is *the* real space, the question whether we are or not makes sense This illustrates in a vivid manner both the possibilities and the limits of such theories Aiming at strong objectivity is undoubtedly a powerful incentive for research, but reality seems to be organized in such a strange way that even the most scientifically conducted efforts in that direction lead to views that partake somewhat of the arbitrariness of traditional metaphysics

It is true that we may feel this unsatisfactory feature of the theory is amply compensated by its conceptual clarity Remarkable indeed is the way in which it explains with but simple, familiar concepts—classical corpuscles and forces—a phenomenon such as that of fringe formation which most authors still believe is unexplainable by such means In Bohm's books, for example, quite illuminating drawings can be found of the trajectories of a bunch of particles after traversal of a diaphragm with two slits It is there clearly seen how these trajectories are "bent" by the quantum potential, in such a manner that, after a while, a fringe pattern gets formed On the other hand, we must also be aware that, actually, the conceptual simplicity of the theory is not as great as this simple example seems to show This is essentially due to the nonlocal character of the theory and is well illustrated by a slightly more complex experiment that of the so-called two-particle interferometry (see Section 9 7, Refs [24–26] and the references given there) A fundamental idea [9] of this recently opened experimental field is to make use of spatially separated, quantum-mechanically entangled two-particle states Four beams are selected from the source of these particles, two of them are interferometrically combined at one locus and two at another When phase shifters are placed in these beams, the coincident count rates at the two loci oscillate as the phases are varied but the single count rates do not The experiment is conceptually similar to a double two-slit experiment in which the two diaphragms and the two screens are symmetrically disposed with respect to the source, with, however, the difference that each pair of impacts must be observed separately, for it is the *difference* of the two abcissas that varies sinusoidally (a histogram of these differences will show that some will be more frequent than others) It is a simple exercise in quantum mechanics to show that this phenomenon exactly agrees with what this theory predicts, and since this is a type of phenomenon concerning which the pilot-wave theory generates the same predictions as conventional quantum mechanics, the mentioned effects must also follow from this theory. Indeed they do, of course. But in this new context it is not any more possible to speak of the guidance of *each* particle by a predetermined field or potential. The configuration space is six-dimensional and it is the representative point *in this space* that is guided by the quantum field (i.e., by the six-variables wave function). In any individual case it is still possible to define the trajectory of each particle as just being the projection of the six-dimensional trajectory on the relevant three-dimensional space. And, if we liked, we still *could* think of the particle as being guided on it by a field. But then this field itself is not determined any more by the general setup of the experiment and the initial wave function. It also depends on what happens to the other particle or, to put it otherwise, it also depends on the detailed, "hidden" initial position of this other particle, so that such an image of guidance of a particle in the three-dimensional space by a field or a potential completely loses, in this case, all of its attractive intuitive power.

As one more drawback let us mention the very general fact that any theory aiming at strong objectivity must of course specify "once and for all" what are the entities that are to be considered "real" in it. And this rigidity can turn to a serious inconvenience when further advances make generalizations necessary. As an example let us consider quantum field theory. Within the conventional view it is just a further step taken in the direction of applying the quantum general axioms to a larger set of phenomena: the formalism is generalized but the basic ideas remain the same. In the case of the pilot-wave or other hidden-variables theories the situation is essentially different, for, there, going over to quantum fields and trying to account for creation and annihilation effects demands giving up the idea that the particles are the basically real things—the "beables" in Bell's language—having real coordinates, real velocities, and so on. Instead, one can declare that the beables are, as suggested by Bohm and Hiley [7] concerning the bosonic fields, the values of the various fields at each point, or, as Bell suggested [8], the fermion density at each point. But seen from the very point of view of the upholders of such theories, namely conventional realism, these modifications amount to advancing altogether new theories. Such radical changes in the whole world picture may make us seriously puzzled at what will come next.

Another rather delicate point in the theory has to do with the notion of *state*. It concerns a distinction which is unquestionably called for between the *full* objective state and the *empirical* objective state. The nature of the problem is most clearly seen if we consider how the theory deals with the question of measurement. Consider first a simple Stern–Gerlach measurement made on a spin 1/2 particle. At a distance of the magnet where the two outgoing beams are well separated, the particle is necessarily in one or the other of the two; but still, the wave function—which, as we saw, is just as real as the particle—is nonvanishing in *both* beams. If, now, we set up a counter in each beam, the situation remains similar: the representative point in the configuration space of the whole system now lies in either one of two regions of this space, a fact that we want to put in correspondence with the one that one or the other counter has "clicked." But still, the wave function of the whole great system has been changed in *both* these regions: so that the just mentioned correspondence actually holds good only if we decide to consider that the real objective state of either counter is determined by the representative point and *only* by this point, which means that it does not depend at all on the wave function (on the so-called quantum state): Two systems that would have the same representative point would be strictly identical in this sense, even if their "quantum states" were different.

This sounds at first quite surprising since, in the case of counters, for example, we are accustomed to associate the fact that a counter has discharged or not discharged with the idea that some of its atoms did or did not make a transition from an excited state to a ground state, in other words, to phenomena primarily involving quantum states and wave functions. Still, according at least to John Bell (1988, private exchange of views with the author), this is *the* correct way to understand the theory. Energy (which is in close correspondence with stationary quantum states), and other so-called observables of quantum physics must be denied—with the sole exception of position—the quality of real physical quantities specifying the real objective state of a system. Within this conception the representative point of, say, a counter is a complete specification of the *reality* of this counter. But—let us be careful here—this holds merely in the sense that we have access only to *this* reality, not to the wave function or "fields," and that therefore, as previously noted, we have to regard as identical two counters, or more generally systems, that have the same representative point (for this reason Bell maintained that the expression "hidden variables" was a misnomer and that these quantities should properly be called "exposed variables") However, still according to this theory, such a reality is not the reality of the world since, to repeat, the wave function of the world is also a reality The conclusion this analysis leads to is that, notwithstanding the fact that the strong objectivity of the theory cannot be denied, still, even in it, the word "reality" has two senses, the "reality of the wave function of the universe and its representative point" which may be identified with the "independent reality" concept and the aforementioned "reality of the counters and other systems," which is the part of the independent reality that is accessible to us *because we are constituted this way* , and which may therefore be identified with "empirical reality" The well-known distinction David Bohm introduces between what he calls the "implicit" and "explicit" orders is of course an idea that is rather close to the one just advanced

13.3 **The Pilot-Wave Theory as a Theoretical Laboratory**

The highly speculative nature of the pilot-wave theory is obvious and acknowledged even by its keenest supporters But many physicists, including those who do not believe it is a true description of what exists, may well find it useful when trying to interpret quantum theory The reason is that in such an enterprise we have to face situations in which the borderline between objectivity and subjectivity is far from clear, and this sometimes creates some muddle In the pilot-wave theory things are clearer since everything is objective

An instructive example of such a use of the theory in question is the description Bell gave [10] of the nonlocal features it involves This he did by making an explicit calculation, in this theory, of the correlations at a distance that occur between the, say, z spin components S_z^U, S_z^V of two spin 1/2 particles, U and V produced in a correlated state at the source (for example in the singlet state) In order to do this he first had to build up a way of describing such spin components in the pilot-wave theory This he did by considering the interaction Hamiltonian

$$H = -ig(t)\sigma\frac{\partial}{\partial r} \qquad\qquad (13\ 9)$$

where σ is the Pauli matrix for the chosen component, S_z, and r is used as the "instrument reading" coordinate For simplicity we may assume that there is no other term in the Hamiltonian and that the coupling $g(t)$ arises from the passage of the particle along a definite classical orbit through the instrument Let the initial state be

$$\psi_m(0) = \phi(r)|m> \qquad\qquad (13\ 10)$$

where $\phi(r)$ is a narrow wave packet centered on $r = 0$ and $|m>$ ($m = 1, 2$) is an eigenket of S_z The solution of the Schrödinger equation

$$\frac{\partial\psi}{\partial t} = -iH\ \psi \qquad\qquad (13\ 11)$$

is

$$\psi_m(t) = \phi(r - (-1)^m\alpha(t))|m> \qquad\qquad (13\ 12)$$

where

$$\alpha(t) = \int_0^t g(t')\,dt' \qquad\qquad (13\ 13)$$

We may now consider an initial state $\psi(0)$ that is a linear super-position of the two possible initial states (13 10)

$$\psi(0) = \sum_m a_m\psi_m \qquad\qquad (13\ 14)$$

At time t this state will change to the corresponding super-position of states (13 12) and, even after a short time t, the two components will separate in r space In order to get from this model a description of the behavior of the particle that, in the pilot-wave theory, essentially reproduces the qualitative feature formula (13.6) then yields, analogously to (13.16)

$$\frac{d}{dt}x_1 = g_1 \frac{\sum_{mn}(-1)^m |a_{mn}|^2 |\phi(x_1 - (-1)^m \alpha_1)|^2 |\phi(x_2 - (-1)^n \alpha_2)|^2}{\sum_{mn} |a_{mn}|^2 |\phi(x_1 - (-1)^m \alpha_1)|^2 |\phi(x_2 - (-1)^n \alpha_2)|^2}$$

$$(13.20)$$

and a similar formula concerning x_2. These expressions greatly simplify when the two spins are uncorrelated, that is, when the a_{mn} factorize: $a_{mn} = b_m c_n$. The factors referring to the second particle then cancel out in (13.20) and vice versa concerning x_2, so that we have just two independent motions of the particles, of the type already discussed. Conversely, when the spins are correlated, when for example, the two particles initially are in an s state, the a_{mn} do not factorize, and then it follows from (13.20) that the detailed behavior of x_1 depends not only on its initial value and the program α_1 of the first instrument but also on the initial value of x_2 and, what is more remarkable, *on the program α_2 of the second, remote, instrument* (and vice versa). In other words, the detailed dynamics is quite nonlocal.

One consequence of this nonlocality is that the strict correlations that are known to occur when, for example, the z spin components of two faraway spin 1/2 particles lying in an s state are measured, are accounted for in totally different ways in classical mechanics and in the pilot-wave theory. In classical mechanics these correlations get established at the source. The initial values of x_1 and x_2 are such that when each particle interacts with an appropriate instrument the first one *must* "go up" and the second one *must* "go down" or conversely, and the fact that the first one must, say, go up does not depend in any way on the structure, position and program of the instrument that will interact with the other one. In the pilot-wave theory, on the contrary, the correlation is not exclusively or even primarily determined by the initial conditions at the source. It results from a whole network of complex causes, including the combined influences exerted by *both* instruments on *each* particle. It is interesting to observe that in a model which, as this one, is fully intelligible in "strongly objective" terms—in terms involving no reference to subjective or intersubjective concepts such as "measurement"— the notion of influences that are nondecreasing functions of the distance is an unavoidable one. In particular, the model may help some physicists to better understand and discuss the substance of the criticism Bohr formulated against the EPR criterion of reality, when he referred to "the very conditions which define the possible types of predictions" on a system and suggested that even conditions defined by "instrument number 2" could have a bearing on the predictions regarding "system number 1." Of course the strong objectivity of the present model would have raised averse feelings in Bohr, but it is interesting to note that even when a point of view totally different from the one of this author is taken up, still results are obtained that are quite consonant with what he said. The only substantial difference is that Bohr merely suggested an influence on the *type* of predictions that can be done (position or momentum or.) whereas here we have to do with an influence bearing on the very *values* of the involved dynamic quantities.

Still on the subject of "understanding Bohr" a remark of a more epistemological nature is here in order. It is that for this purpose a comparison between his approach and that of Bohm or other "realists" or "would-be realists" is helpful. It has to do with the definition and use of the concepts. Contrary to some other thinkers with a realist turn of mind, all these physicists have in common the guiding idea that to be scientifically meaningful a concept must be *operationally* defined, that is, defined with reference to a possible way to measure it. In other words, the difference between Bohr and, say, Bohm is not that one would make use of this touchstone and the other not. It must be looked for elsewhere. But it is easily discovered. It is that when Bohm and other realists have made sure that a concept can be thus referenced to some well-specified measurement process—concepts such as those of position and momentum are obviously in that case—they consider that it is a reliable one "full stop." In other words, they consider that it can safely be used, also in other contexts, that do not involve this measurement process and that may even be incompatible with it. Thus, the pilot-wave theory, for example, does make use of the concept "position" to describe situations in which no position measurement whatsoever is considered. As we know, Bohr—and this is the difference—takes, without really saying it explicitly, a much more restrictive attitude. He considers a concept can safely be used only in situations in which its defining measurement is actually done.

of a Stern–Gerlach experiment, Bell then simply postulates that in this model the density and current are given by (13 4) and

$$j(r,t) = \psi^*(r,t)g\sigma\psi(r,t) \qquad (13\ 15)$$

The motion of x is then determined by an equation analogous to (13 6), that is, explicitly, by

$$\frac{dx}{dt} = g(t)\frac{\sum |a_m|^2|\phi(x - (-1)^m\alpha)|^2(-1)^m}{\sum |a_m|^2|\phi(x - (-1)^m\alpha)|^2} \qquad (13\ 16)$$

As soon as $\alpha(t)$ gets larger than the half-width of ϕ, no value of x can be found that would make both terms in the numerator (and both terms in the denominator) simultaneously different from zero In other words, when the wave packets have separated, given any x one of the two terms is zero, both in the numerator and in the denominator Thus we have essentially the same observable effects as in the wave-packet reduction picture And although, in any individual case, which trajectory is selected is fully determined by the initial x value, still, in an ensemble the observed effects will—in view of the general argument—occur with the same probabilities as in conventional theory

Now let us turn to the two-particle problem we have in mind Let r_1 and r_2 denote the coordinates of these particles If the initial state is

$$\psi_{mn}(t) = \phi(r_1)\phi(r_2) \,|m > |n > \qquad (13\ 17)$$

solution of the Schrödinger equation yields

$$\psi_{mn}(t) = \phi(r_1 - (-1)^m\alpha_1)\ \phi(r_2 - (-1)^n\alpha_2)\ |m > |n > \quad (13\ 18)$$

where $\alpha_1(t)$ and $\alpha_2(t)$ are defined by means of formulas similar to (13 13) involving time-dependent couplings g_1 and g_2 between each one of the two particles and the corresponding instrument (magnet) In the case in which the initial quantum state is a superposition of the type

$$\psi(t) = \sum a_{mn}\psi_{mn} \qquad (13\ 19)$$

But the most interesting feature of the pilot-wave theory is perhaps that it goes quite a long way toward removing the *"and or or "* difficulty mentioned in Section 10 1 This is simply due to the fact that there is only one representative point, so that, when the support of the wave function is composed of several, disconnected, regions of space R_1, R_2, etc , the representative point is necessarily *either* in R_1 *or* in R_2 *or* etc Is the difficulty thereby *completely* removed? We may answer yes, but only if we are careful to remember that, as repeatedly noted, the representative point only describes a part, or an aspect—the "visible" one—of reality, and that reality itself—in the model the wave function of the universe plus its representative point—is quite hopelessly nonlocal and ignores the "or "

• *Remark 1* It is difficult to decide whether or not the Gell-Mann and Hartle theory (Section 12 4) is to be considered as one of those whose purpose is to faithfully describe what "really exists," as is explicitly the case concerning those hitherto reviewed in this chapter Reasons why it should *not* were given above As noted, it seems that its authors did not want to commit themselves too much on realism since, at a place, having to face the necessity of choosing between realism and locality they even made it clear they preferred to keep the latter On the other hand, it is somewhat difficult to understand how a theory whose purpose is to describe the universe could make no ontological claims whatsoever, and, in fact, the very words this one makes use of—in particular the word *history*—seem to indicate that it *does* imply such a claim Indeed it even looks as if its authors implicitly considered that of the several competing consistent histories they introduce, only one *really* takes place If this interpretation of what they wrote is taken up, then, as noted by Bohm and Hiley [7] their theory is similar to the pilot-wave theory at least on one point, namely that both agree that the wave function (or the density matrix) is not a complete description of reality The (very important) difference then is that while in the pilot-wave theory the additional "beables" (to use Bell's expression) are clearly designated and foreign to the Hilbert space (they are the coordinates of the "representative point"), in the Gell-Mann, Hartle theory they emerge as such (as beables) only indirectly and through the notion of coarse-grained projection operators, which, of course, *are* defined in Hilbert space Most of the more mathematically minded physicists will certainly consider this as imparting to the theory in question a formal beauty that the pilot-wave theory is deprived of. But the physicists who yearn for conceptual definiteness more than for mathematical beauty will presumably feel that on the question of realism—especially in connection with the locality or nonlocality problem—the pilot-wave theory is, of the two, the theory the internal consistency of which is by far the least questionable.

• *Remark 2:* To end this section with a remark of a general nature let it be noted that nonlocality does away with the hopes, entertained for some time, of making *events* the building stones of reality. In prerelativistic times this role of "building stones" was predominantly played by the—static—notion of *material objects;* and, in this conception, motion appeared as a mere, momentary and somewhat secondary, property of the objects. Special relativity overthrew this view by making the—basically dynamical— notion of *events* the central one, and lowering the status of the objects to that of mere combinations (world tubes) of events. And for a while physicists and philosophers could consider that this revolutionary change was final, that is, that the nature of the "building stones" of the universe had finally been discovered. However, events are, by definition, strictly local. The discovery that any realist (i.e., strongly objective) theory *must* be nonlocal therefore demonstrates that this conjecture was a delusion.

13.4 Contextuality

Bell's theorem and the corresponding notions, nonseparability, nonlocality, and others, were reviewed in Chapter 9 They are, as we know, general concepts, applying to any theory that reproduces the quantum mechanical predictions, whether or not it is explicitly a hidden-variables theory. The purpose of this short section is to acquaint the reader with the existence of another notion, contextuality, which is more restrictive than the ones just mentioned in that it essentially concerns deterministic hidden-variables theories but which, within this realm, reaches beyond the nonlocality concept.

To define this notion let us consider a physical system S described by a wave function ψ and a representative point x in configuration space (as in Section 13.1), let A, B, and C be three observables defined on S and let B and C be both compatible with A though not necessarily compatible with one another. Let us assume that A is measured on S simultaneously with either B or C. The outcome of this measurement depends of course on ψ and x, but it could prima facie be expected that it does not depend on whether it is B or C that is measured together with A. More generally, its seems natural to expect that in a deterministic hidden-variables theory, measurement of an observable would yield the same value independently of the "context," that is: quite independently of the other measurements that may be made simultaneously with this one. A theory for which this is generally true may appropriately be called noncontextualistic. Conversely, a hidden-variables theory in which this is *not* generally true is called *contextualistic*. It is then a theorem [10–12] that any deterministic hidden-variables theory that exactly reproduces the quantum mechanical predictions is necessarily contextualistic.

This theorem will not be proven here but some comments on its bearings will be presented.

First, let it be stressed that although contextuality is surprising, still it falls into line with one of the "founding fathers" greatest intuition. The point is that while the observables A and B can be measured "simultaneously," that is with *one* complex experimental setup—and same concerning A and C—when B and C do not commute no experimental setup exists that would make it possible to perform the measurements of A, B, and C simultaneously. It is then not even possible to check whether or not a measurement of A yields the same outcome when performed with one of C as when performed with one of B. Contextuality can therefore be considered a verification, and, (remarkably enough!) a *generalization* to theories of a type explicitly rejected by Bohr, of the well-known assertion of the latter that two different experimental arrangements define two different phenomena.

Second, the question may be asked in what sense contextuality appears as a generalization of Bell's theorem and nonlocality. The answer is quite simple. What Bell 1 proves is that the outcome of the measurement of a component of the spin of particle U along a certain direction $\mathbf{n}$ depends not only on $\mathbf{n}$ and the hidden variables of the system but also on what spin component of particle V is measured in the same overall measurement act Clearly, this is just an instance of contextuality, where the role of A is played by the U spin component along $\mathbf{n}$ while those of B and C are played respectively by the V spin components that are measured together with this U spin component within one given experimental setup and within another, different one Hence, the deterministic hidden-variables theories that are non-local in the sense of Bell 1 are necessarily contextual

But the contextuality of deterministic hidden-variables theories reproducing quantum mechanical predictions also reveals itself in problems in which the notion of spatial distance does not play any priviledged role A remarkable example is the problem Kochen and Specker [11] investigated It has to do with the fact that for spin 1 the three operators S_x^2, S_y^2, and S_z^2 representing the squares of the three components of the spin vector happen to commute, which implies that they also commute with $\mathbf{S}^2$ Since $\mathbf{S}^2 = 2$ this implies that one of the $S_i^2 (i = x, y, z)$ must be zero and the other two must be equal to one Within a deterministic hidden-variables theory we might therefore rather naturally expect that for any given complete set of values of the hidden variables, there should exist a set M of directions $\mathbf{n}$ such that

$$S_n^2 = 0 \text{ and hence } 1 - S_n^2 = 1 \quad \text{if } \mathbf{n} \text{ is in } M$$
$$S_n^2 = 1 \text{ and hence } 1 - S_n^2 = 0 \quad \text{if } \mathbf{n} \text{ is outside } M$$

However, Kochen and Specker could show that this is impossible Such a result does not prove that deterministic hidden-variables theories reproducing the quantum mechanical predictions do not exist, but it shows [12] that in such theories the hidden-variables do not in general unambiguously assign a value to S_n^2 for each given direction $\mathbf{n}$ In fact, as it turns out, they do assign such a value only if the two other directions that, together with $\mathbf{n}$, make up a set of three orthogonal vectors have been specified

• *Remark* The preceding discussion deals only with *deterministic* hidden-variables theories S M Roy and Virendra Singh [13] succeeded in generalizing the contextuality notion to non-deterministic (stochastic) hidden-variables theories They gave a natural definition of stochastic, noncontextual hidden-variables theories and they could prove that there exist quantum systems for which some experimentally verifiable quantum predictions cannot be reproduced by any such theory.

Other Approaches

Hidden-variables theories are not the only theories susceptible of being interpreted as faithful descriptions of "what exists" (so-called ontological theories). In particular, several physicists investigated the possibility of building up ontological theories in which the state vector would constitute the complete specification of the physical state of the system. Within such an approach the main difficulty is of course the one linked with the reduction of the wave packet. Since this difficulty has obviously to do with the linearity of the theory, one conceivable way of circumventing it is to slightly modify the Schrödinger wave equation, or Eq. (6.18) for the state operator, by plugging in some nonlinear terms. A number of theories—or "models"—were constructed along this line, which makes it impossible to do justice here to them all. Philip Pearle, for example once proposed a model [14] in which the nonlinear term in question has the following effect. In a case in which the state vector of a macroscopic system is a superposition of macroscopically distinct state vectors this nonlinear term rapidly drives the amplitude of one or another of these state vectors to one and the rest to zero. He suggested that it is the phase angles of the amplitudes immediately after a measurement that determines which amplitude is driven to one. Somewhat later he introduced [15] a different mechanism: a randomly fluctuating source.

Other theories proceed along rather different lines. The one of Ghirardi, Rimini, and Weber (GRW) [16] consists in changing Eq. (6.18) for the state operator describing an ensemble of systems to

$$ih\frac{d}{dt}\rho = [H, \rho] - i\hbar\lambda(\rho - T(\rho))$$ (13.21)

where λ is a suitably chosen parameter and $T(\rho)$ is so defined that Eq. (13.21) is trace preserving but describes an evolution in which a pure case is transformed into a (well-specified) proper mixture. Eq. (13.21) is thus compatible with the assumption that the various elements of a superposition lose some of their coherence when time elapses In the theory it is further assumed that this takes place at random times, through a localization process consisting in the fact that a broad wave function spontaneously makes a random jump to a reduced one More precisely (see also Bell [6 ch 22]) it is assumed that the wave function $\psi(x)$ of a single particle evolves according to the ordinary Schrödinger equation, unless it suddenly makes a "jump" to the "collapsed" (i e , more peaked) function

$$\psi'(x) = [N(z)]^{-1}\psi(x)\exp[-(\alpha/2)(x-z)^2] \qquad (13\ 22)$$

where $N(z)$ is chosen so that ψ' is normalized to unity and the width of the Gaussian is such that

$$\alpha^{-1/2} = 10^{-5}\text{cm} \qquad (13\ 23)$$

The collapse center z is randomly chosen, with probability distribution

$$|N(z)|^2 dz \qquad (13\ 24)$$

and the probability per unit time, λ, of a jump is such that λ^{-1} is of the order of 10^{16} sec In the case in which the wave function ψ consists of several rather well separated wave packets the collapse is most likely to occur with its center within one of these wave packets, due to the fact that, because of the normalization condition on ψ', $N(z)$ is small whenever the center of the Gaussian lies far from the packets The evolution equation of the state operator for an ensemble of particles (with no selection made) can then be shown to be of the form of Eq (13 21)

For the center-of-mass wave function of an n-particle macroscopic system lying in a superposition of several states, it was shown by GRW that the theory naturally produces a state operator evolution that is again described by Eq (13 21), except that λ is replaced by $n\lambda$ This shows that while a single-particle wave function is only affected over a long time scale λ^{-1}, the many-particle wave function of a solid is reduced very rapidly, on the time scale $(n\lambda)^{-1}$ The conceptual problem raised by the quantum superposition of macroscopically distinct macroscopic states in, for example, a quantum measurement process receives thereby quite an attractive solution

Like any other "ontological" theory this one is of course not free from difficulties One of them is that, since the reduction process makes a wave packet narrower, it thereby increases the particle's energy However, with the previously specified values of the parameters α and λ this energy increase turns out to be unobservable Another difficulty is of course that the values in question are purely ad hoc An interesting theoretical suggestion has been put forward by Karolyhasy [17–19], to the effect of explaining state-vector reduction with reference to limitations on the sharpness of the structure of space-time, themselves obtained by combining Heisenberg's uncertainty relations with gravitation In principle, this approach should make it possible to relate the values of parameters similar to α and λ to universal physical constants and in particular to the gravitational constant G The articles quoted above describe inspiring advances in this direction In the GRW theory, however, the state-vector reduction is treated phenomenologically only

Since in the Ghirardi, Rimini, and Weber (GRW) theory the wave function completely specifies the physical state of the system, this theory is quite obviously nonlocal For example, a pair of spin 1/2 particles produced in a singlet state will normally remain in that state for a very long time, hence even at times when the particles have appreciable probabilities of being observed at quite distant places, and all the considerations developed in Chapters 8 and 9 therefore apply also here In particular, the theory violates local causality, as defined in Chapter 8 Together with the fact that the theory is "ontological," this of course leads one to expect that attempts at making it relativistically covariant must encounter serious difficulties

Very interesting such attempts were made nevertheless, not with the GRW proper but with a combination of it and Pearle's previous attempts that was put forward by Pearle and Gisin [20] and developed by Pearle, Ghirardi, Grassi, Rimini, and others [21] under the name "Continuous, Spontaneous Localisation Theory" (CSL) In this theory the state vector reduction is induced by variables that fluctuate randomly The CSL theory has most of the nice features of its predecessor It can, moreover, be given a form that exhibits at least a weakened form of relativistic covariance, called "stochastic relativistic invariance" by the aforenamed authors because, at least the *ensemble* of predicted results of experiments is frame-independent. In our language we would say that the thus built-up theory is *weak-objectively* relativistically invariant. Unfortunately, even this partial result—which does not quite reach to what a realist would expect of a relativistically invariant theory—could be obtained only at the price of replacing the Gaussian localization function of the initial theory by some more local entity (in fact a delta function); and this imparts to the model the unsatisfactory feature that there is now an infinite energy production per unit volume instead of a small finite amount as in the nonrelativistic theory.

If the standpoint is taken up that this difficulty will somehow be removed by further advances (the field is being actively explored by several groups), then [21] such a relativistic CSL may be made compatible with macro-objectivism. But what about micro-objectivism?

The problem at hand is quite similar to the one considered in Section 10.12 and must be studied approximately along the same lines. For example, we may take up again the Aharonov and Albert example of an electron trapped in a double-well potential and lying in a superposition with equal amplitudes of two time-independent wave functions each of which vanishes in one well (we assume the wells are very narrow and quite far apart). If a position measurement is performed in well number I (WI) at a time t_1 in the reference frame F where the wells are at rest and its outcome is *no*, the probability that a position measurement in well number II (WII) would yield outcome *yes* (if it were performed) is $1/2$ at space-time points $\{t < t_1$ & WII$\}$ and 1 at space-time points $\{t > t_1$ & WII$\}$ One would therefore be tempted to say that, due to the system-instrument interaction in WI, the property "the electron lies in WII" emerges at the space-time "point" $(t = t_1$ & WII). However, there exist reference frames F', slowly moving with respect to F, which are such that a hyperplane $t' = $ Const whose intersection with hyperplane $t = t_1$ passes through WI intersects WII at a time-in-F $t_0 < t_1$ (Fig. 13.1). From the point of view of an observer in F' the probability that a position measurement in WII would yield outcome *yes* is $1/2$ at all space-time points $(t < t_0$ & WII) and 1 at all space-time points $\{t > t_0$ & WII$\}$ Hence the same type of reasoning would lead one to say that it is already at time t_0 that the property in question emerges. From this ambiguity, as they called it, Ghirardi, Grassi, and Pearle [21] drew the conclusion that for what concerns micro-objects, no objective local property can emerge as a consequence of a measurement occurring in a spacelike separated region.

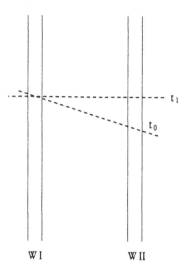

Figure 13.1. A double-well potential, with two spacelike hyperplanes whose intersection passes through well W_l.

The interaction-at-a-distance problem will be analyzed in more general terms below (Section 16.2). Here, let it merely be observed that, in view of the just noted ambiguity the "stochastic relativistic invariance" of the theory does not immunize it from an inconsistency connected with relativity. This inconsistency consists in the fact that, in a theory bearing (as it is claimed this one does) on "reality" as opposed to "intersubjective appearances" no statement concerning a "physical event"—in the relativistic, i.e., local sense of the word "event"—can be both true in one reference frame and false in another one, so that the ambiguity described is unacceptable.[1] It therefore seems likely that, if the capacity of the theory at describing "beables" is to be

1. Considerations similar to these ones but based on a quantitative analysis of a specific thought-experiment were recently put forward by L. Hardy. *Phys. Rev. Lett.* **68**, 2981, (1992).

salvaged, this can only be at the price of admitting (as we noted Bell suggested [8]) that there is but one "real" division of space-time into space and time Within such a conception, since there exists a real Euclidian space and a real Newtonian time, reality itself is not structured according to the relativistic views But the phenomena—the set of what the observers perceive—do obey relativity As we know, length contractions, slowing of clocks, and so on do appear to take place just as relativity predicts they must be observed It seems clear that, at any rate, the theory under review can be *interpreted* this way And if it turns out that this is the only way it can, then this, of course, will confirm the view already expressed in this book, according to which "independent reality" and "empirical reality" are really quite different things, and we must specify whether we speak of one or the other

13.6 The (So-called) Ensemble Interpretations

In the literature on the interpretation of quantum mechanics a great deal of room is taken up by developments that, under the name *ensemble,* or *statistical,* or *stochastic* theory, essentially claim that the conceptual difficulties besetting quantum mechanics can be removed in a rather straightforward way, provided the assumption is made that quantum mechanics describes, not *individual* physical systems but *ensembles* of such systems A comprehensive critical review of this approach recently appeared [22], to which the reader should turn for details Only a short survey of the subject is therefore given here

For clarity's sake, it must first of all be mentioned that in itself the ensemble assumption is still rather vague It must be made more precise and indeed this was done by its supporters As a consequence, it took different forms, which however can schematically be gathered under three headings, corresponding to three different proposed interpretations of quantum mechanics. These are

(a) The *Preassigned Initial Values Assumption* (PIV), as Home and Whitaker called it ([22, 23] *At all times, every physical quantity pertaining to a quantum system has some definite value, which is just the one a measurement of the said quantity would (unambiguously) reveal*

(b) The *Hidden-Variables Assumption:* this is the assumption already considered under that name in several sections above. The PIV assumption is of course just a special case of the hidden-variables assumption. The latter is more general, essentially because it admits of nonlocality and contextuality, as defined here in Section 13.4. Note that, as we saw in Chapter 9, a hidden-variables theory is not necessarily deterministic.

(c) The *Minimal Ensemble Assumption* as P Gibbins called it [24]: *Quantum theory is a theory of ensembles and a wave function is a complete representation of an ensemble. There are no hidden variables.*

Unfortunately, most authors who wrote on the ensemble interpretation did not actually specify—or did not specify accurately enough—whether they meant it in the sense of interpretation (a) or in the sense of interpretation (b) or, perhaps, in the sense of interpretation (c). And, what is worse, they often shifted from one interpretation to another, depending on the nature of the difficulty they purported to remove. Ballentine, for example [25], defined the statistical ensemble interpretation he espouses by writing:

> A state vector describes certain statistical properties of an ensemble of similarly prepared systems. Individual systems possess definite, although perhaps unknown, values of certain fundamental observables (such as position) which are not described by the state vector unless it happens to be an eigenvector of that observable.

As Home and Whitaker point out, this is quite definitely a hidden-variables assumption. Yet, in that same article the quoted author did not put it under the hidden-variables heading and somewhere else [26] he even wrote that any hidden-variables theory that reproduces all the results of quantum theory exactly must, as a consequence of the Bell theorem, be *physically unreasonable.* This makes his argumentation unclear and, on the whole, inconsistent. He seems to have overlooked the fact that any statistical ensemble interpretation of the type he favors must unavoidably fall into the category of those to which Bell's theorem applies.

Similar consistency troubles arise concerning the many attempts that were made at accounting for the *quantum measurement process* by means of the ensemble interpretation. It has been repeatedly claimed that in contrast to von Neumann's scheme, the interpretation in question provides a clear, nonproblematical account of what takes place during the measurement process, and in particular of the fact that the "instrument pointer" always takes up a definite macroscopic position. Of course, this is true, and even quite trivial, if the ensemble interpretation is identified with the Preassigned Initial Values (PIV) assumption, as implicitly or even explicitly (see Ref. [26]) suggested by some proponents of this kind of explanation. But, reconciling the PIV assumption with contextuality and nonlocality is like trying to square the circle. And additional trouble comes from the fact that this PIV assumption meets with serious objections, the substance of which have been known for a quite long time. One of them is the famous argument of von Neumann against hidden variables. This argument is based, as is well known, on two assumptions. The first one, already encountered and discussed in Chapter 7, is that, for some classes of systems at least, any Hermitean operator represents an observable. The second one—call it the v.N. postulate—is that, given any real linear combination of any two Hermitean operators, the expectation value of the observable this combination represents is just the linear combination of the expectation values of the observables these operators represent. This happens to be true for quantum states, because of formula (5.6). For example, the two observables might be the kinetic energy $T = p^2/2m$ and the potential energy $W(x)$ of a particle. The sum $E = T + W$ of these two operators corresponds to the total (nonrelativistic) energy and if ψ is the state the particle is in, the same linear relationship

$$< \psi|E|\psi > = < \psi|T|\psi > + < \psi|W|\psi >$$

obviously holds true between expectation values. Von Neumann's disproof of the hidden-variables hypothesis was grounded on an extension of this linearity requirement to the hypothetical dispersion-free states and, as Bell pointed out [10] there is in fact no basic reason for postulating such an extension to be valid. However it *must* be valid if the PIV assumption is made. Consider again the foregoing example. Within the PIV assumption each element S_i of the ensemble has a potential energy W_i, a kinetic energy T_i, and a total energy E_i that coincide with the values measurements of these quantities would yield if performed, and the total energy is then, of course, $E_i = W_i + T_i$ On a hypothetical dispersion-free state describing an ensemble composed of identical replicas of S_i the mean values of these quantities coincide of course with their individual values, so that the same linear relationship (here, just a sum) holds between the mean values defined on the hypothetical dispersion-free states as between the corresponding operators The v N postulate thus being valid because of the PIV assumption, von Neumann's disproof of the existence of hidden variables—hence in particular of PIV—can be circumvented only by assuming that sufficiently many Hermitean operators correspond to no observables, or (say) that sufficiently many superselection rules are at work in the considered systems—an ad hoc assumption that, even disregarding all the other difficulties, would considerably reduce the plausibility of the PIV conception

Going over to a more general, *contextualistic* hidden-variables theory such as the pilot-wave theory admittedly removes the latter difficulty, but this is, as we saw, at the unavoidable price of having to accept nonlocality Moreover, giving to a hidden-variables theory the name "ensemble theory" would be somewhat misleading since it would amount to emphasize, in it, the *ensemble* concept, whereas in fact the latter concept has but a secondary role in such theories, the primary ones being played by the very notion of hidden variables and that of contextuality

Another approach to measurement theory would be to identify the ensemble interpretation with the *minimal ensemble* assumption, as defined in interpretation (c) above As already mentioned, some proponents of the interpretation in question seem, at places, to favor this view Their standpoint then is that the right-hand side of Eq (10 4) merely tells us that the relative frequency, over the ensemble, of finding the measuring apparatus in the macroscopic state labeled m is $|a_m|^2$ However, a question then arises as to what this statement actually means We can understand it as merely meaning that if, on an ensemble of similarly prepared $S + A$ systems, *we look* at the pointer positions we shall find a proportion $|a_m|^2$ of them in the macroscopic "state m But the thus formulated hypothesis—sometimes rightly or wrongly called the "Einstein hypothesis"—is in fact indistinguishable, as Jammer [27] pointed out, from the conventional Born hypothesis, according to which quantum mechanics predicts the probability of the result m of a measurement performed on a single system to be $|a_m|^2$, since typical quantum probabilities are identical with relative frequencies. Alternatively, we could try to interpret the statement in question as asserting that, in the ensemble described by the right-hand side of Eq. (10.4), there really is, even before we look, a proportion $|a_m|^2$ of apparatus pointers lying in the macrostate m. But it must be seriously questioned whether this view is compatible with the minimal ensemble assumption we are presently considering, according to which there are no hidden variables of any sort. This nonexistence of hidden variables must mean that all the systems that have the same wave function—and therefore, in particular, all the $S + A$ elements of the final ensemble—are identical, since nothing differentiates them. It is hard to see how this could be made compatible with the fact that the pointers of the instruments are not lying in the same graduation interval. To put it otherwise: to say of an individual system that it belongs to an ensemble described by some state vector is, in this context, to make precisely the same statement—as Sudbery [28] stressed—as that the system is in that state. Since we are now considering that some of the initial $S + A$ systems went over in the state described by one term only in the right-hand side of Eq. (10.4), we must consider that the states of these systems, considered individually, have undergone a real, sudden change, as described by the projection postulate. But then we are back to what we called standard textbook quantum mechanics, with an additional "realist" assumption concerning the said final states. But we have seen at length in Chapter 10 all the difficulties such an assumption raises. The words "ensemble theory" are no magic key for removing them.

To sum up: It must be considered that, when all is said and done, the attempts at developing the ensemble interpretation did not, up to now, contribute to a better understanding of the conceptual foundations of quantum theory.

• *Remark:* The remarkable recent developments in the so-called *classical chaos theories* have induced some people to expect that these theories will eventually make it possible to provide quantum mechanics with a realist interpretation entirely stated in terms of the (more or less disordered) motions of *localized* objects. The contents of Chapter 9 and the present chapter show that this is a delusion.

13.7 Dynamical States

In any ontologically interpretable theory elements must be added to standard quantum mechanics in order that objective states of individual physical systems can be defined. In the case of the pilot-wave (alias Bohm's) theory, these supplementary elements—or *variables*, or *parameters*—are the positions of the particles, that is, physical quantities that, in the formalism of standard quantum mechanics, do not, strictly speaking, correspond to any operator defined on a Hilbert space. One subset of the ontologically interpretable theories may be defined as made up of the, rather numerous (see, e.g., Belinfante [12]), theories constructed along these lines. They are conventionally called hidden-variables theories (although, as some of their proponents, notably Bell, as we know, stressed this word is a misnomer since, on the contrary, these variables are those that are directly seen).

In contrast, the supplementary elements of other ontologically interpretable theories—which we may consider as constituting a second subset—are (values of) physical quantities that, in the quantum formalism, do correspond to Hilbert space operators. This would, for example, be the case of the Gell-Mann and Hartle theory if it were tentatively interpreted as a strongly objective theory· for, as already pointed out, the parameters that would then specify *which* history actually takes place would just be particular eigenvalues of projection operators, which, of course, *are* Hilbert space operators. It is also the case in a theory put forward—with some variations—by Simon Kochen [29], Richard Healey [30], Dennis Dieks [31], and Bas van Fraassen [32] and sometimes referred to as the KHDF or "modal" theory. Its essential idea is to introduce a distinction between two notions, called "quantum state" and "dynamical state" by Healey. The former is, in this scheme much as in standard quantum mechanics, a description of the probabilistic dispositions of a system while the latter essentially specifies truth value assignments to sentences of the form "*S has P*" (where *P* is a quantum dynamical property appropriate to system *S*). According to this scheme, if the eigenvalue equation of observable *A* is

$$A|k,r > = a_k|k,r >$$

the dynamical state of a system *S* may (under conditions to be defined) be one in which the proposition $A = a_k$ is true, even in cases in which the quantum state vector of *S* either is nonexistent (if *S* is part of a larger system lying in some entangled state) or is none of the $\{|k,r >\}$ This means that the dynamical state provides a finer description of the state in which *S* really *is* than the quantum state does (if it does). In this sense, Healey's and others' dynamical state is a notion that closely parallels the previously considered notion of objective state. The parallelism with the pilot-wave theory at least in Bell's description of it, is even enhanced by the fact that, also in the KHDF theory, while what we see is the dynamical state, the physical behavior of the system is entirely governed by the quantum state. A difference is, as already noted, that, in a sense, this approach involves no additional nonquantum variables, since, contrary to the physical quantity *position* of the pilot-wave theory, observable *A* does correspond, in strict mathematical sense, to a Hilbert space operator. But the theory involves all the same attributing specific values to quantities that, in the cases of interest and according to conventional quantum mechanics, would not be attributed any. In other words, it is at any rate a "supplementary elements" theory, just as, in Bell's conception, the pilot wave is. Since, in practice, at the price of some lack in mathematical strictness, position and momentum can be treated *as if* they could be described by Hilbert space operators, the thus defined difference between the theories composing the first and second subsets is not conceptually considerable. Of course, all these theories or models are contextual and nonlocal, at least in that they describe a world containing compound systems having irreducible dynamical properties in spite of the fact that they are composed of spatially dispersed components (in other words, they violate nondivisibility by thought). The same, of course, is true of the theories composing what may be considered as a third subset, namely the theories, such as the GRW and the CSL theories, in which independent reality is directly described by the wave function, with no hidden variables added. Incidentally, concerning these theories let us also note that, of course, there are dynamical "supplementary elements" in these theories as well as in the ones here under study They are the parameters specifying which one of the possible collapsed states actually is, just after a "random jump," the final quantum state of the system

The foregoing is just a sketchy description of the qualitative ideas of KHDF theory To be a little more specific and quantitative, let us briefly summarize the account Dieks gave of it in Ref [33] There, the conditions under which a system S_1 has what has been called here a dynamical state are specified by making use of Schmidt's theorem This theorem concerns physical systems composed of two subsystems S_1 and S_2 and whose Hilbert spaces H are therefore tensor products $H_1 \otimes H_2$ of the Hilbert spaces of these systems It states that there exists a *bi-orthonormal* decomposition of every state vector in H

$$| \psi > = \sum_k c_k | \psi_k > \otimes | R_k > \qquad (13\ 25)$$

with $< \psi_i | \psi_j > = \delta_{i,j}$, $< R_i | R_j > = \delta_{i,j}$ (degeneracy indexes dropped) which is unique if no two $|c_k|^2$ are equal In the case of a measurement-like interaction between a system and an instrument, Schmidt's theorem may suggest an alternative way, different from the one explained in Section 10 6, of selecting a "pointer coordinate basis" (this idea was explored, notably by Zeh [34]) But steps can also be taken for trying to go further and use this theorem in order to build up an ontologically interpretable theory along the lines sketched here The idea is to consider that when bi-orthonormal decomposition (13 25) obtains system S_1 possesses exactly one of the physical properties associated with the set of projectors $\{| \psi_k > < \psi_k |\}$ (and system S_2 similarly possesses the corresponding property, associated with the operator $| R_k > < R_k |$ with the same index k) In other words, the physical quantity whose eigenvectors are the $| \psi_k >$ has one definite value on S_1 (but the completeness hypothesis of Section 4 3 is not made the subensemble of systems S_1 having this property is not quantum theoretically described by $| \psi_k >$)

As pointed out by Dieks [33], for the purpose of accounting why the classical description is so successful on the level of everyday experience, the scheme can be associated in a natural way with the environment (alias decoherence) theory The point

is that, as shown in Section 10 6 in connection with the measurement problem and in Section 12 3 concerning the more general
problem of the classical appearances, when a system undergoes
repeated interactions with its environment, factors of the type
of $< \chi_{x'} | \chi_x >$ in Eq (12 15), which correspond to factor $z(t)$ in
Eq (10 35), rapidly become very small, which means that an
approximate bi-orthonormal decomposition of the overall state
vector—the roles of S_1 and S_2 being played by the system and the
environment respectively—obtains in a very short time Hence,
if we consider an ensemble of such $S_1 + S_2$ systems, the ensemble E_1 of the S_1 subsystems, although it is an improper mixture,
still is one on each individual element of which the "dynami-
cal quantity"corresponding to $|\psi_k > < \psi_k|$ has a definite value
At first sight this seems to nicely fulfill a request we expressed
near the end of Section 12 3, and which was that, on the issue of
accounting for classical properties of macrosystems, the theory
should be able to tell us something about the systems *considered
individually* On the other hand, a significant point made in Section 12 3 (Eq (12 33) and below) must be kept in mind here It is
that although the eigenkets $|\psi_k >$ of the reduced-density matrix
describing E_1 are orthogonal (or nearly so), still this does not generally imply the kind of localization that is expected concerning,
e g , instrument pointers

To sum up This idea of making use of Schmidt's theorem is
attractive, and the KHDF scheme may certainly be considered
as one of the most satisfactory of the ontologically interpretable
ones On the other hand, it must be kept in mind that, at the
present stage at least, it is essentially nonrelativistic, that it is not
clear that the problem of unambiguously defining the distinction
between the two systems S_1 and S_2 is solved and that, when the
Schmidt theorem is used, in addition to the difficulty we just
spelled out, there is also the one that the formulas describing
the time evolution of each subsystem are redoubtably intricate
(see [34]) Some other delicate points concerning KHDF schemes
were analyzed by Albert and Loewer [35]

13.8   Far Realism and Outlook

The models reviewed in this chapter are but a sample—even if
a representative one—of what has been proposed in the domain

of ontologically interpretable theories. In fact, an exhaustive review of the great many suggestions that were put forward in this field would be utterly impossible. On the other hand, all of these models—I mean all of those that are compatible with the data as well as internally consistent—necessarily have some features in common, due to the general constraints, such as nonlocality, they must obey. These constraints result in the fact that, in these theories, although by assumption, independent reality exists, it has properties extremely different from the properties we (including most scientists and even most theorists when far from their writing desks) normally ascribe to the physical world. I called "far realism" [36] a version of realism according to which this is the case. In view of what physics has shown, anybody who is not a radical (avowed or crypto-) idealist must accept far realism as thus defined. But within it, at least three different options can be taken. Independent reality may be assumed to be knowable, unknowable, or veiled. Of course, the whole set of the ontologically interpretable theories entirely pertains to the first option (the two other options are distinguished from each other and commented on in the forthcoming chapters).

What makes such theories attractive is the fact that, being strongly objective, they can, like good old classical physics, be interpreted as lifting the veil of appearances and describing what "really is" (independent reality, in our language). But, paradoxically, this is also one of the main sources of their weakness: for there are several of them, among which experiment seems quite unable to discriminate while the *pictures they yield* of (independent) reality are incompatible with one another. Other difficulties they raise, particularly with relativity, are discussed in the next chapter, where general motivations for skepticism concerning them are summarized. Let it be stressed however that, even if we cannot actually adhere to any of these theories, their very existence is most valuable and interesting. Indeed, as will become apparent below, they constitute invaluable "theoretical laboratories" for exploring the meaning and significance of new general ideas—nonseparability and so on—that basic physics imposes upon us and that reach beyond both commonsense *and* scientific usual notions. Moreover, when we mentally contrast these theories with the various versions of standard—that is, weakly objective—quantum physics we better realize how vague and misleading some words commonly used—such as "Nature" or "world"—actually are. For instance, the term "world," which may designate either independent or empirical reality, is sometimes used by philosophers who claim they are pure empiricists in contexts in which it can only refer to independent reality. This may be viewed a symptom of what Einstein branded as "the illness of contemporary philosophy" and saw as the negative side of the, in other respects so highly positive, Humean criticism: the illness that he called the "dread of metaphysics" [37] and that he saw as fraught with serious risks of confusion. The authors of the ontologically interpretable theories do not accept such confusion. They lucidly refuse to pretend that weakly objective theories and concepts are strongly objective and, in this, they set an example to many physicists and philosophers.

References

1. J. S. Bell, *Found. Phys.* **12**, 989 (1982); reprinted in Ref. [6].
2. D. Bohm, *Phys. Rev.* **85**, 165, 180 (1952).
3. L. de Broglie, *J. Phys.* **5**, 225 (1927).
4. J.S. Bell, *International Journal of Quantum Chemistry: Quantum Chemistry Symposium* **14**, 9 (1980); reprinted in Ref. [6].
5. J.S. Bell, in *Quantum Gravity 2*, C. Isham, R. Penrose, and D. Sciama (Eds.), Clarendon Press, Oxford, 1981, reprinted in Ref. [6].
6. J.S. Bell, *Speakable and Unspeakable in Quantum Mechanics*, Cambridge U. Press, 1987
7 D. Bohm and B. Hiley, *The Undivided Universe: an Ontological Interpretation of Quantum Mechanics*, Routledge and Kegan Paul, London 1993.
8. J.S. Bell, in *Quantum Implications, Essays in Honour of David Bohm*, B. Hiley and F. Peat (Eds.), Routledge and Kegan Paul, London, 1987; reprinted in Ref. [6].
9. M. Horne, A. Shimony, and A. Zeilinger, "Two-particle Interferometry," *Nature* **347**, 429 (1990).
10. J.S. Bell, *Rev. Mod. Phys.* **38**, 447 (1966).
11. S. Kochen and E.P. Secker, *J. Math. Mech.* **17**, 59 (1967).
12. F. Belinfante, *A Survey of Hidden-Variables Theories*, Pergamon Press, Oxford, 1973.
13. S.M. Roy and Virendra Singh, *Quantum Violation of Stochastic Noncontextual Hidden Variable Theories*, Tata Institute of Fundamental Research preprint 92–115, 1992.
14. P. Pearle, *Phys. Rev.* **D13**, 857 (1976).
15. P. Pearle, *International Journal of Theoretical Physics* **48**, 489 (1979).
16. G.C. Ghirardi, A. Rimini, and T. Weber, *Phys. Rev.* **D34**, 470 (1986).
17 F. Karolyhazy, "Gravitation and Quantum Mechanics of Macroscopic Bodies," Thesis, *Magyar Fisikai Foloirat* **22**, 23 (1974).

18. F. Karolyhazy, A. Frenkel, and B. Lukacs, in *Physics as Natural Philosophy*, A. Shimony and H. Feshbach (Eds.), MIT Press, Cambridge, Mass., 1982, and in *Quantum Concepts of Space and Time*, R. Penrose and C.J. Isham (Eds.), Clarendon Press, Oxford, 1986.

19. A. Frenkel, *The Reduction of the Schrödinger Wave Function and the Emergence of Classical Behavior*, KFKI preprint 17/A, Budapest, 1988.

20. P. Pearle, *Phys. Rev.* **A39**, 2277 (1989); *Sixty Two Years of Uncertainty*, A.I. Miller (Ed.), Plenum Press, New York, 1990, p. 193; N. Gisin, *Helv. Phys. Acta* **62**, 363 (1989).

21. G.C. Ghirardi, R. Grassi, and P. Pearle, *Symposium on the Foundations of Modern Physics 1990*, World Scientific, Singapore, 1991, p. 109; *Found. Phys.* **20**, 1271 (1990).

22. D. Home and M.A.B. Whitaker, "Ensemble Interpretations of Quantum Mechanics, a Modern Perspective," in *Phys. Rep.* **210**, 225 (1992).

23. D. Home and M.A.B. Whitaker, *Phys. Lett.* **A115**, 81 (1986).

24. P. Gibbins, *Particles and Paradoxes*, Cambridge U. Press, 1987

25. L.E. Ballentine, *Am. J. Phys.* **54**, 81 (1973).

26. L.E. Ballentine, *Rev. Mod. Phys.* **42**, 385 (1970).

27   M. Jammer, *The Philosophy of Quantum Mechanics*, Wiley, New York, 1974.

28. A. Sudbery, in *Quantum Concepts in Space and Time*, R. Penrose and C.J. Isham (Eds.), Clarendon Press, Oxford, 1986.

29. S. Kochen, "A New Interpretation of Quantum Mechanics," in *Symposium on the Foundations of Modern Physics*, *'85*, World Scientific, Singapore, 1985.

30. R. Healey, *Philosophy of Quantum Mechanics, An Interactive Interpretation*, Cambridge U. Press, 1989.

31. D. Dieks, *Synthese* **82**, 127 (1990).

32. B. van Fraassen, in *Current Issues in Quantum Logic*, E. Beltramelli and B. van Fraassen (Eds.), Plenum, New York, 1981.

33. D. Dieks, "Continuous Wave Function and Discrete Physical Properties," in *Erwin Schrödinger· Philosophy and the Birth of Quantum Mechanics*, M. Bitbol and O. Darrigol (Eds.) Editions Frontières, Gif-sur-Yvette, 1992.

34. H. Zeh, *Found. Phys.* **3**, 109 (1973).

35. D. Albert and B. Loewer, in *Symposium on the Foundations of Modern Physics 1990*, P. Lahti and P. Mittelstaedt (Eds.) World Scientific, Singapore, 1991.

36. B. d'Espagnat, *In Search of Reality*, Springer-Verlag, New York, 1983.

37   A. Einstein, in *The Philosophy of Bertrand Russell*, Northwestern U. Press, Evanston, Ill., 1944.

Other Reading

D.J. Hoekzema, *Contextual Quantum Process Theory*, Found. Phys. **22**, 467 (1992).

Open Realism

14 Concerning the title of this chapter, two preliminary remarks are in order.

The first bears on the very word "realism." Although most people would willingly define physics as being "the science of physical reality," such words as realism, reality, and so on are in fact more philosophical than scientific. This chapter must therefore, of necessity, have some kind of a philosophical admixture in it. It must also, of course, take the content of the preceding "purely scientific" chapters into account. So that it will offer a combination of philosophy and physics.

The second remark is of a more practical nature. It is that, in order to make the thread of the ideas proposed here more apparent, we proceed below in four steps. In Section 14.1 the *structure* of the argumentation is outlined. In Sections 14.2–14.6 the main features of each separate element in the chain are explained. Sections 14.7–14.11 go into complementary details of each of the said elements that are not essential for understanding the overall argument but an account of which philosophers are entitled, for strictness' sake, to demand. Finally, in Section 14.12, conclusions are derived.

14.1 Structure of Argumentation

The argumentation of this chapter consists in answering in succession three questions:

1. What are the main riddles?

Our answer will be, they are two in number: *weak objectivity* and *nonseparability*. We already know what these expressions refer to. Let us just remember that neither of these two riddles, or enigmas, exists in classical physics; and their presence in quantum physics is essentially what accounts for the fact that while the quantum mechanical predictive rules work remarkably well, their interpretation in terms of what the philosophers sometimes call "external reality" raises considerable problems. Hence the second question arises:

2. Actually, do we need a reality?
This will constitute the philosophical part of this chapter. Our answer will be yes, and this will lead us to our third question:

3. Is this reality intelligible?
To answer this question there is no better recipe than to try and determine the credibility level of the claims of the physicists who maintain they can describe things "as they really are" without violating any of the observational predictions of quantum mechanics. Essentially such claims are grounded, as we know, either on some alleged "realistic" explanation—by means of quantum mechanics—of the classical features of macro-objects (including instrument pointers) or on some so-called ontological reformulations of general quantum mechanics. Basing our argument on our foregoing detailed examination of these two families of schemes, we shall eventually reach the conclusion that, when all is said and done, the credibility level in question is low.

The inference from this is that, most presumably, reality is either unknowable or veiled. A further examination of the question will lead us to the conclusion that it is veiled.

14.2 Main Riddles, a Schematic Review

Let us here sum up a few things we know.

Weak Objectivity

All statements in physics are, of course, objective. But this assertion is, to some extent, deceptive for, (Section 1.6), there are two distinct types of objectivity, *strong* and *weak* objectivity. Strongly objective statements are those that a conventional realist can (and therefore would) interpret as bearing *on* reality—as describing *as they really are* some features of an "external reality" the existence of which—this the conventional realist takes for granted—is independent of *our* existence and knowledge. With the exception of some statements in statistical thermodynamics, all the basic statements of classical physics are strongly objective ones. And, at a time, it could reasonably be expected that this state of affairs would go on, that even thermodynamics would eventually be reformulated in strongly objective terms, and that all future theories would be stated in such terms as well. But this is not what actually happened. In particular, some of the axioms of standard quantum mechanics (Chapter 3) are not strongly objective. They refer in a basic way, for example, to "measurements"— to the very *notion* of measurement, irrespective of who actually performs the operation. Should such statements be viewed as "subjective"? No, since they are valid for everybody. So we have there a category of statements that are neither strongly objective nor merely subjective. These are the statements we called *weakly objective* in Chapter 1, where a somewhat more elaborate analysis of the distinction between strong and weak objectivity was outlined. From a philosophical standpoint such an analysis should, however, be carried on further, and this will be done in Section 14.7 In this same section, the crucial question whether more elaborate forms of quantum theory—or of measurement theory—such as those we met with in Chapters 10–12 can restore strong objectivity will also be examined—and answered in the negative.

Nonseparability

Classical physics (like practically all other sciences) was built up on the extraordinarily fruitful notion of *analysis*. I mean, not just mathematical analysis but also *conceptual, physical* analysis: the view that the physical systems we normally encounter can be conceived of as composed of distinct parts and that, once the "laws of force" are known, an exhaustive knowledge of these parts yields, ipso facto, a complete knowledge of the whole. This was called the idea of *complete divisibility by thought* in Section 8.1. Now, as Schrödinger was, presumably, the first to remark (or at least to explain neatly), the idea of complete divisibility by thought does not fit well at all with quantum mechanics. This is most clearly apparent within the elementary formalism in which, for the description of several particle systems, the many-body wave function plays the basic role. For there, even if we start with a wave function that is just a product of— approximately localized—one-particle wave functions, the time evolution induced by the Schrödinger time-dependent equation has the effect that after some time the overall wave function will in general *not* be a product any more. Thus the individual particles cannot any more be attributed a wave function (nor therefore can be thought of in general as occupying some specific region of space). As we know, this effect was (appropriately) called *entanglement* by Schrödinger.

Some physicists thought they could nevertheless salvage the notion of divisibility by thought. They proposed to go over to the density matrix (or "state operator") formalism. They pointed out that, in the final situation just considered, even though the particles do not each have a wave function, each one has a density matrix of its own. This of course is quite true, but it does not restore divisibility by thought. The reason simply is that, as we saw in Section 8.1, a knowledge of all these one-particle density matrices does not (even if we have full knowledge of the Hamiltonian and so on) provide us with a knowledge of the total, composite system, since it comprises no knowledge concerning the *correlations* between the particles (although these correlations are, of course, experimentally observable).

This is a situation that is altogether different from what had been known until then. Indeed, as Herman Weyl is credited to have said, quantum mechanics is thereby revealed as "the first holistic theory that really works."

Among the conceptual riddles aroused by quantum theory, mention must of course also be made of the formidable problem of accounting for what really takes place in a *measurement process*, when we try to describe the instruments as ordinary physical systems. Here a special line is not attributed to this problem, just because, in a way, it is (as we saw in Chapter 10) a special case of the entanglement, or nondivisibility by thought, problem. Contrary to what Einstein seems to have, at a time, expected, the so-called ontologically interpretable theories (Chapter 13) do not get rid of nondivisibility by thought. As the Bell theorems have shown, these theories merely shift the riddle. In them, it takes the form of a violation of local causality (Chapter 9). The generic name "nonseparability" was found convenient for covering the riddle in question in any one of its two forms.

To sum up: Weak objectivity and nonseparability are two basic features of quantum mechanics that do not fit at all with our usual views, and force us therefore to modify them. But it is to be noted that, in fact, they are true enigmas only for he who takes seriously the notion of "external reality" (to use the standard term, in spite of its defectiveness). They do not weaken or alter the operational power of the theory (it has been shown that nondivisibility by thought cannot be used to send signals, see Section 8.2). So that recognizing their existence and weight naturally leads us to ask the question: After all, *do we need* an external (or, better to say, an "independent") reality?

14.3   Do We Need an "External" Reality?

Doubts concerning this may be said to originate from Berkeley's idealism, although Descartes may be seen as a forerunner. From this seed the idea gradually developed, as we know, that such general notions as Being, things-in-themselves, Reality per se, and so forth are redundant, useless, and "therefore" invalid, concepts; and that, when all is said and done, it is only on the basis of human experience that we can define anything that can properly be called real. This is what was called "radical idealism" or "radical phenomenism" in Chapter 1. Indeed it may be said that the modern and contemporary philosophical doctrines that were (or now are) the "greatest hits" in their field—we refer to phenomenalism, positivism, antirealism, internal realism, and so on—partake of radical idealism in a way that is in fact more fundamental and basic than, perhaps, most of their proponents would willingly grant. Let us just very briefly summarize once more the argument that a supporter of this general line of approach may put forward.

At the basis of it we find a fact: that much of the very content of our sensations depends in part on our sense organs. Hence, the argument goes on, it is natural to consider that what we immediately perceive is not the physical object that we say we perceive, but a mental image of it.

But then (continuing along this line): Why believe at all in the existence of external objects? We should only assign meaning to operationally defined entities. The very word "existence" has therefore no meaning in itself. It gets one only in specific cases, by referring to our experience. And correlatively the concepts, such as *space, time,* and *causality,* we use for describing our experience do not correspond to structural elements of a reality per se: they are a priori modes of *our* sensibility and understanding (Kant).

So runs the argument of these philosophers, whose doctrine, as already noted, partakes, in fact, very much of idealism. It is interesting to note that quite a number of physicists did entertain ideas very near these, at least for a time. Among the most prominent of them let us merely quote the young Einstein and the young Heisenberg. On the other hand, noteworthy is also the fact that in their more mature age many of them changed to a more balanced position. This was, for example, the case for Einstein, who became a true realist, and, to some milder extent, it was also the case for Heisenberg, when he introduced his *potentia*. This may be considered as an indication that the radical-idealist, neo-Kantian approach is not totally satisfactory. And indeed, there *are* forceful objections to it as we began to see in some of the foregoing chapters. It is claimed here neither that these objections are new (there are presumably no strictly new arguments to be found in this very well explored field) *nor* that they can be considered as valid without any further detailed scrutiny. But they are, all the same, significant. At this stage let us just mention what some of them consist of, and let us postpone the finer analysis and discussion of them all until later (Sections 14.8–14.10). They are as follows.

1. Kantianism and neo-Kantianism quite naturally lead to the view that, since the physical laws are, when all is said and done, regulated by the a priori modes of our sensibility and understanding, the laws in question must be constructed in terms of "visualizable" concepts only. Some science historians, such as A. Miller [1], have stressed the fact that trusting Kantianism or neo-Kantianism too much on this point may well have hindered some of the physicists who, during the first quarter of this century, took up the task of building up a theory of the atomic processes. And indeed, it is conceivable that Schrödinger's longlasting insistence that everything should be describable in terms of the (visualizable) *wave* concept had no other cause.

2. We sometimes build up quite beautifully rational theories that experiments falsify. Something says no. This something cannot just be "us." There must therefore be something else than just "us."

3. The approach in question openly makes the concepts of knowledge and experience logically prior to the concept of existence. However, it seems quite impossible to impart any meaning to the very word "knowledge" without at least implicitly postulating the *existence* of somebody or something, or whatnot, who knows. In other words, the logical ordering postulated in this approach does not seem to be compatible with normal requirements concerning meaningfulness.

4. Intersubjective agreement seems hardly explainable without *some* reference to something existing outside us.

As already mentioned, these arguments—as well as a few others—must be subjected to detailed critical analysis before they can be finally accepted. To some extent this was already done in Chapter 1 and additional developments along similar lines will be produced in the complement Sections 14.8 and 14.9 For maximal clarity's sake let us, however, anticipate the outcome of this delicate and unavoidably somewhat tedious analysis. It is on the whole a positive one: it seems hardly consistent to dispense completely with any notion of a reality logically prior to knowledge. It is appropriate that this standpoint be given a name. Let it be called *open realism*.

Open realism. There is "something" (is this "something" the set of all the objects, that of all the atoms, that of all the events, God, the Platonic Ideas?—this, at this stage, is not specified: one just says "something") the existence of which does not proceed from the existence of the human mind.

Some will consider that the arguments sketched above make open realism extremely plausible. Others will call it a postulate. Although the issue is important it is not what matters most in our overall argumentation. What is essential is that anyhow, *no postulate except this one (if it is one) will prove necessary for the developments that follow.* This means: contrary to many conventional realists we do not postulate anything concerning the nature of "what exists", not even that it is knowable. Nor do we postulate that it is unknowable! The matter is just left open: it is to be decided on the basis of what factual knowledge, and in particular scientific knowledge, reveals.

Under such conditions it is difficult to find for "what exists" a name "neutral" enough not to suggest, concerning it, features or properties that, at this early stage in our inquiry, it would be preposterous to attribute to it. The least compromising name that could be thought of is "Independent Reality," meaning, not that we cannot act on this reality but that its existence is independent of our own.

14.4 Is Independent Reality Atomizable?

In the (admittedly quite unrefined) sense we attribute here to the word "atomism," the latter means the theory according to which ultimate—alias independent—reality consists of myriads of simple entities. The question that forms the title of this section could therefore be reformulated as "Is it true that Being is 'dispersed' in a multitude of such primitive elements?"

As physics is described in elementary or popular books, it seems to strongly support an affirmative answer. But, clearly, the information gathered in the foregoing chapters points, on the whole, toward quite a different picture. Without going into any systematic review of what has already been noted on this issue, let us just simply observe here that also *relativistic* quantum physics, and more precisely *quantum field theory*, indicates atomism is basically wrong.

This is perhaps not immediately clear even to the concerned "professionals"—to wit, the high-energy physicists—since the language their science is couched in is that of an implicit atomism and the tools they use the most—the Feynman graphs—seem to refer to particles as to entities existing per se. But on further reflection they must concede: first that Feynman graphs are but diagrammatic descriptions of calculation procedures aimed at predicting with what probabilities such and such measurement results will be obtained (in such and such circumstances); and, second, that these calculation procedures follow from quantum field theory· so that if we ask for information of a conceptual nature, if our query concerns the "real entities" conceivably underlying the calculations in question, it is to the very structure of quantum field theory that we must turn our attention.[1] Now

---

1. In particular this approach sheds light on a subject sometimes considered as puzzling: the "epistemological status" of the virtual particles. Virtual particles

what, from a philosophical standpoint, is by far the most remarkable feature of quantum field theory is that it reduces the
(scientifically unmanageable) notion "creation" to the (scientifically tractable) notion "state change." And the point that is
relevant to the here considered issue is that it succeeds in doing
so by making primary some concepts of a *general* nature—such
as fields associated with types of particles—and secondary the
concept of individualized particles. Consequently, if we are on
the lookout for some concept, or "mathematical algorithm," that
in this theory could be identified as referring to the "basic stuff,"
we can find none except, conceivably, the element the state of
which changes when a particle gets "created" or "annihilated."
This element is the state vector in Fock's space. Now, in the theory, there are not myriads and myriads of such elements. Indeed,
there is just one! Which means that, conceptually speaking, the
theory is as far from atomism as it is conceivably possible for a
theory to be.

To the, thus broadly sketched, reasoning it could prima facie be objected that even if the basic stuff is just *one*—a kind
of *materia prima*—still it materializes into myriads of particles
and there is no reason not to consider every one of these particles as being "real," in the commonly accepted sense of the
word. However, when we look at the detailed structure of quantum field theory we discover that in fact this objection is not
tenable. The point is that quantum field theory takes over all
of the basic axioms of standard quantum mechanics, including
in particular (a) the correspondence between operators and observables and (b) Rules 5 and 6 of Chapter 3, which associate the
eigenvalues of some operators, not with attributes actually *possessed* by quantum systems but with outcomes of—possible or
actual—*measurements* of the corresponding observables. Since,
in quantum field theory, the quantity "number of particles of
such and such a type in such and such a state" is nothing else
than such an observable (associated with an operator built up
from basic field operators), consistency requires that it should
not be considered as an attribute possessed by some underlying reality but just as the possible outcome of some conceivable

---

in Feynman graphs are nothing but pictorial representations of denominators
that appear in some quantum field theory formulas expressing probabilities of
observation.

measurement. In other words, quantum field theory induces us to consider that the (contingent) *existence* of such and such a particle is not something intrinsically different from a mere dynamical property; that, indeed, it is *nothing more basic* than just an eigenvalue of an observable, and that, as any other eigenvalue in standard quantum mechanics, this one characterizes, not a determination reality *has* but the possible or actual outcome of a conceivable *measurement* (i.e., in the general case it is an entity of the same kind as, in elementary quantum mechanics, the not-yet-measured position of a particle when this particle is represented by an extended wave function). Hence, to repeat, there is, to say the least, a considerable *mismatch* between atomism as defined above and conventional quantum field theory.

It goes without saying that neither the Feynman "zigzag account" of pair creation-annihilation nor the Feynman path integral technique open the way to a possible refutation of the above. Both are computational techniques "internal," so to speak, to quantum field theory and derive part of their efficiency from the fact they are allegorically expressed in a pseudorealistic language that makes them evocative. But the images they call to mind cannot be interpreted literally. For example, in the path integral technique, to get the probability of certain observational outcomes we must sum over, not the probabilities (as the literal interpretation would force us to) but the probability amplitudes. And, moreover, what determines which one of the two operations is to be performed (summing on probabilities or amplitudes) is the fact that *we* can, or, as the case may be, *we* cannot, build up (through some measurement) a "knowledge of which path was taken" a "rule of the game" that, quite typically, is but weakly objective.

14.5   Is Independent Reality Knowable?

This is a question that, as already noted, should be investigated in the light of physics. Do we have at present, or can we entertain reasonable hopes of building up in the future, a theory that would be both ontologically interpretable and scientifically convincing? Let it be explained here why the answer no is the more reasonable one, by having a look at the currently available proposals.

1. It is appropriate to mention briefly two approaches that are sometimes understood, though erroneously, as admitting of an ontological interpretation. One of them is a tentative—but, I claim, faulty—interpretation of Bohr's approach (see, e.g., Feyerabend [2]). The idea is to identify this approach with the ontological theory often called "macro-objectivism," according to which macroscopic objects, contrary to microscopic ones, obey classical physics and are therefore just as "ontologically interpretable" as were the physical systems in the times in which classical physics was believed to be *the* true theory. This idea is not tenable, as we saw (Chapter 11), in particular because of the fact that in, for example, thought experiments involving movable diaphragms, Bohr [3] sets parts of the setup—just these movable diaphragms—either on the side of the instrument or on the side of the (quantum) object, not according to the physical properties they may have or not have, but just according to whether they *are used or not* as instrument parts. As Bohr himself put it, a *logical distinction* [4] is thereby introduced between instruments and quantum objects; and, clearly, such a *logical* distinction has nothing to do with the difference in complexity of these two types of systems. This, of course, merely confirms Bohr's general statement, according to which: "The description has. a perfectly objective character *in the sense* (emphasis added) that no explicit reference is made to any individual observer" [4]. This phrase exactly characterizes what we called "weak objectivity," and shows that this—and *not* strong objectivity—was what Bohr really aimed at. In other words, Bohr's approach was not meant to, and does not, reconcile quantum mechanics with realism (for more details, see Sections 11.1–11.3). To the extent that theories such as, for example, that of Omnès may be considered as formulations of Bohr's conceptions stricter than Bohr's own one (see Section 11.4), the same remark should hold good also concerning the aims of the authors of these theories, even if—same as Bohr—they do not always find it necessary to expatiate on this point.

The other idea that is sometimes erroneously interpreted this way is the environment theory of Joos and Zeh [5] and others, here investigated in Chapter 12. To be sure, the energy levels of macroscopic bodies are so tightly packed that these bodies never can be considered as isolated. To be sure the environment variables whose overall correlations would, if measured, reveal the inconsistency of attributing well-defined classical properties to these bodies are nonmeasurable in practice But undisputable is also the fact (see Section 10 6) that such variables and such correlations could in principle be measured (well-defined experimental programs can be specified and the correlations between the outcomes of distant measurements can in principle be read *later,* in the notebooks of the experimenters, even if they operated in spacelike separated space-time regions) This (through the "Schrödinger cat" argument—section 10 6)) precludes any *ontological* interpretation of the theory in terms of objects having in all circumstances well-defined dynamical properties of their own Admittedly this argument has to be elaborated on, but to develop it further implies going into details that, if described here, would lead us somewhat astray We shall therefore have to bring up the subject again, and this will be done in the complementary Section 14 10 Anticipating on the conclusion of this section we may however assert already that no better summary exists of the epistemological status of the descriptions yielded by the theory than the one provided by the very title of one of Joos's articles *"the appearance of a classical world"* [6]

2. We are thus left with some more unconventional approaches Of them, admittedly, it can be said that they are genuinely ontological They are the pilot-wave theory, some other hidden-variable theories (KHDF and others) and the spontaneous-reduction theories, GRW, CSL These theories are extremely interesting, mainly because of the fact that they do restore strong objectivity (this is why they are "ontological"), which is not a small achievement On the other hand, this feature, which is their strong point, turns out to be at the same time a *weak* point because of the fact that in them there is an equivalent to nondivisibility by thought, namely nonlocality (Bell's theorems) and this feature is much more disturbing in their case than in the case of conventional quantum mechanics The reason stems from the fact that (see Section 8 2) neither nondivisibility by thought nor nonlocality allows for superluminal signaling When we have to do with a nonlocal, *weakly* objective theory this makes it possible to—if not completely forget about such things—at least speak, in Shimony's and Redhead's words, of a "peaceful coexistence" between it and relativity theory (also conceived of as weakly objective only). Within the realm of a theory the strong objectivity of which is stressed, this kind of escape is of course no more admissible since, for consistency, also relativity must then be conceived of as a strongly objective theory· so that it is not only the "signals" that should travel no faster than light but also any "influence" whatsoever, which is hardly or not at all reconcilable with nonlocality (see the remark on this issue near the end of Section 16.1). Hence what makes the interest and the value of these ontological theories is also, ironically, the source of their main weakness (some more on this end of Section 14.10).

Of course, the foregoing remarks do not amount to asserting that these theories are falsified. Nevertheless, they lower very much the degree of confidence one feels inclined to attribute to them. And all the more so as there are several different such theories as we saw, that experimental discrimination between them seems, even on theoretical grounds, something very difficult to achieve and that extending such theories to the realm of quantum field theory (creation, annihilation) implies a considerable upheaval within their ontological attributions since particles can then no more constitute the basic stuff. The conjecture that therefore seems the most reasonable is that the real business of physics is not to construct a scientifically well-established ontological theory, that this is probably quite an inaccessible goal, and that physics should be content to describe and synthesize the phenomena.

14.6 Is Independent Reality in Space and Time?

The question whether or not independent reality is embedded in space and time (or in space-time) is, clearly, a momentous one: the answer we give to it commands the whole of our apprehension of the world as well as of the place we have in this world. Unsurprisingly, it is also quite a difficult one. To state as we just did that the role of physics is but to describe the *phenomena*— in a more or less Kantian sense—does not provide us with any ready-made response. It would, of course, if we took Kantianism for granted, but, as we have seen (Section 1.1), there are nowadays good scientific reasons for not accepting this doctrine in its entirety.

One point, at any rate, is clear: current scientific knowledge—special and general relativity in particular—does not make us inclined to think that independent reality lies in ordinary space and time or in flat space-time. It remains true, however, that relativity-inspired remarks cannot answer basic questions of the type: Is independent reality lying *in* some kind of—perhaps distorted—space-time? Is it evolving *in* the so-called cosmic time?

The sets of arguments favoring an answer yes and an answer no to such questions are both impressive. Unquestionably the whole of astrophysics presently develops under the implicit assumption that the yes answer is correct; otherwise said, that stars, galaxies, and so forth exist by themselves, quite independently of our knowledge concerning them (thus being elements of *independent* reality), *and* are localized *in* space and time roughly as we see them to be. In fact, it is often said that denying this would amount to claiming that stars, galaxies, and so forth are created by us and it is stressed that this is quite plainly absurd since, obviously, Earth existed before human beings existed and so on. In Section 1.4 we saw that idealists have a ready reply to this. But we also saw that this reply is rooted in some quite basic tenets of idealism so that, for it to be consistent, it is necessary that idealism should be accepted up to and including some of its most disconcerting features: which may make us waver accepting the reply. But on the other hand, such "data" as nonseparability, weak objectivity of all truly productive versions of quantum theory, and so on, seem to constitute quite strong arguments against an embedding of reality in space-time and/or cosmic time. In fact, the power of these arguments is such as to make us critically reconsider those pleading *for* an embedding. Let us try and find out whether it is possible to counter the latter (the argument for embedding) even within the realm of open realism, that is, without going over to *radical* idealism.

As a first step in this direction, let us indulge in a kind of "first year course in philosophy" discussion. Let us compare the two concepts "localization of the moon in the sky" and "taste of a fruit." It is usually considered, in particular by realists, that there is a qualitative difference between the two. Our apprehension of the direction where the moon lies at a given time is considered as informing us directly of an objective fact (the moon really lies in this direction, full stop), whereas our apprehension of the taste of the fruit is conceived of as resulting from a complex combination of "objective" facts concerning external reality (the existence of some specific molecules in the fruit) with facts of an essentially subjective or intersubjective nature, concerning our sensory system and brain. However, it is theoretically conceivable that the difference should be more apparent than real: that *both* apprehensions should result from complex processing, by our sensory system and brain, of "remote" factual elements, pertaining to an "independent reality" extremely different from what we see. Of course, such a conjecture would be totally arbitrary—and, to a scientist's eye, not worth a moment's attention—if a straightforward interpretation of the whole of contemporary physics in strongly objective terms were available. But this is not the case, as the content of the foregoing chapters has amply shown. In other words, it is conceivable that the "direction in which the moon lies" should be something related just as indirectly—through our sensory and mental structures—to the underlying reality as the taste of a fruit is: and should therefore be just as much "created by us" as the latter. Indeed, in some ontologically interpretable theories such as Bohm's [7], the "explicit" order—the order we observe in astronomy as well as in daily life and in our laboratories—far from being a faithful reflection of the underlying "implicit" order, is on the contrary thoroughly different from it.

In the exploration of how open realism could possibly be reconciled with the thesis that independent reality is *not* embedded in space and time, a further step consists in remembering what we noted in Section 10.8 concerning the two roles, the mathematical symbols, ρ, E, B, r serving to designate physical quantities (be they vectors, scalars, or whatever) all have in classical physics. As noted, one of these, which I called *Role a*, is that they serve for writing down the general laws, such as the Maxwell equations, considered as good candidates for describing the *general structure* of the world. The other one, I called *Role b*, is that they designate the values the quantities they refer to actually have *in some particular instance*. In quantum mechanics the Heisenberg-picture operators have Role a only, and Role b is played by no algorithm endowed with a strongly objective sense whatsoever. This, as shown in that section, results in the fact that if we assume the structure of independent reality is adequately and fully described by quantum mechanics (thus parting from idealism, in the direction of realism), we must conceive of independent reality as a highly entangled whole, with the consequence that it is impossible to conceive of parts of this whole that would each occupy definite places. So that indeed, as is the case concerning the taste of the fruits, it is *we* who *create* (or collectively *imagine*) the localizations of the objects. It is true that, as just formulated, this argument holds for space only, not for time. But the relativistic (partial) equivalence of space and time cannot here be ignored. To the extent that *we* create space, there must be a sense in which *we* also create time. It is well known that there are mathematical difficulties in formulating quantum mechanics, even in its Heisenberg-picture form, in a quite general (interactions included) *and* manifestly covariant way. In view of these difficulties, the fact that the foregoing argument does not directly show we do "create" time as well as space does not come much as a surprise, and is not an indication that we do not.

At this stage, we already have in our possession all the information that allows for a reasonable conclusion to be drawn concerning the reality concept. This conclusion is stated in Section 14.12 and, if pressed for time, the reader is urged to turn straightaway to it. The following complementary sections (Sections 14.7–14.11) are only meant for the benefit of the "perfectionist" readers who, on points they consider as delicate, are prepared to go into detailed discussions.

14.7 Complements on Weak Objectivity

As mentioned earlier, this concept is one that, from a philosophical standpoint, should be analyzed in more detail than has been done above. As explained in Chapter 1, one delicate point is that weak objectivity (alias *intersubjectivity*) is a feature either of statements or of concepts. A statement is but weakly objective if, while being true for everybody, still it basically refers to what human beings actually do, or can do, or observe. In parallel, we may form the idea of calling "weakly objective" the concepts defined by specifically referring to some human procedure. But, as already hinted (Section 1.6) such a definition of the weak objectivity of *concepts* is still, in fact, too schematic. Indeed, it lays itself open to the criticism that, if taken at face value, it makes *any* concept, or at least any scientific one, weakly objective, since, if we had to define any concept whatsoever—be it even such an apparently simple one as the position of a macroscopic object—to somebody not yet possessing it, we would, in some way, have to refer to human observations. However, there exists a philosophical standpoint, namely conventional realism, within which strongly objective concepts can be defined in quite a straightforward way. Following Dummett we may characterize realism as the belief there are statements "that possess an objective truth value, independently of our means of knowing it: they are true or false in virtue of a reality existing independently of us" [8]. Of course, there are different types of realism, (a classical example, which however needs not directly concern us at this point, is *realism concerning the universals*, the philosophical standpoint that, in the Middle Ages, was the subject of hectic controversy). But within any of them there are, by assumption, at least *some* statements (which, in our language, are the *strongly objective* ones) of the kind Dummett specifies; and it is natural to consider that the terms in which these statements are couched are, by definition, strongly objective concepts. Here, realism about material objects is the type of realism we start from (although we shall have to greatly modify its views so as to adapt them to modern physics). In it, although, as noted, the position—for example—of a macroscopic object is a notion whose meaning must ultimately be explained by mentioning observations, still it is a concept normally considered as referring to a state of affairs not dependent on our possession of evidence concerning it, and therefore strongly objective.

At least within such a type of realism, a definition of weakly objective concepts more restrictive than the one above can therefore be formulated, and this allows for a clearcut distinction to be made between the notions of strongly and weakly objective concepts as well as between those of strongly and weakly objective statements. Let us begin with strongly objective statements. To take over the essentials of Dummett's approach again: we may define them as those that, within a Dummett-type realism, could and would be considered as being true or false in virtue of a reality existing independently of us, and as possessing therefore an objective truth value, independently of our means of knowing it. This makes it possible to define the strongly objective concepts as—to repeat—just the entities or attributes the terms used in such strongly objective statements refer to. With regard to the weakly objective statements, the already considered definition remains appropriate: they are those that explicitly refer to human actions or observations and hold true for anybody. But within some theories there exists a third category of statements. These statements resemble the strongly objective statements of classical physics in that they apparently adequately describe—or refer to—*objects as we see them to be* (they make no explicit reference to actions or observations); however, reasons of internal consistency of the theories in question prevent us from interpreting these statements as possessing strongly objective truth (for instance, they are inconsistent with predictions derived from the theory and bearing on "conceivable but practically unfeasible" experiments: we met with such a case in Section 10.6). Within such a theory we may decide to call "weakly objective" (i.e., *merely weakly* objective) the concepts to which the statements belonging to this third category refer. A typical example of such concepts is that of the position of a pointer after a measurement has been performed (see Eq. (10.5)) and before anybody looked. Such a definition will, of course, have to be refined (see Chapter 15) but as it stands, it already makes clear that, within some theories aiming at realism, to say a concept is "but weakly" objective is a meaningful, nontautological statement.

Incidentally, in order to prevent possible confusions a semantic remark is appropriate at this stage. It consists in the observation that, as just defined, the class of the weakly objective concepts is, contrary to that of weakly objective statements, a rather restricted one, since these concepts refer to objects. We should therefore keep in mind that there are many concepts—the one of a mathematical function, for example, or the one of truth and so on—that are quite "objective" in the usual sense of the word but that are neither "strongly" nor "weakly" objective in the limited sense conventionally imparted above to these expressions.

A theory should be said to be weakly objective if its basic statements or "axioms"—the statements from which the theory deductively follows—involve at least *some* weakly objective statements or concepts. Conversely, a theory classifies as strongly objective if all of its basic laws are exclusively formulated in terms of strongly objective statements and concepts. Needless to say that, given a strongly objective theory, we are not *forced* to understand it in realistic terms, that is, as bearing

on objects themselves and describing them "as they really are."
Cogent philosophical reasons led prominent physicists, even in
the "classical age," to understand such theories otherwise. The
strongly objective character of a theory merely consists in the
fact that, when considered in itself, independently of any external, philosophical considerations, its basic statements *can* consistently be understood in such a way.

Setting aside, as already mentioned, the questionable case
of thermodynamics, all the various theories that composed classical physics were strongly objective according to the definition
above. Indeed, this is also true regarding relativity theory in spite
of some appearances to the contrary. Rosenfeld for example [9],
rightly claimed that, although, in relativity, reference is made to
observers (to the "things" observers in such and such reference
frames observe and so on) this does not imply that relativity
theory is incompatible with realism. His argument was that it is
the *formal invariance* of the laws of relativity theory with respect
to changes of reference frames that imparts objectivity to the
theory. In fact, a better argument, or at least one that is more in
the line of Dummett's conception of realism, can be thought of.
It consists in pointing out that, as already noted in Chapter 1,
within relativity theory space-time and events can consistently
be considered as elements of a reality existing quite independently of us. Since, in this theory, events are knowable, it follows
that the theory can consistently be interpreted as describing "reality as it is," even though this reality manifests itself to human
observers in ways that depend on the reference frame in which
they are (in other words, the "basic statements" of the theory are
those that bear on events and space-time and these statements
are strongly objective).

From his own argument Rosenfeld inferred that, with respect
to such general features as the use of the notion of observers and
the like, there is no basic difference between (standard textbook
or Copenhagen-like) quantum mechanics and special relativity and, accordingly, he considered quantum mechanics just as
objective (i.e., strongly objective, in our language) as relativity
theory. Such a conclusion can, however, not be upheld any more
when the second argument—the one just explained—is taken
into due consideration since (contrary to the relativity case) in
(standard textbook or Copenhagen-like) quantum mechanics, *no*
concept whatsoever, neither that of particle, nor that of wave, nor

that of state operator, and so forth can be interpreted as referring to contingent *elements* of a reality existing independently of us. While, if we so wished, we could unfold special relativity theory in terms just of myriad point-sized events existing per se in some per se existing space-time (details of the connection with experience only being stated later), a similar procedure cannot be applied to standard quantum mechanics, as many data analyzed in the foregoing chapters clearly show. It therefore comes as no surprise that such "normative" books as those of von Neumann [10] and Dirac [11] should, as done above in Chapter 3, express the basic quantum mechanical axioms in terms of measurement outcomes. As stressed by John Bell [12] the phrases "any result of a measurement of a real dynamical variable is one of its eigenvalues," "if the measurement of the observable   is made a large number of times the average of all the results obtained will be   ," "a measurement always causes the system to jump into an eigenstate of the dynamical variable that is being measured   " are statements of *basic axioms* taken literally from Dirac's book. It is true, of course, that this does not make the theory *subjective*, as some philosophers claimed, because reference is not made there to such and such *individual* observer getting conscious of the outcome. But it remains true that the theory involves in a *basic* way the very notion of measurement, so that—to quote Bell [12] once more—"it would seem that the theory is exclusively concerned about 'results of measurements' and has nothing to say about anything else."

While this fact is not in the least of a surprising nature for the phenomenalists,[2] it is deeply unsatisfactory for a realist, since the latter does, by assumption, believe in the existence of an intelligible reality that physics describes. He then cannot but ask such questions as those Bell put forward in the paper quoted above: "What exactly qualifies some physical systems to play the role of 'measurer'? Was the wave function of the world waiting to jump for thousands of millions of years until a single-celled

---

2. At least for those of them who lean toward operationalism. Right from the start, others claimed (as Berkeley did) that what we have to do is just change the meaning of the word "matter" and apply it to the sense-data: And that therefore giving up philosophical realism should entail no difference whatsoever in our effective ways of arguing and dealing with the given physical world. For *these* phenomenalists the conventional quantum mechanical axiomatics should, of course, be just as unpalatable as it is for the realists.

living creature appeared? Or did it have to wait a little longer, for some better qualified system   with a Ph.D?" It is therefore natural that a number of theorists should have strived to clarify this question and get quantum physics out of what they could not but consider an "awful mess."

As we know, most of them took up what may be called a "direct" approach to the problem, in that they tried to define physically and theorize the disputed notion "measurement." Considering that the Hilbert-space formulation of quantum physics is unsurpassable in every respect, most of these physicists aimed at removing the difficulty without introducing in the theory any element foreign to the said Hilbert-space. In the spirit of the above developments we may understand their program as having been to reconcile Hilbert-space quantum theory with a kind of macro-objectivism in which some statements concerning macroscopic objects could consistently be considered as strongly objective ones, with, of course, the corollary that the dynamical quantities on which these statements bear should enjoy the status of strongly objective concepts. As we said, of all the theories of that type, the one that centers on the interaction of macrosystems with their environment and on such notions as decoherence is one of the most interesting and can serve as a paradigm for the others. We already know that it accounts remarkably well for the *appearance* of a classical world (and so do also, to a good extent, other, rival, classical-limit theories and, in particular, measurement theories). But, if we are to have a realism in Dummett's sense, the material-object statements we make must be supposed to reveal attributes as they really are. For example, the observation of a pointer position must reveal what this position *is*, in the ontological sense of the word, that is, independently of what human observational and technical ability limits happen to be. However, this condition is not fulfilled, as we know and as emphasized in Section 14.10. And this precludes the possibility of interpreting the statements in question as referring to strongly objective concepts.

These physicists' endeavor bears, of course, some relationship with the quest a number of philosophers engaged in quite independently of quantum mechanics and that consisted in trying to find a middle way between conventional realism and what was called "radical phenomenism" in Chapter 1. Dummett [8], for example, considers that the systematic reduction, proposed by the phenomenalists, of all physical statements to statements about sense-data might conceivably be avoided—and hence a kind of moderate realism might be consistently restored—just by rejecting the law of excluded middle for material-object statements. More precisely, the idea he sketches there, but does not pursue, is based on the fact that a material-object statement (say, a statement about a physical property of some system), when not asserted as a direct report of observation, reduces to a subjunctive conditional whose constituents are still couched in material-object language.[3] Such a conditional should, of course, be judged true on the basis of observations and generalizations from them. But, in the view Dummett suggests here, it can be considered true only if there are observations that have actually been made, which would serve as grounds for its truth. If this is not the case the statement is neither true nor false. We should say that regarding the particular case at issue, it is meaningless.

There is some appreciable similarity between this idea, of which Dummett asserts that it is "neither realist nor phenomenalist," and Bohr's main "guideline" idea that, for defining a phenomenon, the role of the instruments of observation is essential. Consequently, this is a proper place for recalling that the whole discussion of Bohr's views carried out in Sections 11.1–11.3 amply demonstrates Bohr cannot be classified as a realist in the sense of the Dummett's definition of realism: and that it is not by turning to Bohr that we could hope making quantum physics a strongly objective theory.

More generally, let it be stressed that the whole detailed investigation carried out in this book points to the following picture: due to recent, important advances in the fields of environmental theories, theories of complex systems, calculation procedures making use of unfamiliar algorithms such as quasi-classical operators, we have gained a much better—indeed almost completely satisfactory—understanding of why, in a, presumably, *quantum* universe, so many phenomena that are of overwhelming importance to us *look* classical. But none of the developments carried out in this direction has truly met the—admittedly quite demanding—conditions that (conventional) realism sets forth. None of them, in other words, leads to a

---

3. The statement "there is a table in the next room" reduces to "if we were to go into the next room we would ascertain the actual presence of a table there."

strongly objective theory. So that it seems necessary that those of us who really want one—at all costs—should give up strict, exclusive adherence to standard quantum theory, and turn to such conceptions as those explored here in Chapter 13: nonlocal hidden-variables theories or nonlinear quantum physics.

14.8 Complements on "Do We Need a Reality?"; The Notion of Explanation

In Section 14.3, some arguments in favor of open realism were stated. It is appropriate that they should here be supplemented by some more detailed considerations.

The Notion of Explanation

As recalled in Chapter 1 and again in Section 14.3, a number of philosophers have for a long time challenged realism. It so happens that even pieces of physical knowledge of a rather elementary kind may reinforce this questioning. I refer especially to what some of them suggest concerning the notion of *explanation*. At first sight it may seem that this notion is quite tightly linked with that of elements of reality existing "out there." However, as recalled in Section 1.4, explaining some phenomenon is not necessarily tantamount to discovering "how things really are." Indeed, the history of physics indicates that such a goal may well be overambitious: Explaining the return of Halley's comet by referring to a *really existing* gravitational force proved incorrect when general relativity showed this force not to exist. In fact two more modest definitions of the "explanation" concept seem more suitable.

Definition (a). It consists in asserting that "to explain" means "to reduce to what is already known." It often happens that this is indeed possible, and we then *do* have the feeling a genuine explanation has been discovered. Unfortunately, there also are many instances in which such a procedure is deceptive. The idea of explaining the sensation of heat due to contact with, say, a hot hydrogen gas by referring to the more or less energetic impacts of gas molecules on our skin is an example. It does reduce heat to a known fact (we all have seen impacts of tennis balls on nets) and looks so simple and straightforward that it is difficult to renounce it: and that, therefore, explaining *heat itself* as due to the existence of small classical bodies endowed with more or less energetic disordered motions seems convincing. However, the hydrogen molecules are quantum objects, to which quantum theory does not attribute definite trajectories, so that this explanation and similar ones cannot quite be taken at face value. Indeed, if present-day statistical mechanics "explains" in a very general way most phenomena involving heat, it is by means of such concepts as that of Gibbsian ensembles, which are abstractions precluding "literal" descriptive interpretations in terms of "the known."

In a way, the hidden-variables theories may also be considered (setting any yearning for realism aside) as attempts at reducing what is unknown to what is known: specifically to the familiar concepts of matter-points, force, position, velocity, and so on. But with the discovery of such facts as nonlocality and contextuality it became clear that even the hidden-variables theories did *not* actually perform the reduction in question. So that there are some grounds for falling back on the even more modest definition that follows.

Definition (b). It consists in asserting that "to explain" just means to connect up a great many different facts, by showing that they come under the same law. According to this view, to explain the return of Halley's comet is to link it up to the general gravitation law, conceived of not as an element of a description of how nature is constituted but merely as a relationship associating some phenomena together.

Clearly, the basic quantum facts discussed in detail in this book and summarized in the foregoing sections point in the same general direction as the just mentioned ones: they constitute, when considered together, a strong indication that these definitions of the "explanation" concept—and particularly the weaker definition (b)—are by far more secure than those centered on realism, and this observation must considerably reinforce our doubts concerning realism, as understood as Dummett defined it.

Why Not "Radical Phenomenism?"

We may then feel tempted to go to the other extreme, consider that this notion of reality is, when all is said and done, a useless one, and adhere to radical idealism (possibly giving it some other name, but keeping its theses intact). Such was, for example, Schrödinger's position. In particular this author severely criticized the Kantian notion of things-in-themselves. "No single man, he wrote [13], can make a distinction between the realm of his perceptions and the realm of things that cause it. The story is occurring once and not twice. The duplication is an allegory "

Even if it is granted that Schrödinger had some grounds for speaking of "duplication" when specifically referring to Kant's views—since the latter *did* often use the expression "things in themselves" in the plural—it must be observed that we now have to do with quite a different state of affairs. This is due to the nonseparability of any tentative description of independent reality. Phenomena appear to us as multiple and localized, while we now know that any attempt at describing independent reality along these lines is bound to fail. It is therefore quite impossible to describe the phenomena as a *duplication* of independent reality or vice versa. But there are many other facets to the question. In fact, it is difficult—for anybody and quite especially for physicists—to part altogether with realism. Let us now try and make quite explicit why this is so.

First, if not Schrödinger, at least most scientists would willingly agree with the following quotation from Pierre Duhem [14]:

However much the physicist may try to convince himself that theories have no capacity to grasp the Real, that they only serve to give a condensed and classified representation of the experimental laws, he cannot force himself to believe that a system able to order so simply and easily such a huge number of laws, which seemed to be so disparate, could be a wholly artificial system. Through an intuition, in which Pascal would have recognized one of the reasons of the heart "that reason does not know," he puts his faith in a real order, of which his theories are an ever clearer and more exact picture.

However, there is more to say. Admittedly Duhem is right when, as implied by his reference to Pascal's "reasons of the heart," he grants there are no *logically binding* arguments in favor of realism. This, at any rate, is unquestionably true concerning the type of realism that Dummett's definition refers to. But it does not entail that there are no arguments whatsoever for any form of realism. In what follows we list a number of arguments that, if not totally binding, still have quite a strong plausibility value. Some were already sketchily mentioned in Section 14.3. Most of them are definitely *not* the traditional arguments of realism since, in view of present-day quantum physics, many of the latter appear shaky (to say the least!). They are openly philosophical and, as we shall see, they do not point to quite the same "kind" of reality as the traditional ones. But they are worth considering nevertheless.

1. Already mentioned was the argument that whoever, literally accepting the teaching of Kant, assumes all our concepts—including the scientific ones—basically are a priori forms of our sensibility must, by this very assumption, logically be led to expect that the concepts to be used in any theory whatsoever form a subset of those we naturally have. In other words he must expect all our basic scientific concepts to have the quality some philosophers call "visualizability" (*Anschaulischkeit*). And indeed, Kant's own concepts of Euclidian space and universal time do have, to quite a good extent, this character. We noted that just such too great a belief in Kantianism—or neo-Kantianism— may well have hindered the physicists who, during the first quarter of this century, had the task of building up a theory of the atomic processes. Anyhow, we know quite well now that a distinction has to be made between the concepts with which we describe our experience, which, as Bohr kept saying, must be visualizable (expressible in "plain language") and those that serve in formalizing our theories, *most of which must be abstract*. But then, that is, if Kant's accounting for the set of all the concepts we use is not valid, a question arises as to where these concepts "come from." Why do we use such and such concepts instead of others? Clearly, the (obvious) answer "It is because they work" is not totally satisfactory, for it immediately calls for the question already encountered above: "Why do *they* work, while others do not?," which Kant, with his scheme, did not even have to consider. To sweep this last question away just "by decree" ("no question beginning with a 'why' shall be considered") is somewhat of a poor answer. A slightly better one would be to point out that these new concepts are borrowed from mathematics, which is also a form, if not of our sensibility, at least of our understanding. But then, why precisely just these—curved spaces, tensors, Heisenberg-picture projectors and so on—instead of so many other ones, that swarm in textbooks on mathematics? When all is said and done, the idea that, even though the form of our scientific descriptions admittedly owes much to the mind structures, still it does not owe *everything* to it, seems to force itself upon us.

2. We also noted the well-known but still quite impressive argument based on the remark that we sometimes build up quite beautifully rational physical theories that experiment falsifies. Experiment cannot falsify the rules of the game of chess—nor those of any other game—because these rules are just *created* by us. In this case, therefore, there is nothing "external" that could say "no." But in physics it sometimes (and even quite often!) happens that something *does* say no. How could this "something" still be "us?" It seems that the degree of intellectual contorsion necessary for answering such a question in any positive way exceeds what is acceptable. As Bonsack [15] puts it: "The world constructed by the subject lacks an important dimension, that which gives a sense to error. Somehow it is its own measure."

3. Another previously noted argument is of a more "metaphysical" nature. The approach of radical phenomenism is based on the view that we only know the phenomena, combined with the wise maxim that we should only speak of what we can possibly know ("Whereof we cannot speak, thereof we must keep silent," as Wittgenstein put it in his *Tractatus* [16]). Radical phenomenism extrapolates this maxim to the idea that only the phenomena have a meaning. But, reasonable as it may look at first sight, this extrapolation openly makes the concepts of knowledge and experience logically prior to the concept of existence, and *this* is a standpoint the internal consistency of which seems questionable. Should not, in fact, the tables be turned? It seems quite impossible to impart any meaning to the very word "knowledge" without postulating, implicitly at least, the *existence* of somebody, or something, or what-not, who knows. As we said, the logical ordering implied by the approach of radical phenomenism does not, when all is said and done, seem compatible with normal requirements concerning meaningfulness.

4. A fourth point, already hinted at in Chapter 1, has to do with the discovery of the finite age of the universe, a fact unknown to Berkeley, Kant, Henri Poincaré and practically all the great advocates of (radical or moderate) phenomenism. In a way this discovery strengthens the message of evolution theory. In a noneternal, hence basically evolutionary, universe, more and more complex systems gradually appeared, starting with atoms, molecules, and other nonliving systems and culminating in the emergence of animal and human mind; and mind, in turn, gradually increased its abilities, in particular by forming good concepts, which, for the very reason that they were thus gradually *formed*, can hardly be considered as being the ultimate building blocks of the "outside world." Admittedly this argument is partly circular, hence not a decisive one (for the phenomenist, time is not an arena *within* which mind evolved). Still, to some extent it does lessen the a priori credibility of radical phenomenism.

5. Finally, these arguments must be supplemented by referring again to a fact pointed out in Section 10.8 in connection with the Heisenberg picture. As stressed once more in Section 14.6, the mathematical symbols such as ρ, E, B, r that are made use of in physics have two different roles that classical physics tends to confuse: one being of referring to the contingent value of a dynamical quantity pertaining to this or that particular system, and the other one being to serve in writing down the general laws of physics, such as the Maxwell equations, the Lagrangian of this or that field theory and so on. As long as these two roles are identified with one another it is impossible to uphold realism without ipso facto postulating that the basic concepts of the theory (say the electric field at any given space-time point, for example) describe elements of a reality existing independently of us. This kind of realism is not very far from material-object realism and, because of this, is sometimes called "mathematical realism" (although this expression is used by the pure mathematicians in a somewhat different sense). It is, as we saw, definitely at variance with what is implied by standard textbook quantum mechanics. In quantum field theory, for instance, we already noted that the

Heisenberg field operators essentially appear in combinations, such as the "particle number" operator, whose eigenvalues are nothing more than the possible values that would be observed *if* we measured the corresponding observable. But as soon as the two aforementioned roles are distinguished it becomes possible, without falling back into an incompatibility with quantum mechanics (even considered in its standard, weakly objective, version) to uphold the view that the general laws do indeed reflect, in some way, structures of a reality that exists quite independently of us. This makes it somewhat less surprising that general laws such as the Maxwell equations should have remained so efficient in spite of the many radical changes the underlying concepts underwent. And this kind of up-to-date version of Poincaré's *structural realism* then may be viewed as *accounting* for the regularities of the observed phenomena instead of just recording them, which, we claim, is an advantage.

• *Remark:* In quantum field theory the Heisenberg field operators $\psi(x, t)$ etc. are defined *in* space-time. For the reason explained in Section 14.6, this fact cannot be understood as meaning that independent reality is *immersed in* space-time. But, to repeat, such Heisenberg operators can consistently be attributed the role named "Role a" there, that is, it can be assumed that the Lagrangians, Hamiltonians, and so on written down with the help of these fields do reflect some general structures of independent reality. If this reasonable assumption is made, then, pursuing in the same vein as above, the fact that these operators are functions of space-time coordinates seems to imply that space-time is, in a way, something more basic than just an a priori mode of our sensibility, as contemporary Kant followers might well believe it to be. Even if it is a *construction*—in the sense that, as just pointed out, we cannot think of it as a preexisting arena in which independent reality would be immersed—still it seems that this construction does capture "some truth" about independent reality.

To sum up: There are good reasons for not accepting radical phenomenism, that is, for answering *yes* to the question "Do we need a reality?" On the whole, the detailed considerations of this section thus confirm the standpoint of open realism, as stated at the end of Section 14.3.

14.9 Intersubjective Agreement Revisited

Our most common, ever recurring experience as members of the human species is that of an intersubjective agreement between us concerning contingent facts. Unquestionably this experience—together with that of the permanency of many objects and phenomena—is one of those that lie at the basis of our strong, intuitive belief in the existence of a "real world out there." In other words, this intersubjective agreement, as noted in Section 1.4, seems prima facie to be another argument, and an extremely powerful one, in favor of some sort of realism. However, in this case as in the other, above reviewed, ones, the question deserves further, detailed scrutiny. As we shall see, it is quite a delicate one.

As just pointed out, the naive, commonsense explanation of the intersubjective agreement relative to contingent facts is grounded on realism: if we all agree that such and such an object is red, or is at such and such a place, it must be because it *really is* red, or at that place. As noted in Section 1.4, while the idealists reject this explanation on the ground that it is metaphysical, most of them do not put forward any substitute to it. On the other hand, it was also pointed out there that in the cases in which quantum objects are concerned such kinds of explanations—for example, explanations of the type "both of us (each one with the help of his own instruments) observed the electron in the upper beam *because, before we observed it, it really was* in the upper beam"—are, most of the times, flawed. The question is therefore worth some more study.

Agreement with Respect to Secondary Qualities

This case is the least controversial, so let us begin with it. Concerning these qualities (or "properties") it was in fact pointed out at a rather early stage in the history of the investigations on such matters that the "reality" supposed to lie behind the intersubjective agreement is in fact something comparatively evanescent. It is even doubtful that Peter has the same sense feelings as Paul when Paul shows him an object he says is red, and questionable that the words "the same" have a meaning in this context. What is unquestionable is only that, in an overwhelming majority of cases, Peter agrees and says "yes, it is red." In the spirit of these observations it is nowadays generally considered that concerning colors—and other secondary properties as well—the aforementioned realistic explanation of the agreement, based on the idea of a property possessed by the object and observed by us "as it really is" does have something too naive.

But this observation never persuaded anybody not to look for an explanation of some sort. And of course, even a very schematic theory of vision can in principle yield one, by pointing out that the agreement in question is a consequence of general laws of physics. For example, in the case of color the laws of classical electromagnetic theory do readily provide a suitable such explanation.

It should be observed that, at this stage, two philosophical options still remain available. One (O_1) is to stop at the point just reached. Explicitly, this option consists in observing that we have reliable predictive physical laws that do predict a correlation between Peter's and Paul's sensations and in claiming that *this* reference to independently known physical laws constitutes by itself the explanation looked for (see the definition (*b*) of the explanation concept in Section 14.8). The other one (O_2) is conceived for and by the people who demand more. It resorts to realism. It consists in observing that the general laws in question (the laws of classical electromagnetism in the example considered above) deal with primary qualities (field strengths, etc.) and in claiming that the explanation consists in that these laws describe the fields as *having* such and such values at such and such times. Clearly, within option (O_2) the problem concerning the secondary qualities merges into the one concerning primary qualities.

Primary Qualities

Let us then turn to these. If we are content with option (O_1) we are entitled to consider that, concerning them, quantum mechanics completely solves the intersubjective agreement problem; that, in other words, it explains the intersubjective agreement about primary qualities in a fully satisfactory way. It even yields several possible and equivalent ways of couching this explanation in words, all of them based of course on the fact that only measurements—in the broad sense of the word—can yield sense-data. One of these ways is to resort to the notion of wave-packet collapse: the person who performs the first measurement—for example, a position measurement—gets a certain outcome and thereby reduces the wave packet to the eigenfunction corresponding to that outcome A person who, immediately afterward, measures the same quantity has then to do with the reduced wave packet and is therefore bound to get that same outcome This description must of course be refined in order to account for relativity effects and, in Section 10 12, we saw how this can, at least formally, be done A second way consists in not resorting explicitly to collapse and in considering the quantum states of the two measuring instruments instead When both measurements have been performed, the overall quantum state of the system *plus* the two instruments is in general a quantum superposition of states that are eigenstates of the measured quantity corresponding to different eigenvalues, but in each of these states the two instruments are in the corresponding eigenstates of their respective "pointer coordinates," so that these eigenstates are strictly correlated Under these conditions the probability is zero that, when each observer looks at his instrument (i e , measures its pointer coordinate) the outcome they read should differ Finally, a variant of this is Everett's "relative state" conception, in which even the observers' consciousnesses are included into the overall quantum system Again, there is a zero probability that the states of consciousness of the two observers—here identified with the eigenstates of these two consciousnesses—should differ from one another

• *Remark 1* Note that within option (O_1)—or equivalently within the realm of Definition (b) (Section 14 8) of the notion of explanation—the famous question whether or not animals participate to "our" intersubjective agreement gets an unambiguous positive answer For ascertaining this it is not necessary to go as far as to unreservedly subscribe to Everett's conception As soon as we attribute to animals the ability to observe—for example, to become conscious of the position of an instrument pointer— and associate ordinary quantum probabilities to such "events," the quantum rules predict the same correlations between their sense-data and ours as between just ours Dogs, cats, and so on therefore see, and live in, the same empirical reality as we

• *Remark 2* One of the reasons why the notion of a hidden-variable "position" is intuitively attractive is that we are accustomed to account for intersubjective agreement about positions of macroobjects by means of the commonsense "realist" argument that "we both see the object at the same place because it *is* at that place" (option O_2); so that we have a natural tendency to extend this explanation to micro-objects. At least in the case of the latter, the standpoint may consistently be taken that this explanation is a delusion. Think again of the, theoretically indisputable, fact that if two of us measure—one immediately after the other and each one with his own equipment—the position of an electron we both get the same outcome. As we know, the set of the standard quantum rules *accounts* for this, in spite of the fact that, in the standard picture, the electron did not have beforehand any definite position: Hence, on the issue, the fact that, in the pilot-wave theory, the electron has, at any time, a definite position that we detect may be said to bring nothing new. It may be conceived of as redundant. For it to be considered as yielding a genuine explanation (*the* explanation) of our agreement it is necessary that our whole outlook on physics be changed, and the so-called "hidden" variables be considered as ones and the sole ones of which we have *direct* knowledge—a far-reaching reinterpretation of which it is generally doubted that it has sufficient justification.

The foregoing illustrates the fact that, as long as we adhere to option (O_1) ("to explain" is to show that different facts come under the same general law) intersubjective agreement is satisfactorily accounted for by quantum mechanics. And of course this observation holds true in general, that is, it is not restricted to "quantum" objects. To the extent that quantum mechanics is considered a universal theory, it applies to intersubjective agreement concerning any contingent material fact. This renders quite questionable the, aforementioned, prima facie rather "obvious" view that the observed intersubjective agreement about contingent facts is an indication in favor of the existence of an "external" or "independent" reality (or, better to say, of the meaningfulness of such a concept). However, the issue cannot be considered as finally settled through this remark. It is true that this agreement follows from the predictive rules of quantum mechanics. But what is the nature of these rules? Clearly, when applied to them, the often used expression "rules of the game" is deceptive, for, actually, they are not of the same nature as those of a game. As we noted, the rules of a game are quite obviously freely *decided* by human beings. By contrast, not even the most "internalist" among the epistemologists would claim the rules of physics are just whimsically decided by men. Hence, to the philosopher who would maintain that intersubjective agreement "proceeds" from the quantum rules, another philosopher could respond: "But then, whence do the rules in question proceed from?" This would be just another way of questioning the view that option (O_1) is really tenable. It opens on basic questions concerning admissible extensions of the causality concept, that will be touched upon in Chapter 16. Briefly, if the remarks about the Heisenberg picture and the dual role of symbols such as **r**, **v**, **E**, **B**, and so on developed in Section 10.8 (and referred to again in Sections 14.6 and 14.8) suggest something, it is that the concept of causality between the phenomena—the only one the philosophers consider as meaningful—should be supplemented by a concept of causality *from* the general structures *to* the phenomena. But for such a notion to make sense it seems necessary that a reality of some sort be attributed to the said general structures.

14.10 Complements Concerning Limited Knowability and Realism

In Section 14.5 the main reasons were listed why it seems hopeless to strive at building up an "ontological" physical theory, describing independent reality "as it really is." These critical arguments are developed here in more detail.

The Environment Theory and the Conventional Measurement Theories

The environment theory is aimed at being more than a purely "phenomenalistic" theory since, (contrary to, e.g., Bohr's approach) it imparts a meaning to the notion of the quantum state of the instrument independently of whether or not the latter is used as such. It is based as we saw on the quite crucial remark (Zeh [17], Baumann [18]) that macroscopic systems of even quite a small size (such as dust specks) have such a dense energy spectrum that they never can be considered as totally isolated from environment.

There are several reasons why this beautiful theory—which so nicely accounts for the classical *appearances*—cannot be interpreted as reconciling quantum mechanics with conventional realism. Let us here analyze three of them that could only be outlined in Section 14.5.

1. The first one concerns the description of measurement based on this theory. It was pointed out in Section 10.6 and mentioned again in Sections 14.2 and 14.5. It essentially goes as follows. Within a Gibbsian ensemble of systems and instruments (and environments) let a measurementlike process be considered, in which the measured system S initially lies in a quantum superposition of several—say, two—eigenstates of the measured observable and let t be some time after this process has come to a close and the interaction between the instrument and the environment has produced the effects analyzed by the theory. There then exist some operators that correspond to in-principle-observable quantities and have the property that the quantum predictions concerning the outcomes of possible measurements of these quantities, to be performed at time t or later, are incompatible with the assumption that the instrument pointers *are* in definite macrostates. It is true that these quantities—the "sensitive observables," as we called them in Section 1.6—involve the environment in a way that makes them unmeasurable *in practice*. But this expression "in practice" clearly refers to limitations of the human abilities, and cannot therefore be part of the wording of any condition that makes a statement strongly objective.

To grasp somewhat more in detail the conceptual problems that are at stake here we may follow Zurek [19] and use a simplified model which he puts forth as an attractive implementation of the ideas of the theory in question. Essentially it consists in representing the whole $S + A$ system (where A stands for the apparatus) by just a particle, free to move on the $x'x$ axis. When the system S initially is in a state such as the one specified above, the state of the $\Sigma = S + A$ system at a time t_0 immediately after the interaction between its two parts has taken place is pictured in the model by a state of the representative particle that is a coherent superposition

$$\psi(x) = 2^{-1/2}(\psi_1(x) + \psi_2(x)) \qquad (14.1)$$

344   14. *Open Realism*

of two strongly peaked wave packets, $\psi_1(x)$ and $\psi_2(x)$ (for simplicity's sake the two amplitudes are here taken to be equal) which (also for simplicity's sake) we may take as being

$$\psi_1(x) = \phi(x - a), \qquad \psi_2(x) = \phi(x + a) \qquad (14.2)$$

a being the position of the peak of ψ_1 The corresponding state operator ("density matrix") at time t_0 is then $\rho_\Sigma(t_0) = \psi\psi^*$ Obviously, in the $x'x$ plane this ρ_Σ has four peaks, two on the main diagonal, at the points $x = x' = \pm a$ and two on the second diagonal, at points $(x = a, x' = -a)$ and $(x = -a, x' = a)$ But from time t_0 on, the representative particle interacts with the environment, in a way that can be described by an interaction Hamiltonian qualitatively somewhat similar to the one described by Eq. (10.30). When this environment is traced out, as done in Section 10.6, the resulting $\rho_\Sigma(t)$ is equal to $\rho_\Sigma(t_0)$ multiplied by a function of coordinates and time that, much as the exponential factor in Eq. (12.16) has the effect of rapidly decreasing the off-diagonal parts of the state operator, that is, in the present case the two peaks lying on the second diagonal. To the extent that these two peaks may, after some time, be considered as having vanished, the resulting state operator is then identical to one describing a proper mixture of states ψ_1 and ψ_2, which, if we were to take Zurek's assertions in the quoted article completely at face value, would imply that an ensemble of such representative particles may then be considered as composed, in approximately equal proportions, of particles *in* state ψ_1 and particles *in* state ψ_2. However, the assertions in question cannot really be understood this way. The reason (as we know) is that no general rule of physics implies that the environment should not be observable, so that there is no element in the theory that would make it meaningless (or "unphysical") to consider physical quantities of the type called "sensitive observables" with the consequence that the argument developed in Section 10.6 can be taken over here and a discrepancy follows between the quantum predictions and the literal form of the assertions in question. The only reason that could be invoked for discarding such sensitive observables would, again, be that they cannot be measured *in practice* (perhaps because the limited lifetime of the universe will anyhow prevent "us"—i.e., humanity—from doing so). But, as stressed earlier, while this is a very good reason for considering that the theory accounts extremely well for the *phenomena*, it is not one that can be seen as making it strongly objective

• *Remark 3* Incidentally, remember that, even assuming we could disregard the sensitive observables question, the proposition—call it Q—according to which, under the above stated conditions, $\sum$ is either in state ψ_1 *or* in state ψ_2 (either the pointer position has one of its two possible values or it has the other one) could not be considered as tautological As remarked in Section 4 3, a condition for statement "*A* has value a_1 on $\sum$" to be true is that we can imagine somebody who would know beforehand that a measurement of quantity *A* on $\sum$ would yield outcome a_1 Symbolically[20], this can be written $D > P$, where D stands for "an instrument suitable for measuring *A* is made to interact with $\sum$", *P* stands for "outcome a_1 is registered" and > is the symbol for *strict* implication Taking *A* such that ψ_1 and ψ_2 are its two only eigenkets, with eigenvalues a_1 and a_2, Q then has the form

$$Q = (D > P) \vee (D > \sim P) \qquad\qquad (14\ 3)$$

where $\sim P$, the orthogonal complement to *P*, stands for "outcome a_2 is registered" and $\vee$ is the usual symbol for "or" The fact that, while $(P \vee \sim P)$ is a tautology, Q is not one is then a well-known theorem in modal logic

Clearly the existence of the sensitive observables, if they are at all measurable, implies the existence of some gross differences between the quantum predictions and the localized-pointers worldview It is true that, as pointed out in Section 10 7, also in classical physics far-fetched "sensitive" observables may be thought of, the measurement of which, if at all possible, would grossly invalidate the simple statements about "localized" objects that are good enough for common practice However, as also pointed out there, while in classical physics such sharp statements can always be replaced by unsharp but strictly exact ones bearing on the *individual* objects under study, this is not possible in quantum physics This constitutes quite a critical difference For example, the remark is sometimes made that taking sensitive observables into account implies conceiving of super-instruments—instruments as much larger and more complex than ours as ours are compared to atoms—and this remark is used for claiming that the gross differences alluded to above are such only when "seen" with the help of the magnifying glasses these superinstruments are. So that, it is claimed, *in reality* (that is, relatively to the scale of the actually used implements) these differences are vanishingly small. This inference is valid within the realm of classical physics. But it is so, only because of the fact that, there, the sharp, approximate statements can be replaced by unsharp, strict ones bearing on individual objects; that is, by statements of the type: "Such and such a *physical* quantity is at least x percent smaller than such and such an other one." As we saw, no such replacement can be made concerning the quantum mechanical sensitive observables here at issue. The remark in question cannot therefore be used for justifying the aforementioned claim.

Still on the subject of the sensitive observables a final remark—one of a general character—may be in order. It concerns the fact that among physicists there are some who believe that the purpose of physics is to describe *what really is*, and there are others who, while discarding such ontological objectives because of their "metaphysical" undertones, nevertheless consider that, concerning what science can say, this "setting aside" of ontology has no implication whatsoever since physics anyhow must describe the phenomena in a realist *language*. When the latter claim that their disagreement with the former is merely formal they are mistaken. It is true that it has no implication in the realm of "normal" physics, but it does induce a difference as regards the problem of the sensitive observables, for which an experimental measurement procedure is well defined but practically unworkable. *They* can claim that such observables are void of any physical meaning whereas the realists cannot do so.

2. The second reason why environment theory does not reconcile quantum mechanics with (conventional) realism has to do with the case when it is applied to the more general problem of accounting for the classical properties of the macroscopic objects. In measurement theory the $S + A$ (or Zurek's "representative particle") state considered at time t_0 has, by virtue of the Schrödinger time-dependent equation, naturally the form of a quantum superposition of several more or less strongly peaked and weakly overlapping wave packets This is not normally the case concerning the wave functions of macroscopic objects such as dust grains, the classical features of which are also well accounted for by the environment theory, as noted in Chapter 12 It is quite remarkable indeed that this theory does account for what is observed, also in this more general case, and, since the problem there dealt with is more general, it is not surprising that it should bring in an *additional* reason preventing the theory in question from being interpretable as a "macro-objectivist" theory As we saw in Chapter 12, this reason basically is that the whole argumentation of the environment theory is grounded on the notion of ensemble, and that the density matrix ρ_S relative to the ensemble of the considered macrosystems (after the environment has been traced out) is one that correctly describes not one but an infinity of proper mixtures, most of which—including the "most natural" one, the one composed of eigenvectors of ρ_S—do *not* correspond to macroscopically localized macroobjects With the consequence that although, as shown there, *one* of these ensembles does indeed exhibit that "localization" feature, an individual who, contrary to us, would not have, ingrained in his mind, the preconceived notion that macroobjects *must* be localized could not derive from the theory the idea that in fact they are

　　3. To make this analysis more complete, let a third argument be developed here that goes in the same direction as the two foregoing ones It has to do with the way the Wigner phase-space distribution function $f(x,p)$ defined in Section 6.9 is applied in the environment theory As shown in that section, the mean value of any observable A on an ensemble described by a statistical operator ρ can be expressed by means of a formula, (6 42), that resembles very much the *classical* expression of the corresponding mean value Indeed, to an A it is in general possible to associate a function $a(x,p)$ in phase-space by means of formula (6 41); and formula (6 42) is then formally identical to the one that, in classical mechanics, yields the mean value of $a(x,p)$ when the ensemble over which this mean value is defined is described by the joint probability distribution $f(x,p)$

The main, or at least the most conspicuous, difference between $f(x,p)$ and a classical probability distribution in phase-space is that, for some statistical operators, f can take up negative values.

The most interesting case is of course the one in which $f(x,p)$ is nonnegative for all values of its arguments. This is for example what takes place when the wave function is a Gaussian wave packet, for then $f(x,p)$ is also a Gaussian (in x and p). Some authors seem to suggest that in such cases the x and p coordinates of the system both simultaneously have definite values. This, for example, is the most natural way of understanding Zurek's statement [19] according to which, a system described by a Wigner function corresponding to a minimum-uncertainty Gaussian wave packet "is localized in both x and p." As pointed out in Section 6.9, because of the Heisenberg indeterminacy relationships these values cannot then both be the values that would be observed if the corresponding measurements were actually made in succession on one and the same system. On the other hand, we might still surmise that they are "supplementary" or "hidden" variables, possessing values that differ from one element of the ensemble to the next. As we know, the hypothesis that such variables exist does not, by itself, conflict with the view that all the *experimentally verifiable* predictions of quantum mechanics are exact. However, if the assertion that the system is localized in both x and p is to be understood as an implicit acceptance of the hidden-variables hypothesis, it must immediately be observed that in fact it implies more: it implies acceptance of a special form of the hidden-variables hypothesis. This is the hypothesis that, following Home and Whitaker, we called the preassigned values assumption in Section 13.6 and which consists in assuming, not only that the values a measurement of x or one of p would yield are predetermined, but that even if no such measurement takes place, x and p *have*, on each individual system, the very values these measurements would reveal (and, of course, that these values are distributed within the ensemble according to the distribution laws $|\psi(x)|^2$ and $|\phi(p)|^2$, ψ being the wave function and ϕ its Fourier transform). It is known (see Section 13.6) that when this assumption is considered as a general one, valid for all quantum systems whatsoever, it is incompatible with the hypothesis that the experimentally verifiable predictions of quantum mechanics are all correct. But the question may be considered as to whether, by any chance, consistency could be recovered for the special case of the systems that are here of interest, namely for all those whose Wigner function is positive definite. Fortunately an argument put forward a long time ago by Schrödinger [21] (and mentioned in the Home and Whitaker report) yields additional information on the issue. This is because it concerns, not the general case of all possible systems and states but just precisely one specified state of one particular type of system—namely the ground state of the harmonic oscillator—and this ground state is a Gaussian so that its Wigner function *is* positive definite. Its energy is sharply defined and equal to $hv/2$, which means that within an ensemble of harmonic oscillators in their ground state all the elements have exactly the same energy $hv/2$ (no fluctuation). Since the kinetic energy is necessarily nonnegative this implies that the potential energy $kx^2/2$ of any of these systems cannot exceed $hv/2$. This consequence, however, is incompatible with the idea that, in the ensemble in question, the (sharply defined) x coordinates of the systems are distributed according to the distribution law $|\psi|^2$ since this function is of unlimited support, that is, is nonzero even for values of x that are large enough for the potential energy to exceed $hv/2$. The answer to the question at issue is therefore a negative one: even the systems whose Wigner function is positive definite cannot generally be considered as being localized[4] in both x and p.

The preceding remark should not prevent us from appreciating what the environment theorists have to say about the more complex situations where $f(x,p)$ is negative for certain values of its arguments. This, for example, is the case concerning the two-peaked wave function (14.1). In such a case, of course, it is not

---

4. This is not to say that ensembles whose Wigner's phase-space distribution function is everywhere nonnegative have no special features of their own. Indeed, it was pointed out, notably by Bell (Ref. [22: Ch. 21]) that on EPR pairs whose quantum states correspond to such nonnegative phase-space distributions, nonlocality problems are not to be expected, at least when only positions are directly measured (momenta being derived from positions at different times).

necessary to refer to the foregoing argument. It is immediately clear that $f(x,p)$ cannot be considered as a probability distribution. As noted, what then turns out (this is clearly shown, for instance, in Zurek's article that was quoted) is that the interaction with the environment—when, as above, the environment variables are finally traced out—has the effect of suppressing the negative values of $f(x,p)$ Consequently, the physicists who feel they can somehow assert that a system described by a nonnegative $f(x,p)$ is localized in both x and p can say the same concerning, for example, the $S + A$ system after the interaction with its environment has operated. This may induce them to state that the pointer of each instrument in the ensemble is in one and only one definite graduation interval and so on. But of course the foregoing Schrödinger-like objection to such a standpoint cannot be forgotten. It makes it clear that a distinction we, at first sight, might think Zurek [19] proposes between the systems that are "truly quantum" (those whose Wigner function is not positive definite) and those that are not (those whose Wigner function *is* positive definite), presumably does not correspond to what this author actually means. What, in fact, these developments show is that *for all practical purposes* we may do *as if*, once the interaction with the environment has operated, the system was localized in both x and p (as if the pointer of each instrument was in one definite graduation interval). Hence, as already noted, the theory "saves appearances," as was said in Galileo's century. In other words, the theory shows that quantum mechanics provides us with a fully consistent description of *empirical reality* in the sense defined in Section 10.6. But, to repeat, the objection based on the Schrödinger argument may not be overlooked here any more than it could be in the previously considered simple case. So that, when all is said and done, the theory cannot be considered as providing us with a strongly objective description of the "real" state of affairs. The same is true of the cosmological quantum theory of Gell-Mann and Hartle, as we saw in Section 12.4.

As we know, there have been, especially in the field of quantum *measurement* theory proper, a great number of very ingenious attempts at building up "realistic" descriptions not departing—or at least not departing in any too conspicuous way—from the completeness (alias no-hidden-variables) hypothesis (see Chapter 10). Although they are rather different from one another, these proposals typically show, when adequately scrutinized, the same, basic general feature as the environment theory. They account quite nicely of the phenomena as phenomena, but they cannot be interpreted as descriptions of "reality as it is."

A consequence of this which, obvious as it is, still is worth mentioning is that when proponents of such a theory claim it yields a physical interpretation of the measurement process, the qualificative "physical" is to be understood, not in an *ontological* but in a strictly *operational* sense. In other words, within this context such a qualificative cannot mean that the process in question has been identified with one taking place within a physical reality conceived of as existing quite independently of us and the way we understand it. On the contrary, it can only mean that there is no basic difference between this and other processes because *all of them* are mere *phenomena,* in the philosophical sense of the word. In this respect, it must be considered that an assertion frequently made by the proponents of the environment theory, namely that "the environment continuously measures the system" is more than just a metaphor: it adequately stresses the implicit reference to "mentality" (weak objectivity) that underlies the whole of standard textbook quantum mechanics, including its "measurement theory" extension.

The Explicitly Ontological Approaches

Do the foregoing analyses imply the impossibility of any strongly objective theory accounting for the experimentally verifiable predictions of quantum mechanics? The answer is no for, as we know, unconventional approaches have been put forward that succeed in doing just this. These theories have been reviewed in Chapter 13: concerning them let us therefore keep here to remarks of a general, conceptual nature that are mere extensions of those already put forward in Section 14.5.

The first of these remarks is that, as we saw, any theory that leads to the same observable predictions as quantum mechanics in the domain of measurements performed on particle pairs must violate local causality (in the broad sense, see Chapters 8 and 9), that is, it must violate either outcome independence or parameter independence or both. This means that none of these theories can restore anything conceptually resembling the classical worldview, for all of them will exhibit a form of what we generally called "nonseparability" in Section 8.5. For example, the pilot-wave theory (alias Bohm's hidden-variables theory) involves, either a *quantum mechanical potential* that does not decrease with increasing distances (Bohm), or alternatively (Bell), highly nonlocal fields, coinciding with the real and imaginary parts of the wave function in ordinary quantum theory and which are just as "real" as the electric and magnetic fields of classical Maxwell theory. The GRW theory (Section 13.5) involves no hidden variables but it attributes an ontological status to wave functions that, for microscopic systems, are practically identical to the wave functions of ordinary quantum mechanics and are therefore—for composite such systems—also highly nonlocal.

The second remark is that, as we also saw, since all these theories violate local causality, it could, prima facie, be thought that the ontologically interpretable quantum theories and the conventional quantum theory stand on equal footing concerning their relationship with special relativity, that is, that they all violate it equally severely. But, actually, we observed this is not the case, because of the conjunction of two facts, one of them being that (as shown in Section 8.2) no theory exactly reproducing the quantum predictions allows for superluminal signaling, the other one being that conventional textbook quantum mechanics is weakly objective only. A consequence of this is that if we cling to conventional textbook quantum mechanics we ipso facto give up strong objectivity in physics, so that we have no ground for demanding that relativity theory should be strongly objective. But then, giving up local causality does not prevent us from keeping a formal, operationally defined (Einsteinian) causality, consisting in the rule that quantum field operators localized in spacelike separated space-time regions commute. And, in fact, this is just the principle defined under the name "causality" in quantum fields textbooks. As we noted, it is in this way that a kind of "peaceful coexistence"—to use Shimony's and Redhead's expression again—is salvaged in practice between quantum theory and special relativity. But, clearly, this result is reached at the expense of giving up realism, so that, by definition, it cannot be applied to ontologically interpretable theories that *are* ontologically interpreted. In other words, a lasting conflict between these theories and special relativity seems unavoidable. Compared to this defect, the other shortcomings of these theories seem relatively minor, although they are objectively significant.

Essentially they consist in the facts that (a) there are several such theories, many of which can hardly, or not at all, be distinguished experimentally from one another since they essentially all yield the same verifiable predictions (namely just those of ordinary quantum mechanics) and (b) they are "unfruitful," in the sense that they never yielded any experimentally verifiable prediction that could not be derived from ordinary quantum mechanics in a simpler way and was verified by experiment. Note this state of affairs may well be related to the fact that the most distinctive feature of these theories is they "materialize," so to speak, nonlocality, while nonlocality is "nonoperative" (it opens no possibility for altering the phenomena at a distance). For these reasons, it is not exaggerated to consider them, more as the beautiful metaphysics that fit a scientific age than as truly scientific proposals.

14.11 Status of Nonseparability

As we defined it (Chapter 9) the word "nonseparability" refers to the violation of concepts that had hitherto been considered as valid: divisibility by thought and locality. The discovery of such violations may remind us of previous ones, such as, for example, that of the inadequacy of the concept of universal, Newtonian time (retardation of moving clocks, etc.). The similarities are obvious. Both discoveries originated from theoretical developments: relativity and the Bell theorems. Both were then experimentally verified: through many relativistic experiments in one case, through the experiments of Clauser, Fry, Aspect, and others in the other case. Considered from this angle, relativity of time and nonseparability equally deserve the appellation: "scientific discoveries."

It is clear, however, that there is some epistemological difference between the two. To try and make it precise, let it be suggested that two levels be considered in scientifically established pieces of knowledge. The first (lower) level is the one just made apparent on our examples: the piece of knowledge at issue must have some experimental confirmation and (preferably) have a place in some general, valid theory. The other (higher) level is reached by the pieces of knowledge that, in addition, must be taken into account for explaining data *other than those collected for the specific purpose of testing their validity.* This, of course, is the case concerning relativity of time, which must be explicitly taken into account in predicting the behavior of unstable relativistic particles and in many other experiments, not in the least intended for verifying it. And indeed, the generality of the scientific laws has the effect that most pieces of knowledge that reach the lower level automatically also reach the higher level. But it does not seem likely that this will ever be the case concerning nonseparability, and the reason for this is clear: it is to be found in *parameter independence* (see Chapter 8) and in the fact that, correlatively, nonseparability cannot be directly made use of for transmitting information. Clearly, the circumstance that in the usual (operationally formulated) textbook descriptions of such branches of physics as quantum field theory no reference to nonseparability need be made is linked in some way to this state of affairs.

14.12   Outlook, Veiled Reality

To sum up the content of the foregoing sections: we have forceful arguments against conventional—that is, traditional—realism. Essentially·

1. Conventional realism assumes strong objectivity, which is at variance with the standard—by far the most efficient—formulation of quantum mechanics.

2. Admittedly, there are other formulations (the ontologically interpretable ones, see Chapter 13). But within these, as we saw, conventional realism can be made to agree with the experimentally verifiable, and well-verified, quantum mechanical predictions only at the expense of giving up the kind of *peaceful coexistence* between quantum mechanics and relativity that was described in Sections 14.5 and 14.10. Under these conditions, holding on to strong objectivity implies giving up *some*, at least, of the essential features of special relativity. This is certainly a high price to pay for realism. And, even if we accept paying the price in question, we are faced with the problem alluded to that several models are available, that more of them are probably to come, that, in seventy years time, none of those that existed has yielded original and experimentally verified predictions and that to start trying to discriminate between them by experiment seems to be to launch into a task without an end.

3. On this issue the review, in Chapters 10–12, of important theories of measurement and classical appearances built up within the conventional quantum framework is quite instructive. It is all the more so because most of the authors of these theories—reluctant as they, apparently, are to go into philosophical considerations—seem to have genuinely believed they were constructing theories capable, like old classical physics, of being understood as describing physical reality as it really is. And yet, by scrutinizing the premises of these theories we found out that, at the present stage in their development, if no new ingredient is added, not a single one of these descriptions—be it the environment-based theory, the formalism of operations and effects, the Griffiths theory, the Omnès theory, the Gell-Mann and Hartle theory, or any other of the conceptions we met with in these chapters—is ontologically interpretable.

However, quite strong arguments were also enumerated (Sections 14.3 and 14.8) against radical idealism or radical phenomenism, to the effect that making knowledge logically prior to existence (a) looks basically inconsistent and (b) (even if we forget about this basic inconsistency) has misleading aspects concerning the very development of scientific research.

But then, if radical idealism (neo-Kantianism) as well as all the cryptoradical-idealisms that proceed from it are to be rejected, that is, if the notion of "something" (perhaps, but not necessarily, a set of "things," for language betrays us here: perhaps events, or minds, or Mind, or gods, or God, or what not; let us just say "something") the existence of which is not dependent on (does not boil down to) *our* existence is considered as logically necessary, and if, on the other hand, it is realized that all the ontological theories are too speculative, that is, that the detailed features of this "something" are beyond our reach, then only two possibilities remain: either this "something" is altogether unknowable, a "pure X," or it is such that we can get, or guess, some knowledge about it, *but merely general or merely allegorical.*

Our claim in this section is that rather convincing arguments favor the second branch of this alternative over the first. Among these arguments are some of those we already met in Sections 14.3 and 14.8 in the argument that we need a reality· for indeed, when properly looked at, these arguments also indicate that the latter is not totally unknowable.

This is the case concerning argument 1 in Section 14.8, which led us to the view that even though the form of our scientific descriptions owes much to the structures of our own minds, these descriptions do not owe everything to it. To put it bluntly· it certainly would be mathematically simpler if the radiation field were just a scalar, but it is not. And though we cannot take any definite ontological commitment as to its ultimate nature, still we do have there a piece of knowledge that cannot be thought of as being exclusively our own construction.

This is also the case concerning the argument 2 same section, which consists in the observation that some beautifully rational theories are falsified by experiment. Such facts not only convincingly show that there is "something" outside us—that "something" that says no—but at the same time give us some knowledge about that "something." It is true that it is merely *negative* knowledge, but it is knowledge all the same. Nonseparability, for example, is a negative knowledge of this kind.

• *Remark:* In the foregoing section it was noted that nonseparability, while being a scientifically established piece of knowledge, still does not enjoy quite the same status as most other pieces of theoretical knowledge since it cannot directly be made use of in various areas of physics. This backs up the view that nonseparability is hardly—presumably even *not*—relevant to empirical reality, that is, to the subject-matter of physics proper, and refers in fact to the "deeper layer" of reality that independent reality is.

Similarly, argument 5 in Section 14.8 favors, as we saw, a kind of up-to-date version of Poincaré's structural realism, implying that the general structures of independent reality are not totally obscure to us.

Hence, when all is said and done, the conception that seems most favored by modern physical data is that [23] of a "veiled" independent reality. Let a first, preliminary explanation of what this expression means be given here (a more elaborate one is described in Section 15.1).

1. Independent reality is not in space-time (see Section 14.6). For a long time, indeed since the discovery of entanglement by Schrödinger and even earlier, it has been known that quantum mechanics does not harmonize well with the locality concept.

That this mismatching is not a temporary defect but something much more basic has become clear, now that we know that any "ontological" reformulation of quantum mechanics necessarily involves nonlocality (Bell). This is why, to repeat, we must consider that while space is indeed the arena where the *phenomena* are rightly viewed as taking place, it is not (nor is space-time) an arena in which *independent reality* evolves. Rather it (and time and space-time) are primarily allegorical, human modes of apprehension (as Kant might have said) of independent reality.

2. This veiled reality partly exceeds human power of intellection, and *on this basis* (not just on the basis of indeterminism, which does not seem to be a convincing argument) we may agree with Pauli, when (at several places in his correspondence) he spoke of the "irrationality" of Reality.[5]

3. Still, what science says has "something to do" with it, but this information seems to be limited to some general structures of Reality, and in this sense it can hardly be thought exhaustive.

Hence the object of the factual descriptions of physics, the set of phenomena, should be given another name: *empirical* or *effective* reality, for example. And what emerges from all this is that both notions of (veiled) reality-per-se and empirical reality must be considered as significant. It is worth noticing that this conclusion is not derived here from any particular theoretical model such as, for example, the bootstrap theory. Views partly similar to those here expressed were sometimes derived from such models. The inconvenience of this procedure is, of course, that the validity of the views in question then apparently hinges on that of the reference model. It must be stressed that the derivation presented here of the veiled reality idea cannot be objected to on this basis.

As a final observation, let it be pointed out that although the veiled reality idea has, as previously noted, some rather close similarities with Henri Poincaré's "structural realism," still it differs from it on at least one important point that I repeatedly alluded to. When Poincaré mentioned "the images substituted

---

5. That the word "irrationality" should have appeared in the writings of a physicist of such a great ability as Pauli is not quite so astonishing as it may seem. As pointed out by K.V. Laurikainen (private communication), in Pauli's view the theory is rational; it is reality that is not.

to *the real objects that Nature will hide from us for ever"* (emphasis added) [24] he, as we see, wrote "objects" in the plural. It seems that nondivisibility by thought, or nonlocality, which, in ontological theories, plays the same role, now precludes any reasonable use of the plural when referring to independent reality and that we should therefore speak of independent reality in the singular, rather than of the real objects. The reason why this difference between the veiled reality concept and Poincaré's views is significant is of course that, according to the latter, knowledge of the "relationships between the objects" *is* accessible and indeed constitutes the very essence of science while the very *notion* of these relationships between objects becomes obscure if we can no more think of such objects as being separate entities.

For any kind of metaphysical extrapolation that may be ventured on the basis of such developments, this shift from the plural to the singular induces momentous differences.

References

1. A.I. Miller, "Imagery, Probability and the Roots of Werner Heisenberg's Uncertainty Principle Paper," in *Sixty-Two Years of Uncertainty*, A.I. Miller (Ed.), Plenum (NATO Series), New York 1990.
2. P. Feyerabend, "On the Quantum Theory of Measurement", in *Proceedings of the Colston Research Society*, Butterworth Scientific Publications, London, 1957
3. N. Bohr, *Atomic Physics and Human Knowledge*, Science Editions, New York, 1961.
4. N. Bohr, "Quantum Physics and Philosophy—Causality and Complementarity," contribution to *Philosophy in the Mid-Century*, R. Klibansky (Ed.), La Nuova Italia Editrice, Florence 1958; reprinted in *Essays 1958–1962 on Atomic Physics and Human Knowledge*, A. Bohr (Ed.), printed by Richard Clay and Co., Bungay, Suffolk, 1963.
5. E. Joos and H.D. Zeh, Z. *Phys.* **B59**, 223 (1985).
6. E. Joos, "Quantum Theory and the Appearance of a Classical World," in *New Techniques and Ideas in Quantum Measurement Theory*, Annals of the New York Academy of Sciences **480**, 242 (1986).
7 D. Bohm, *Wholeness and the Implicate Order*, Routledge and Kegan Paul, London, 1980.
8. M. Dummett, *Truth and Other Enigmas*, Duckworth, London, 1978.
9. L. Rosenfeld, in *Louis de Broglie, physicien et penseur*, Albin Michel, Paris, 1953.
10. J. von Neumann, *Mathematical Foundations of Quantum Mechanics*, Princeton U. Press, Princeton, N.J., 1955.
11. P.A.M. Dirac, *Quantum Mechanics*, 3rd ed., Oxford U. Press, Oxford 1948 (1st ed., 1930).

12. J.S. Bell, "Against 'Measurement'" in *Sixty-Two Years of Uncertainty*, A.I. Miller (Ed.), Plenum (NATO Series), New York, 1990.
13. E. Schrödinger, *Mind and Matter*, Cambridge U. Press, 1958.
14. P. Duhem, *La théorie physique*, 2nd ed., Paris, 1914.
15. F. Bonsack, "Prolegomena to a Realist Epistemology," in *Dialectica* **43**, 157 (1989).
16. L. Wittgenstein, *Tractatus Logo-Philosophicus*, Routledge and Kegan Paul, London, 1961.
17 H.D. Zeh, *Found. Phys.* **1**, 69 (1970).
18. K. Baumann, Z. *Naturforsch.* **A 25**, 1954 (1970).
19. W.H. Zurek, "Decoherence and the Transition from Quantum to Classical," *Physics Today*, October 1991, p. 36.
20. B. d'Espagnat, "Nonseparability and the Tentative Descriptions of Reality," *Phys. Rep.* **110**, 203 (1984).
21. E. Schrödinger, *Naturwissenshaften* **23**, 804, 824, 844 (1935). (English transl.. *Proc. Am. Philos. Soc.* **124**, 323 (1980).)
22. J.S. Bell, *Speakable and Unspeakable in Quantum Mechanics*, Cambridge U. Press, 1987
23. B. d'Espagnat, *In Search of Reality*, Springer-Verlag, New York (1983).
24. H. Poincaré, *La science et l'hypothèse*, Flammarion, Paris, 1902.

Other Reading

K.V. Laurikainen, *Beyond the Atom: the Philosophical Thought of Wolfgang Pauli*, Springer, Heidelberg, 1988.

Veiled, Independent
Reality, Empirical
Reality

15 The developments of the foregoing chapter favor the view that independent reality is veiled. In the first section of this chapter this conception is compared with ideas that were recently developed in philosophy of science on purely philosophical bases. In the next ones the propounded notion of empirical reality is made precise.

Philosophical Developments; Bonsack's Approach

It is interesting that ideas showing affinities with the veiled reality concept were independently developed by thinkers working in the field of pure philosophy of science, which means that they did not base their conceptions on detailed physical knowledge of any sort. There can, of course, be no question of presenting here a balanced account of these subtle and often quite sophisticated theories. Indeed it is only possible to mention briefly two or three examples, concentrating only on one of them.

The two examples it is appropriate to mention first are Bas van Fraassen's Constructive Empiricism and Hilary Putnam's Internal Realism. Both are so elaborate that a description of either of them would call for developments that cannot take place in this book. In fact, we must be content with a mere glimpse of each of them.

Concerning *Constructive Empiricism* let us merely take note of one of its major claims, which is that "the aim of science is not truth as such but only *empirical adequacy*," that is, truth with respect to the observable phenomena [1]. It is true that, considered in isolation, such a statement might induce us to wonder whether constructive empiricism should not be identified as a variety of what was called "pure phenomenism" in Chapter 1. But this would grossly oversimplify van Fraassen's subtle standpoint and, when the ways this author applies his views to specific problems are considered, it turns out that, in fact, the standpoint in question is *not* to be identified with pure phenomenism. It is because of this that it does not seem utterly improper to speak of some affinities between constructive empiricism and the veiled reality concept, it being, of course, understood that the concept "affinities" leaves place for significant differences.

With regard to Putnam's *Internal Realism* the situation is not very much different. This author also takes a stand against the "correspondence theory of truth," that is, against the idea that there should exist some mapping of concepts onto the (mind-independent) world, his argument being that a great many such mappings can, a priori, be conceived of, and that in order to pick out just *one* correspondence between words or mental signs and mind-independent things we would have already to have referential access to the mind-independent things [2: p. 72]. On the other hand, Putnam also emphasizes that internal realism is not a facile relativism. "Knowledge," he writes, "is not a story with no constraints except *internal* coherence." In fact, what he denies in conventional realism is essentially the view "that there are any inputs [to knowledge] *which are not themselves to some extent shaped by our concepts, by the vocabulary we use to re-port and describe them, or any inputs which admit of only one description, independent of all conceptual choices*" [2: p. 54]. And at another place he opens a parenthesis to grant that "perhaps Kant is right: perhaps we can't help thinking that there is *some-how* a mind-independent 'ground,' for our experience." All this again shows some resemblance with the veiled reality concept, although, again, this resemblance is but a very partial one: within the same parentheses, Putnam immediately added: "even if at-tempts to talk about it [the mind-independent 'ground'] leads at once to nonsense," a phrase with the spirit of which the approach of the foregoing chapter distinctly parts.

Neither one of the two authors whose views I have much too rapidly sketched would classify himself as a realist (Putnam formerly was one, but he changed). Since physicists are, most of them, naturally inclined toward realism and since a large proportion of the potential readers of this book are physicists, it seems adequate to enter into some more details concerning the affinities the veiled reality concept seems to have with the conceptions of a philosopher who *does* consider himself, on the whole, as a realist. To this end, let us analyze in some detail François Bonsack's approach to realism [3].

Bonsack's starting point was the observation that many, perhaps most, philosophers consider the idealist arguments as impossible to counter, and therefore feel forced to grant that our irrepressible proneness at believing in an external world is, when all is said and done, a delusion. In the views of those among them who nevertheless have a soft spot for realism one would be a realist by instinct, against reason that would demonstrate the contrary. Bonsack makes a stand against this, for, he points out, either realism is a right position and then we must be able to defend it rationally, or it is a wrong one, and then to rely on instinct to uphold it is a wretched line of retreat. In a way that resembles some of the analyses put forward in the foregoing chapter, this philosopher is on the look for a ground on which a realist epistemology would be philosophically unobjectionable.

Anxious not to develop arguments with a metaphysical undertone, Bonsack starts with the only sure and reliable pieces of information we have, namely the uninterpreted "given to consciousness." This, he points out, may, on the first move, give rise to naive realism (things exist exactly as we see them) and, on the second move, once naive realism has been transcended, may lead to a kind of naive, first step idealism: a confinement to what is accessible to us, that is, to the conscious given. He grants that this is a defensible option but claims that, all the same, "the job isn't done so far." For it then remains to explain why and how, starting with this conscious given, the subject is led to postulate a real world and to discern that some of the data he has refer to this world while others (dreams, etc.) do not. "Even if everything goes through representation," he wrote, "the distinctions between an object and its representation and between external world and internal subjectivity have to be restored *within this representation*." For a philosophically minded *scientist*, what constitutes the main interest of Bonsack's approach—and its originality as compared to that of most philosophers—is that, in it, this problem is explicitly stated (in the just quoted sentence) and rationally investigated.

To this end, Bonsack first stresses that the sensory given has a *structure*—since the various perspectives we get on an object are connected in computable ways to one another as well as to our actions—and claims that the aim of knowledge is to give an account of this structure [he quoted Poincaré [4] as having written "objects are not only clusters of sensations, but clusters cemented together by a persistent bond. It is this bond and only this bond that is object in them and this bond is a relation (F.B.. a net of relations)]." The level of the sensory and motor flows and the links between them Bonsack calls level-S (S for subjective or subject). It is the level that some call phenomenological. At this level, he stresses, there is neither world nor space, and therefore neither inside nor outside.

Clearly, to account for this structure by trying to grasp laws directly linking sensations or actions to other sensations and actions would be virtually impracticable (think of what it would be to try and connect directly the orders we give to various muscles for walking around with the resulting changes in the appearances of the world without resorting to such concepts as space and location in space), while, on the contrary, such an account becomes easy if, starting from what we see, we do construct three-dimensional objects, locate them in space, and so forth. In other words, for accounting for the structure in question, the only practically available procedure consists in "going through the instrumentality of building an objective world-O, structured in space and time and furnished with objects."

At this point, although Bonsack is at heart a realist he does not—not any more than other philosophers—mention in favor of realism the (apparently obvious!) argument that if this procedure is so efficient perhaps (some would say· "probably") it is just that it is actually more than a mere trick: a genuine discovery of the *good* concepts; of how the world *really is*. For some reason, this argument, which sounded so reasonable to many physicists of the classical age (in an age of quantum physics, and with the difficulties raised by these very notions of location and objects it does not sound so reasonable any more!) was never taken seriously by philosophers. For the sake of the argument, we shall here follow them—and Bonsack—on this matter. All the same, Bonsack appropriately stresses some points that make his scheme somewhat close to conventional realism. One is that his world-O includes the subject: not, actually, the "subject S" to which the level-S belongs but an objectified subject-O, who is nothing else than an object interacting with other objects and whose resulting sensations and experiences are interpreted by referring to the physical laws of world-O Moreover, still according to our author, the subject-O must be objectified *including his subjectivity*, so that the relation of consciousness and the external world should not fall outside the realm of ordinary causality.

In the model, the objects-O should, for consistency, be considered as having an "existence-O" *not* proceeding from the existence-O of the subject-O and of the knowledge this subject-O is capable of. Obviously this fulfills one of the main requisites of classical realism, and Bonsack is right in stressing this point. But at the same time he lucidly grants that the traditional realists may well object to the whole scheme. They may protest that if the matter develops so easily it is just because a deception underlies it, in that, under cover of defending realism, the model in fact "sails in the thick of" idealism, since world-O is nothing but a representation.

To this possible criticism Bonsack offers a balanced answer. He grants that, at this stage at least, what he has achieved is just to give a sense to realism within an idealist framework. But, he points out, this, already, is significant since idealists stop halfway and forget to meet the real again. Within his own approach this, he claims, is not the case since, in it, as he explains with some detail, the duality between appearance and reality is saved, as well as that between what represents and what is represented. In the same line, Bonsack even claims that, contrary to idealism, his scheme does have, in a way, the dimension of heterogeneity that gives sense to error: though its original is not accessible, world-O is not its own measure, the sensations-O resulting merely from the objectification of the actual sensations S can be compared with the sensations-O predicted by means of the postulated world-O, account being taken of the relevant scientific laws-O, so that discrepancies are conceivable, should be remedied, and so on.

However, Bonsack continues, the realist may still protest that (a) everyone knows the representation the subject forms of the world—that is, the subject's own world-O—is very incomplete and relative to what he knows and (b) the classical difficulties of realism would arise anew if we were to ask *what* is represented. He grants these are serious objections, which he made to himself and to which an answer must be found.

Bonsack considers two conceivable ways of meeting the challenge. One, which he merely sketches, is to remain within a positive and reachable aim but add to world-O a metaprinciple claiming that this world-O is neither complete nor definitive and must be modified according to our experience. He does not criticize this solution but passes over it so quickly that he somehow gives the impression he is not convinced it meets the demands.

The other way is, as he writes: to "go some steps towards metaphysics," but make these steps cautious and rationally justifiable. It consists in postulating, beyond our individual and historical worlds-O, the existence of a kind of limit-world-O that he names world-Ω. World-Ω, he writes, is neither substantially different from our worlds-O nor in principle inaccessible: it is even possible that on certain points some worlds-O coincide with it, but we neither know which ones nor where. Indeed we have directly access only to our worlds-O so that it is out of the question to measure the adequacy of our worlds-O by comparing them to world-Ω. The latter is postulated merely as an ideal toward which we tend. Nevertheless it meets the realist's demands and answers his last objection: for *it*, at least, as Bonsack specifies explicitly, is *not* a representation. It is truly settled as independent of the knowledge of the individual subjects. It is an ontological whole, and, because of this, any discussion of its nature is "vain and metaphysical," the only verifiable thing concerning it being (here again Poincaré's influence is clear) the "net of relations which links its events."

This scheme, to which Bonsack was led by developing purely philosophical considerations, turns out to be remarkably similar to the one a detailed analysis of contemporary physics led us to earlier, in which basic reality is neither a "pure X" nor something truly and scientifically knowable. Above, this state of affairs was expressed by stating that basic reality is *veiled*. To some philosophers this word may prima facie sound rather vague and ill defined. Bonsack's scheme, on the other hand, is philosophically elaborate and should therefore not give them the same impression. Because of the close similarity between it and the "veiled reality" scheme let it therefore be suggested that the philosophers who would request a more precise specification of the last named one should, at least in a first approximation, understand it as coinciding with that of Bonsack. It goes without saying that the matching between the two cannot be considered as quite an exact one it is sufficiently accurate to be very significant nevertheless The correspondence between Bonsack's language and the one used in this book is, of course, clear Bonsack's world-Ω corresponds to our "independent reality" and Bonsack's worlds-O correspond to what we call here "empirical" or "effective" reality Bonsack's view that "it is possible that, on certain points, some worlds-O coincide with world-Ω *but we neither know which ones nor where*" (emphasis added) may, in "first approximation," be considered as explaining adequately what is actually meant in this book when it is asserted that independent reality is veiled This parallelism is all the more remarkable as, of course, the philosophical and the physical *approaches* are different At the start, the philosopher essentially *doubts*—for good, philosophical reasons—that the testimonies of both our senses and science do correspond to independent reality, and, later on, his reflection induces him to circumscribe his doubts, without eliminating them At the start, the physicist, on the contrary, is inclined to believe that science does really lift the "veil of appearances," and, later on, his analysis of scientific facts makes him water down to a great extent this belief of his Consequently, contrary to the philosopher, the physicist will willingly conjecture that the features of our knowledge of which it has not been proved they have no one-to-one correspondence with independent reality do have such a correspondence Here this slight difference reflects in the fact that, since the *general structures* of independent reality may consistently be thought of as being well approached by physics (as explained in Section 10 8), the advocate of the veiled reality conception will not feel averse to the idea that these structures *are* indeed well approached by this science, it is only the *contingent features* of the phenomena that he will consider as basically shaped by us, while, on the contrary, the philosopher has no motivation to distinguish between the two But, compared to the similarities, this difference may be thought of as not being an essential one

Incidentally, it is nevertheless interesting to observe that according to the present analysis, those of our beliefs of which it is most doubtful that they correspond to any elements of independent reality, essentially consist in the contingent features we observe and argue about, both in daily life and in our laboratories This is quite the opposite of the commonsense attitude, which essentially consists in believing in the reality per se of the contingent facts while questioning our possibilities of knowing the basic structural features.

15.2 Analysis of the "Empirical Reality" Approach

The foregoing shows that critical epistemology and quantum physics go hand in hand. Epistemology brought about a soundly argumented questioning of the idea that the things we encounter in daily and scientific life exist per se, lie within some space and time also existing per se and are knowable by physics. It showed that this idea, far from being the obvious truth commonsense believes it to be, is just a mere hypothesis. Quantum physics then brought out the additional piece of information that, in fact, the hypothesis in question is most presumably erroneous ("most presumably," not "surely" because in some ontologically interpretable models objects or events can be conceived of as existing per se; but we explained in Chapter 13 why it seems impossible to scientifically firmly believe in any one of these models in particular). This, we noted, gives credence to the idea that what physics can be expected to describe is not *Reality-per-se* in its totality and its details but primarily the phenomena as they get manifested to the community of all human beings.

It is true that the basic status of this veiled reality view is that of a realist theory (in the sense of "open realism") since, in it, the notion of an independently existing reality (of a world-Ω in Bonsack's language) is considered as meaningful. But on the other hand, according to it, independent reality is not scientifically knowable as it is (not even in its gross, contingent features). From the definitions set forth in Section 14.5 it then follows that although, within the view in question, the notions of strongly objective statements and concepts are *meaningful*, still, such statements and concepts must form in it but a highly restricted set—a set, in fact, that is either totally void or, at best, so scarcely populated that, with its elements, no accurate scientific description of "how the world *is*" can be produced. In particular, it cannot be expected that it includes our detailed scientific statements and concepts concerning *dynamical properties of systems*. In fact all the latter are weakly objective only. What they describe is not independent reality itself, but independent reality as seen through the forms that the mind imposes on the data of the senses, which is just what we called "empirical reality" (realists should carefully note that empirical reality is *no reality at all* in *their* sense; mentally replacing everywhere this expression by the equivalent one "empirical view" may help them avoid getting confused on this issue).

But is this, all by itself, a valid solution to the difficulties that were analyzed in this book, concerning the problem of "understanding quantum mechanics"? The question we address to in this section is whether or not it can reasonably be hoped that it is.

Admittedly, such a hope would be preposterous if the difficulties in question consisted in some inconsistencies internal to the set of the basic operational rules of quantum mechanics. For then, changing the referent of physics from "independent" reality to "empirical" reality would obviously boil down to making a mere, ineffectual change in semantics. But a prejudicial objection of such a type cannot be substantiated since the quantum rules *are* mutually consistent. The difficulties that are encountered in this field—be they those linked with the pointer "really lying" within some definite graduation interval, or those concerning the "real factual situation" of some physical systems, or other difficulties—all involve in some way or other the notion of reality as a referent, and it can therefore prima facie be hoped that some philosophical change in the nature of this referent can remove, or at least alleviate, them. Incidentally, let us note that while such changes can indeed be contemplated, it seems impossible to do completely without any notion whatsoever of a physical reality of some sort—independent or empirical—as some idealist thinkers might think we could. This is because of the fact that the quantum rules do not *directly* relate sense-data and human actions with one another. It is clear (as Bonsack also noted, see above) that the program of constructing such direct connective rules would be unrealizable in practice and that when we formulate physics as a set of "rules of the game" (as, in a way, we did), we still picture to ourselves at least the macroscopic impedimenta used for preparing systems and performing measurements as lying in three-dimensional space, *having* such and such positions and forms, *being* in such and such macrostates, and so forth. The accounts we thus give of empirical reality therefore fulfill the condition that, in them, *at least the dynamical properties of the macroscopic systems used as such impedimenta* are described in terms of (weakly) objective concepts (as this notion is defined in Section 14.7).

All this being said, let us start our program by considering one after the other the various difficulties we met with when trying to understand quantum mechanics as a theory of "reality" in the usual sense of the word—that is of "independent reality," in our language.

We had a preliminary contact with them in Chapter 4, at the place where we had to acknowledge that in important cases it is impossible to speak of dynamical properties of quantum systems as having "definite though unknown" values. In fact, however, this appeared at this stage more as a strange feature of quantum physics than as a basic, conceptual difficulty. It really revealed itself as being one only at the place where, when studying quantum measurement theory, we realized that it prevents us from considering the pointer of an instrument as quite generally lying in some definite state, even if this state is only defined macroscopically. In this context it was labeled 4a and we saw it is associated with other conceptual difficulties, 1–3 and 4b in Section 10.1. It is therefore appropriate that we should consider these five difficulties in this order.

Difficulty 1

As may be remembered, it consists in the fact that in the quantum mechanical description of the measurement process sketched in Section 10.1 [Eqs. (10.1)–(10.5a)] the composite quantum system must necessarily be considered as composed of two distinct parts, the "measured" system S and the instrument A, and that it is not clear what physically defines the borderline between the two. We know, however, that no operational ambiguity follows for, as soon as a definite measurement process is fully specified, it is quite clear what the minimal extension of the "quantum system" must be chosen to be, in the sense that it would make no difference (except with respect to computational difficulties) if we chose this extension to be more than minimal by including parts of the "instrument" into the "system." Under such conditions we are just free to choose any of these allowable extensions and describe the complementary part of the overall system by means of weakly objective concepts. Consequently, within the program of constructing an empirical view—what we call a description of empirical reality—difficulty 1 does not arise. This follows from the fact that, in this approach, there, clearly, is no point in requiring that the distinction between S and A should be physical, that is, that it should refer to the physical structure of these systems. Within the program in question it is natural that the distinction should essentially refer to human features: and to specify it then raises no difficulty whatsoever for the just explained reason.

Difficulty 2

In contrast, this is a difficulty that has genuinely physical aspects. It had to do with the determination of what is conventionally called the "pointer basis" (Sections 10.1 and 10.6). However, it may be considered that the theory has nowadays gone a good way toward removing this difficulty. This is one significant achievement of Zurek's approach to measurement theory. As we saw on a simple model in Section 10.6 [a few lines after Eq. (10.36)], the environment theory does indeed remove the difficulty in question, provided only we assume some prior split of a physical system into microscopic system, detector, and environment. Of course this solution applies without any change to the corresponding problem in the empirical reality approach. Indeed, it applies even better since, for the same reasons as above, the question of the borderline between the instrument and the environment does not arise in this approach.

Difficulty 3

This is the difficulty that has to do with the "von Neumann chain" and the necessity of the so-called quantum-classical cut (that of cutting this chain somewhere). The same argument as that removes difficulty 1 also removes difficulty 3.

Difficulties 4a and 4b

Difficulty 4a was already mentioned in connection with the developments of Chapter 4. It is one facet of the so-called Schrödinger cat paradox, and it may be said to constitute the central riddle of quantum measurement theory, nay, even of the whole of quantum mechanics. Difficulty 4b, the "and-or" difficulty, though often overlooked, is also intriguing. We shall analyze each of them again in detail, with, this time, the aim of precisely determining whether and to what extent they are still there within the present perspective, centered, to repeat, on the problem of describing empirical reality.

Let us begin by observations valid for both; and let us, once more, stress that, by definition, empirical reality is a set of phenomena, that is, that its description is one, not of how things really are, but of how they appear to the collectivity of mankind. This has important implications, in particular concerning the notion of wave function of a system—or ensemble of systems. Admittedly, within the "veiled reality" conception, an Everett-like *wave function of the universe* may be considered—competitively, so to speak, with the time-dependent Heisenberg operators commented on in Section 10.8—as the backbone element of a *conjectural* description of *independent* reality, providing (perhaps!) some glimpses on it. But the notion of the wave function of this or that particular atomic system—or ensemble of the same—is neither a strongly nor a weakly objective concept. As long as we keep to the purpose of describing empirical reality (the set of the phenomena) it must be considered as a mere tool, useful just as an element for making predictions concerning a given set of prospective experiments. It is when—and only when—we conceive of the wave function this way that, after an observation on a system has been performed, we can without qualms attribute a reduced wave function (the eigenfunction corresponding to the outcome) to the system, and worry neither about the "other branches" (in Everett's sense), nor about the systems with which, before the measurement, the considered system interacted.

But then the question arises: Should the completeness hypothesis be kept—does it even still have meaning? The completeness hypothesis ("there is no finer descriptions of quantum systems than those provided by kets") goes beyond mere operationality. It can hardly be interpreted otherwise than by considering that all the elements of an ensemble described by a ket are identical. But this is, in fact, a statement with an *ontological* content, that is, a statement going beyond what can meaningfully be asserted in the realm of a description of empirical reality. For this reason it is appropriate that, when trying to build up a description of empirical reality based on quantum mechanics, we should either drop the completeness assumption altogether or attribute to it a meaning carefully restricted to operationally definable concepts, thus changing it to the weak completeness assumption referred to in Section 4.2: Remark 3. Remember, however, that, in such a weakened sense completeness does not bar hidden (or "supplementary") variables.

In connection with this, and still in the same spirit of keeping close to what is operationally meaningful, it is quite consistent to assume the validity of an axiom that does make sense within the empirical reality conception—since the latter is centered on a reference to human possibilities—and which is, in fact, a necessary ingredient (although it, unfortunately, is most often kept implicit) in all the measurement theories reviewed in the foregoing chapters (the Zeh–Joos and Zurek environment-based theories as well as the Gell-Mann and Hartle cosmological theory, the formalism of operations and effects, etc.).

Axiom of empirical reality. [5]. A theoretical systematization of the empirical view must involve either one or both of the following postulates: (a) replacing large times by infinite times and/or very large particle numbers by infinite numbers is a valid abstraction; and (b) the possibility of measuring observables exceeding a certain degree of complexity is to be considered as not existing, not even theoretically; it being specifically stated that the latter position must be taken even in cases in which, according to quantum mechanics, such a possibility, in principle, actually exists.

Clearly, with postulate (b) the sensitive observables referred to in Sections 10.6 and 14.10 can be considered as void of physical meaning, so that, as explained already in Section 1.6, they can be set outside the realm of the observables we must consider when applying the truth criterion of Section 1.6.

Let us now investigate whether or not these two steps—removing, or at least alleviating, the completeness assumption, and postulating the validity of the axiom of empirical reality—can help us removing difficulties 4a and 4b.

Removing Difficulty 4a

As may be remembered, difficulty 4a consists in the fact that the statistical operator describing the ensemble of the composite $S + A$ systems after the interaction has taken place is a pure case, and is therefore unavoidably different from any statistical operator describing an ensemble in which the instrument pointers each have some macroscopically well-defined position. The environment-based theory reported on in Section 10.6 may be considered as a significant step toward removing this difficulty since, concerning all the observables attached to the $S + A$ system and not involving the environment—that is, all the observables that can be measured in practice—the statistical predictions of measurement outcomes are practically the same, whether they are derived from the overall state vector (10.33) or from the density matrix describing a proper mixture of states $|s_+>$ and $|s_->$ in proportions $|a|^2$ and $|b|^2$ respectively. However, as pointed out in Section 10.6, there are physical quantities—those, such as the quantity M [Eq. (10.37)], we called "sensitive observables"— a measurement procedure of which can in principle be defined *and* concerning which the statistical predictions of measurement outcomes derived by the two just mentioned methods are grossly different. When part (b) of the axiom of empirical reality is taken into consideration these sensitive observables drop out of the picture and difficulty 4a is removed. Incidentally, it may be noted that within the realm of the environment-based theory of Section 10.6, only part (b) of the axiom is necessary for salvaging consistency (with other measurement theories, those, in particular, that are based on the *algebra of observables* theory (Primas [6]), part (a) must also be resorted to).

• *Remark 1.* While this simple, qualitative axiom thus indicates the general line to be followed, it is true of course that, all by itself, it is not sufficient to make quantum mechanics a precise, difficulty-free theory of empirical reality. More elaborate developments in this direction are described in Section 15.4.

• *Remark 2:* At this stage, questions may be asked concerning the time evolution of the overall system. The point is that if we do consider the state operator $\rho'(t)$ [Eq. (10.5a) of the general theory with label t made explicit] as describing—at some time t after the measurement interaction has taken place—the "empirical reality" of the overall system (more precisely of an ensemble of such systems), epistemological consistency would seem to require that, for describing this same empirical reality at a time $t_1 > t$, we should use the state operator—call it $\rho'(t_1)$—obtained by applying the time evolution law of quantum mechanics to the said $\rho'(t)$ (which means: to the proper mixture that $\rho'(t)$ describes). However, in the general case we have no guarantee that, in situations where the differences between $\rho'(t)$ [Eq.

(10 5a)] and $\rho(t)$ [Eq (10 5)], both taken at time t, only have effects on predictions concerning sensitive observables, this will also be the case concerning differences between $\rho'(t_1)$ and $\rho(t_1)$, and if this is *not* the case some ambiguity in the predictive rules of quantum mechanics is seen to arise

To this it may be replied that the program of constructing a general, precise theory in which such observable differences should not appear—at least within times on a human scale—is not blocked by any no-go theorem, and it can be considered that theories based on ideas that implicitly involve the empirical reality axiom—as, for example, the Gell-Mann and Hartle theory—fulfill this program satisfactorily Here we show in Appendix 2 that within simple "decoherence" models, no differences of the considered type occur (this, indeed, is an element of answer to the sometimes asked, question "after all, what do decoherence theories basically bring in, over and beyond the simple Jauch-like argument referred to just before Eq (10 5a)?")

• *Remark 3* Questions that, prima facie, seem quite natural ones may be asked concerning the definition of the borderline between the nonsensitive and the sensitive observables "How is it defined?", "Is it sharp?" etc Indeed, in Appendix 2 a sharp such borderline—defined by the number q of involved environment spins—is used However, let it here be stressed that our sole reason for proceeding this way at this place is to simplify the calculations In the empirical reality approach it is, in fact, neither necessary nor advisable to specify what physical quantities are physically observable by means of a criterion bearing, as this one does, on the complexity of the physical systems they belong to It is, to repeat, in the nature of this approach that it merely aims at synthetically describing human communicable experience and that, correlatively, it is avowedly centered on the human possibilities It suffices therefore to define the sensitive physical quantities by stating they are those that are presently outside our reach Of course the thus defined set is bound to change as time passes But, same as with the von Neumann chain (the parallelism is clear), this is not a difficulty as long as the set in question remains nonempty

Removing Difficulty 4b

On the other hand, taken by itself, the axiom of empirical reality is not sufficient for removing difficulty 4b (the "and/or" difficulty). At first sight it might perhaps be conjectured that it is, for it could be observed that the von Neumann theorem described in Section 7 4, establishing the impossibility of splitting an ensemble described by a ket ("pure case") into two or more *different* ensembles, is explicitly based on the assumption that all the Hermitean operators correspond to observables. Within the realm of the empirical reality axiom it therefore obviously does not apply. That this, however, is not sufficient for removing the "and/or" difficulty can best be seen by considering a simplified model. For example, let us consider the one described in Section 10.1, where, for simplicity's sake, we may suppress the degeneracy index r and let index m merely run from 1 to 2. In this model, the role of the sensitive observables is taken up by observable correlations between quantities pertaining to S and instrument observables such as G' [defined just below Eq. (10.6)] that do not commute with the pointer coordinate G, and we know that as long as measurements of such observables are considered as being possible (and the completeness hypothesis is kept), it is inconsistent to consider the ensemble E of the final-state overall systems as being a proper mixture of N such systems, $N|a_m|^2$ of which ($m = 1, 2$) *have* $L = l_m$ and $G = g_m$. If these "observables" are assumed void of physical meaning (as we assumed is the case of the sensitive observables in more realistic situations) this difficulty (which is just 4a) is removed, as pointed out. But this is at the price of considering E as composed of elements that are not all identical. Now, initially, the ensemble of the overall systems is described by the state vector

$$(\sum a_m \psi_m >) \otimes |0 >$$

If we *did* keep to the completeness hypothesis, as, when discussing quantum mechanical problems, we have an intuitive, natural tendency to do, we would normally consider that all the elements of the said, initial ensemble are identical to one another. Since, in the final ensemble, they are no more identical, we would then have to say that a diversification spontaneously takes place between them during the interaction time, due to the Schrödinger-like time evolution process. This sounds all the more disconcerting as we cannot attribute such a diversification to anything else than just to the circumstance that the sensitive observables (such as G' in the model) are outside human reach. This shows that, indeed, by itself the axiom of empirical reality alone is insufficient for removing the "and/or" difficulty, as long as the completeness hypothesis is kept.

However, as explained previously, this completeness hypothesis is in fact crypto-ontological, and so is the notion of an "identity" between systems. When constructing a theory of *empirical* reality we therefore can consistently keep neither of them. And—this is the point—dropping them removes the foregoing objection and therefore the "and/or" difficulty as well. A simple way of making this clear is to argue as follows. We do not know and cannot scientifically know what independent reality is. Maybe it involves hidden variables of some sort. Maybe it so happens that, for example, some refined version of the pilot-wave theory faithfully describes it. We have no truly scientific way of knowing which one if any, of the available, ontologically interpretable models actually describes independent reality, but we may *conjecture* one does. And the very possibility that this is the case suffices for removing the "and/or" difficulty since in any one of these models (including also models of the GRW type) an individual system ends up by being in one *or* other of the possible final states. In other words, the very existence of such models removes the difficulty *because we may conjecture (or even fancy) independent reality is described by one of them.* But, on the other hand, we are not allowed—and in my opinion we *never* shall be allowed—to go around claiming it has been proved that this or that well-specified model describes independent reality. Our *scientific* discourse must therefore keep to empirical reality, which releases us from the need of explicitly introducing, at the roots of our scientific descriptions, hidden variables, or similar concepts.

At the same time it must be granted that, by giving up the completeness hypothesis, not only are we in fact accepting that the supplementary variables concept should not be banned, but we even extend it to the realm of the descriptions of empirical reality. Relatively to the $S + A$ wave function, the values g_m of the pointer coordinates and the corresponding values l_m of the measured observable indeed are values of supplementary variables describing the phenomena perceived by the observer

Of course, the necessity of giving up the completeness hypothesis is consonant with one of our previous remarks, namely that within such a description wave functions and state vectors should be considered as being nothing else than tools for making predictions And it should even be specified that these tools have specific domains of efficiency The wave function of an ensemble E of systems S is effective only for making predictions concerning the outcomes of a set K of conceivable experiments, which are just those involving but the elements S of E It does not have any meaning outside this realm

• *Important remark* For a genuine understanding of the "veiled reality" and "empirical reality" conceptions, the proposal for removing the "and/or" difficulty described above is quite significant Let us therefore analyze it further by examining a question likely to come to the mind of anybody who considers seriously the said proposal It is as follows Each of us has his own mind and therefore, it we opt in favor of the idea that the "supplementary variables" (definite pointer positions, etc) are *fancied*—that is, somehow, created by the mind—we may wonder why it is so that these appearances, these values of supplementary variables, are the same—or *seem* to be the same—for all of us Admittedly, this problem was, formally, already solved through the observation (see Section 14 9) that the quantum rules do account for intersubjective agreement, in the sense that they yield zero probability for two observers to get different measurement outcomes concerning the same physical quantity But to this it could, at a first stage in the analysis, be objected that this "formal" explanation is insufficient, that, like any quantum mechanical probability, the probability in question is merely that of "observing" something and that, therefore, referring to it in the present context amounts to implicitly supposing some "superobserver," looking at the minds of each of us just as we look at instrument pointers It must be granted that such a "metaphysical" assumption sounds alien to the spirit in which the very notion of empirical reality is constructed

If we took this objection at face value, that is if we actually wanted to build an explanation of the identity, for all of us, of these appearances—or "supplementary variables"—more or less similar in form to the explanations we are used to in "normal" physics and in daily life, we could do so by referring to some models. In fact two models are available for this purpose. In one of them these variables are more than just imaginary. I refer here to enlarged Bell's version of the pilot-wave—or Bohm's—theory, in which the configuration space has as many additional dimensions as there are distinct "minds," and the projection of the "representative point" onto any one of these new axes describes the state of the corresponding mind (see Section 16.6). The other model is the suggested amendment (d'Espagnat [7]) to Everett's relative state theory explained in Section 12.2. In both these models the intersubjective agreement is automatically realized, without the need of referring to any "rule of the game" as in the conventional theory.

However both these models describe the universe, including minds, as evolving *in* time. In fact they were conceived of as models, not of empirical reality but—to repeat—of independent reality. And, as explained in detail in Chapter 14, there are cogent reasons to beware, not only of any firm belief that this or that model of independent reality is "true," but also of the opinion that time is a strongly objective arena in which independent reality lies or evolves. All this may well make us suspicious concerning the pertinence of explanations of this kind. And indeed, upon reflection, the very appearance of the word "explanation" in this context is such as to put us on the alert. The point is one already mentioned at several places in this book. It is that, except perhaps for some "naive realists," "to explain" does not necessarily mean "to describe how things really are." As pointed out in Sections 14.8 and 14.9, another definition of "to explain"—one indeed that is very much favored by the philosophers—is "to reduce to some general law," and when, as is the case in the present chapter, one's interest is focused on "explaining within the realm of appearances," this obviously is the appropriate definition. Hence, to sum up, when all is said and done, the misgivings concerning the formal explanation of the intersubjective agreement that were formulated at the beginning of the present remark are not well founded.[1] When one gets to the root of the

---

1. This, of course, is especially clear for anyone who, without taking the conception described in Section 12.2 at face value, nevertheless considers its basic

matter and takes up the assumption that Being is a-temporal he must, in the last resort, have recourse to an explanation of such a type. As for the two models alluded to above, their interest in the last resort, lies elsewhere. Even if we do not believe any one of the two is true, their existence shows that the concept of an "ontological" basis to what is observed is not logically inconsistent: a piece of information that is important in itself since it suffices for dispelling apprehensions of definitive aporias. In particular, since neither the Schrödinger cat nor the "Wigner's friend" [8] paradoxes are *true* paradoxes in these models, the mere existence of the said models shows that the paradoxes in question vanish within a description that explicitly claims to be one of *empirical* reality exclusively.

Finally, let us note that the empirical reality approach also reconciles measurement theory with relativity. Indeed, it is within *this* approach that, centered as it is on Bohr-like conditions, the Fleming hyperplanes argument (here reported on at the end of Section 10.12) is consistent and does remove the mismatching. So, to sum up the content of this section: it shows that the "veiled reality" conception can indeed be made to remove the difficulties relative to the measurement process. To this end, it is necessary that any aim at producing an ontologically interpretable description to be accepted as "the true one" be set aside. Physics has to be interpreted as a description of the *phenomena*—what we call *empirical reality*—that is, of the intersubjective appearances, together with a set of predictive rules making it possible to connect, statistically at least, these appearances with one another. The description of the said appearances must necessarily be made using a realist language, but it is understood that this is only a *language*, implying no commitment whatsoever to a realist *interpretation*. The elements of discourse that are borrowed

---

idea as substantially correct: for, within this conception, no super-observer is needed. The same holds if, along somewhat similar lines, one goes as far as to consider, as Rovelli [27] does, that *any* correlation between two physical systems may be regarded as information that each one has about the other one: for this is a view in which observers are such relatively to other systems or observers, thus making less unpalatable the notion of a hierarchy of observers. Of course, this conception also makes the notions of *facts* and *truth* (the truth of statements of the "*A* is equal to *a*" type) relative to that of observation. In this it is akin to the *empirical reality* conception and its corollary, the *E-truth* notion (Section 15.4).

from the realistic way of speaking were called weakly objective concepts in Section 14.7.

Still, a closer, more quantitative inspection of the questions raised by this conception is necessary. The next few sections take care of this.

15.3 Empirical Versus Ontological Senses of the Verbs 'To Have' and 'To Be'

The verbs 'to have' and 'to be' are essential elements of discourse and, as already noted, we are at a loss how to imagine a text or a talk that would use neither of them. But on the other hand, both have quite a strong realist connotation, so that it is natural that in the empirical reality approach their use should raise questions. Of course, this use must have some strong relationships with that of weakly objective concepts since, as just noted, it is with the help of these concepts that we describe our collective experience in the *language* of realism. In the foregoing section, the only examples of weakly objective concepts that came in were such macroscopic ones as pointers and pointer positions. And it is a fact that, as a matter of principle, the whole of physics might be described in terms of weakly objective concepts of just this type, together with weakly objective statements (statements such as "if a synchrotron is made to operate in such and such a way, there is such and such a probability that the pointer of the measuring instrument will be observed to lie in such and such a graduation interval"). It must, however, be granted that in practice, describing the whole of physics this way would be tedious and cumbersome. It is therefore tempting to extend somewhat the realist language, with its verbs 'to have' and 'to be', in the direction of the microscopic. Let the extent to which this can be done be investigated in what follows.

To begin with, let us try and get a concrete view of the magnitude of the challenge. For this purpose, modes of description of microscopic reality that prima facie seem to be quite strongly justified at the epistemological level in that they comply with most strict epistemological standards will be compared with the description of this same reality yielded by a model constructed so as to be ontologically interpretable (which, as we know, is not the case of standard quantum mechanics). Let this model be Bell's version [9] of the pilot-wave theory, here reviewed in Sections 13.1–13.3.

As for the just mentioned strict epistemological standards, they are that for being allowed to ascribe an attribute of any sort to a system (to say· "*A* has value *a*"), we should be able to refer to the procedure by means of which we *actually* ascertain (without disturbing the system) the truth of the ascription in question. We know (Section 4.3 and Chapter 9) that a less stringent condition, the one merely stipulating that we *could* ascertain the same though some imagined measurement, cannot be stated as a sufficient one. But, as we also saw there, this one can, and this is why we take it up. The procedure referred to can then be considered as, in fact, defining the *meaning* of the attribution in question. For example, if we are to strictly adhere to this rule, the position of a particle is defined by referring to the way in which, in the circumstances at hand, we happen to know of this position. Let us show that even this (which would seem to guarantee ontological interpretability) may lead to definitions that are extremely different from those the chosen ontological theory leads to.

To this end, let us consider the special example of the pair correlation experiment as discussed within the realm of the pilot-wave theory in Section 13.3. Let us assume the instrumental setup is such that the measurement on particle U—with coordinate x_1—takes place at a time t_1, earlier than the time t_2 at which that on particle V, whose coordinate is x_2, takes place. In this case, knowing the outcome, say +1, of the S_z^U measurement, we know in advance and with certainty, due to the strict correlation, the outcome of the future measurement of S_z^V This is typically a case in which the kind of epistemological definition just mentioned (and which is nothing else than Reality Criterion 1, Section 9.2) applies. According to it we must therefore say that, immediately after t_1, S_z^V already *has* value -1. However, this picture in no way reflects what the ontological theory implies, for, according to Eq. (13.20), the variable x_2 whose value after t_2 expresses the outcome of the measurement of S_z^V keeps its initial (vanishingly small), negative or positive value until time t_2, and takes its final, negative, value only then. If the considered ontological theory is right, the mental process by which we attribute to S_z^V, already just after t_1, the value -1 is therefore misleading, at least if we consider this value as *actually possessed* by S_z^V On the other hand, it is not misleading if this attribution is thought of as being that of just a potentiality Thus, as we see, taking ontological concerns into account forces us, in such cases, to distinguish between an actual dynamical property and the potential dynamical property with the same name, and consider both of them as *real* On the other hand, within the epistemological approach (which is appropriate for describing empirical reality) nothing of the sort is possible We have no alternative but to give up completely the ontological description (which was identified with actuality) and correspondingly raise the aforesaid potentialities to the status of actualities (which, of course, does not signify we should consider them as ontologically real) In the considered example, this means we have no other choice than to state that S_z^V has the value −1 already just after t_1 But we must then add "in a weak (or epistemological) sense" (d'Espagnat [10]) And we must not believe that all the consequences normally derived from the mere presence of the verb 'to have' in this assertion or similar ones are necessarily valid

Which ones are and which ones are not? A clear-cut general answer to this question can only be obtained from a precise, quantitative theoretical description of empirical reality, such as the one outlined in the next section But even at the present stage we may note that when, in some proposition, the verb 'to have' is taken in its ordinary—that is, implicitly "ontological"—sense, some extension of the domain of validity of the proposition to the counterfactual domain seems natural which, on the contrary, is definitely *not* natural when the verb 'to have' is taken in its epistemological sense only An important application of this remark concerns the fields of applicability of the two reality criteria—Reality Criterion 1 and Reality Criterion 2—to which, as we saw in Section 9.2, we could a priori consider having recourse for replacing the Einstein, Podolsky, and Rosen (EPR) criterion of reality, thereby removing an ambiguity in the latter As noted above, within a tentative description of *microscopic empirical reality*, Reality Criterion 1 can be made use of for justifying the validity of a statement of the form "on system *S*, the physical quantity *A has* value *a*," but this holds good only with the specification that the verb 'to have' is there to be taken in its epistemological, not in its ontological, sense It then must be observed that this reservation bars any possibility of widening Reality Criterion 1 in the form of to Reality Criterion 2 The reason is that, as we saw in Chapter 9, the arguments that might support such an extension make use of counterfactuality and/or

Postulate R, which means they are cogent only within an ontologically oriented way of thinking The argument cannot therefore be applied to the situation under study For this reason, use of Reality Criterion 2 should be restricted to systems close enough to classical ones to render harmless the fact of ontologically interpreting their properties, which is the case when this criterion is associated with macrosystems only, as explained below

• *A side remark on epistemology and ontology* Within classical physics the rule of only using concepts defined in an epistemologically sound way—that is, with reference to (actual or possible) measurements and so forth—was commonly used But at the same time it was taken for granted that the idea of attributing ontological significance to the thus defined concepts could be objected to only on philosophical grounds (such as the "danger of metaphysics"), *not* on the basis of empirically grounded theories The advent of quantum theory modified this view to some extent But it may well be said that, on the whole, the changes were not radical enough and that, in particular, the necessity was not always clearly felt of distinguishing between epistemology and ontology Strict epistemological criteria were used for defining physical quantities but, most of the time, these quantities were still interpreted as somehow describing elements of "external" (independent) reality One of our general claims in this book is that the distinction in question must be quite carefully kept in mind and that implicit ontological interpretations should either be banned (in a description of empirical reality) or else made explicit (in tentative descriptions of independent reality)

15.4 Theoretical Description of Empirical Reality

In the foregoing section the need was noted of some precise, quantitative general description of empirical reality In this section an outline is given of one, tentative, such description

To build it up we, fortunately, have a ready-made formalism at our disposal This is the one that underlies the updated version [11] of the Omnès theory, where a deficiency of the original version, namely the lack (see [12] and above) of a criterion for truth, has been corrected and a restrictive truth criterion is given A formalism is not a theory, and the fact that we do make use here of this one does not imply that we take up the whole array of *concepts* that Omnès has set at the roots of his theory (see Section 11.4). Indeed, the very circumstance that we apply his formalism to the description of *empirical*—not *independent*—reality suffices to show that we do not. In fact, it even implies that we must change the language in which, within the Omnès theory proper, the formalism in question is couched. So that it may equivalently be said, either that the theory to be now explained is a new one using the same formalism as Omnès's or that we are here "at last setting Omnès's theory on appropriate conceptual footing" (a bit as Marx thought he had succeeded doing, concerning Hegel's theory[1]). In any case a debt to Omnès on this matter is here acknowledged.

Concerning the just mentioned change in language, it bears primarily on the word "true." While, in Omnès's theory, the values assigned to such macroscopic quantities as 'localization of an as yet unobserved pointer within some given graduation interval' are considered as being "facts," and consequently are "true" or "not true" (this, in the theory in question, is related, as we know, to decoherence, which Omnès considers as capable of generating such "facts"), here they merely correspond to weakly objective concepts (see, in Section 14.10 in particular, the argument based on the "sensitive observables" notion, showing these values cannot be full-fledged facts referring to independent reality). Hence, to assert it is *true* that, in general measurementlike processes such as those considered in Chapter 10, the pointer always is well localized would (because the sensitive observables cannot then be really forgotten) violate the truth criterion stated in Section 1.6.

On the other hand, when what is aimed at is just a theory of empirical reality, the problem thus raised, fortunately, boils down to a purely semantic, hence reasonably tractable, one. Pointers and the like are macroscopic systems, on which it is practically quite impossible to measure incompatible observables (all their practically measurable observables are compatible) and on which, even if such measurements were possible, we should normally not have the idea of performing them. Let us define macroscopic systems just simply as systems on which we are at present not capable of performing such measurements (or unwilling to take the—considerable!—pain of doing so). Admittedly, this is a human-centered definition, which would, as such, be unacceptable if physics were to describe independent reality all the more so as the set of the thus defined macroscopic systems may well change with increase of human powers. But in the context of the present analysis this does not constitute an objection since the proposed definition is obviously consistent with the empirical reality conception. With this definition, since all the practically measurable quantities pertaining to macroscopic systems are compatible, we can build up logics that are consistent in Omnès's sense (see Section 11.4) just by associating any number of propositions concerning the said quantities. These are essentially those Omnès [11] called *sensible* logics. We may then define a notion of "E-truth" of a proposition (E for empirical) by following exactly the same procedure as this author followed for specifying what, in his theory, should be called true. Essentially this consists (see Section 11.4) in asserting that:

(i) the propositions considered above concerning macroscopic systems all are E-true,
(ii) given *any* sensible logic L and a property *a*, *a* is E-true if L can be augmented by adding *a* (and its negation) to its field of propositions while preserving consistency, and if, in all these augmented logics, *a* is logically equivalent to a proposition of the type mentioned in (i).

Similarly, we can also define the concept of *propositions that are E-trustworthy in some consistent logic* by exactly the same procedure as that which is followed in the final version of Omnès's theory for specifying what propositions should be called "reliable" in such a logic.

When all this is done it turns out (see Section 11.4 and/or Ref. [11]) that the only E-true propositions are those concerning the practically measurable quantities pertaining to macroscopic systems as well as, possibly, outcomes of quantum measurements considered just at the time when the measurement is actually performed. All other propositions of quantum physics are, at best, merely E-trustworthy within some consistent logic. This, in particular, holds good, as we saw (Section 11.4), with regard to some propositions that—in connection with a measurement performed on a quantum system having yielded some definite outcome—we often believe we may safely utter, concerning properties of this system at some earlier or later time, finitely differing from the actual measurement time. Assume, for example, that the z spin component S_z of a spin 1/2 particle is measured at some time t_0, with outcome $+1/2$. Assume moreover that the particle is free at both earlier and later times. Even then, the proposition that at some earlier time t_{-1}, finitely differing from t_0, S_z already *had* value $+1/2$ is only E-trustworthy within some consistent logic containing the proposition "$S_z = 1/2$ at t_0" And so is also the proposition that it *will* have value $+1/2$ at some time t_1 finitely differing from t_0. Similarly, in the standard Bohm example of two spin $1/2$ particles U and V lying in a spin-zero state $S = 0$, if the spin component S_a^U of particle U along direction **a** has been measured, with outcome $+1/2$ say, the proposition that S_a^V has value $-1/2$ is not E-true but merely E-trustworthy within some consistent logics containing the propositions $S = 0$ and $S_a^U = +1/2$.

As we see on this last example, asserting that a proposition constructed with the verb 'to have' is merely E-trustworthy and not E-true is equivalent to stating—using a notion introduced in Section 15.3—that in this proposition the verb 'to have' is to be taken in its *weak, or epistemological* sense only. Consequently the claim we made there that when we have to do with such a proposition—inferred from Reality Criterion 1—we cannot go over to Reality Criterion 2 (that is, we cannot infer from it that even if the S_a^U measurement were not performed, or were replaced by that of some other spin component, the proposition would still be true) is corroborated: there is strictly no ground for believing that a proposition that is 'E-trustworthy within some logic' is E-true, which would here mean E-true independently of whether S_a^U is measured or not.

15.5 Empirical Reality and Nonseparability

In Section 8.2 it was shown that the predictive rules of quantum mechanics obey parameter independence and that therefore nonseparability does not make faster-than-light signaling possible (in fact it cannot be made use of for slower-than-light signaling either, except in some very indirect senses: see, e.g., Refs. [13] and [14]). In Section 14.11 it was pointed out that, because of this, nonseparability—contrary to such physical facts as the nonexistence of a universal Newtonian time—is not, as such, likely to become a notion specifically needed for explaining experimental facts other than those gathered for testing it. And it was stressed that, therefore, nonseparability does not enjoy a scientific status quite as respectable as that of some other "negative" discoveries such as, again, the relativistic "nonabsoluteness" of time. Indeed, both in the highly specialized (and strictly operationally minded) developments of basic physics given in advanced textbooks and in the easily understandable accounts of, say, astrophysics put forward in popular books, explicitly mentioning nonseparability would be both inappropriate and confusing.

It is tempting to try and express the substance of this in a short, condensed manner, by asserting that somehow nonseparability does not apply to *empirical* reality. But is it actually possible to say so, and if yes, in what sense? It is to this question that we now turn. Let it be noted, however, that, interesting as it is, it is definitely less important than the points analyzed in Sections 15.2 and 15.3. The reason is that it does not bear on the consistency of our worldview but, primarily, on semantics. Explicitly formulated it is: Do our stated (freely chosen but natural) definitions of *weakly objective concepts, macroscopic systems, empirical reality*, and so on happen to make nonseparability irrelevant to empirical reality?"

One aspect of this question is: What about the status of local causality within an empirical reality theory? Let us begin with this.

Empirical Reality and Local Causality

As pointed out in Section 15.2 (see, in particular, "Removing Difficulty 4b"), the completeness hypothesis is essentially ontological, and its meaning lies therefore outside the realm of the empirical reality concept. Within an empirical reality theory the said hypothesis is, in other words, meaningless: and, correlatively, in such a theory, the notion of a *full* specification of the events (or variables or what-not) in a given space-time region is meaningless as well. This notion is, however, a basic one in the very definition of local causality (see Section 8.4), and thus, within the realm of an empirical reality description, local causality cannot even be defined; which has the obvious consequence that it would be quite meaningless to speak of its violation. In other words, the theorem labeled Bell 2 in Section 9.3 is one the meaning and bearing of which—while considerable in its field— are restricted to the domain of the theories that explicitly aim at providing a description of "reality as it really is" (as was Bell's explicit purpose). The experimental data do of course establish a violation of the Bell inequalities, showing that no locally causal description of independent reality can be correct. But, by itself, this has no implication concerning our "empirical view."[2]

Empirical Reality and Bell 1 (EPR)

Concerning the status of nonseparability within a description of empirical reality, the foregoing does not completely settle the matter. The reason is that, as we saw, nonseparability can also be proved by means of the Bell 1 or Bell 3 theorems. Let it be observed, however, that in both these theorems Bell's inequalities are derived by applying to microsystems premises that include Reality Criterion 2. Concerning Bell 1 this comes in within the first step in the proof, the step consisting in proving incompleteness by means of the EPR argument, since, as shown in Section 9.1, the latter argument does makes use of Reality Criterion 2. Within the Bell 3 proof, Reality Criterion 2 or some idea equivalent to it is implicitly made use of at the place where induction is invoked and where it is claimed that the information gained about the reality of some attributes of the considered microsystems is also valid concerning microsystems on which no measurements were made. Since, as we saw in Section 15.3, in the empirical reality description this criterion is *not valid* concerning microsystems, the proofs in question cannot be transferred to the realm of empirical reality.

At first sight some apparent mismatching may be felt to exist between the argument in this section and the one that, in Section 15.2, served for removing the "and/or" difficulty. For removing the difficulty in question we there *made use* of the notion of ontological models (we referred to the fact that these models do not have this difficulty), while here, on the contrary, we argue against "nonseparability of empirical reality" by using the idea that in the empirical view ontological considerations should be *barred*. The answer is that both arguments are correct and that,

---

2. For this reason, when a short account of the above is requested, it is not inadmissible to assert empirical reality is *separable*. However, it must be realized that, strictly speaking, this word is here a misnomer since the impossibility of defining local causality in this context entails that neither one of the words "nonseparable" and "separable" has, in it, a well-defined sense. The author acknowledges useful exchanges of views with André Frenkel on this point.

notwithstanding appearances, they are not mutually inconsistent. The point is that independent reality, if it is not just simply "irrational" (that is, exceeding human aptitudes at analysis) may happen to be constituted as this or that ontological model tells us it is, and all these models are indeed "nonseparable"; but, since the possibility of scientifically determining which one—if any—is true seems ruled out, these models themselves partake of metaphysics more than of physics and so does their common, nonseparability, feature.[3] A scientific description of empirical reality may consistently ignore the latter.

The result of the analysis carried out in this shows that, as expected, empirical reality as we defined it may be said free from nonseparability effects.

• *Remark:* Interesting proposals have been put forward, as noted (Stapp [15]; Eberhard [16]) to the effect of deriving the Bell inequalities without any resort to realism. However, the meaning and pertinence of the premises on which such derivations are based are not indisputable. For a discussion of these points see, e.g., Refs. [17–20]. In any case, violation of these premises could not be made a feature of empirical reality as theoretically described in the preceding section.

15.6 A Quantitative Illustration

It is not without interest to analyze in detail, using an example, the precise manner in which the simple, general ideas put forward in the foregoing section apply in a specific case. Let this be done here. For this purpose, let us consider [21] the case of a pair of systems, one of which, S, is a microsystem while the other one, A, is a macroscopic one and, more specifically still, an instrument of observation whose interaction with S takes place within a short time interval $(t_0, t_0 + \varepsilon)$, according to the general pattern (described in Section 10.1) of quantum measurement processes.

---

3. In any one of these models in particular, the impossibility of switching from Reality Criterion 1 to Reality Criterion 2 is due to nonseparability (concerning the pilot-wave model this is shown in Section 15.3). But the quite crucial remark that constitutes the subject matter of Section 14.11 already makes it a consistent and reasonable convention to consider that this nonseparability of the model is not to be counted as "nonseparability of empirical reality." And indeed, the theoretical description of empirical reality shown in Section 15.4 corroborates the consistency of this convention.

Let us imagine that, at a time $t \gg t_0$, measurements, the details of which we do not analyze, are performed on both S and A, at places quite distant from one another And let us assume moreover that A is a macroscopic system in our sense To be more precise on this, let us assume that, on A, only a set of collective macroscopic variables can be measured Following Omnès [11], we define the latter notion by referring to the $2N$ dimensional phase-space of the system, a point of which corresponds to the position q_i and momentum p_i coordinates ($i = 1, 2, \ ,N$) of the particles composing the system Within this space we may consider a set of many nonoverlapping cells, C_i It can then be shown [11] that (i) provided their shape is simple enough and their volume appreciably larger than $\hbar^N$, such cells can be associated with projectors (more precisely, quasi-projectors) onto appropriate subspaces of the overall Hilbert space of the system, and (ii) the commutator of two such projectors is vanishingly small as soon as the two corresponding cells are clearly separated By definition, the "collective, macroscopic variables" are then the observables represented by such commuting projectors (together of course with the observables the spectral decomposition of which involves only these projectors) and a macroscopic system is one on which only such collective variables can be measured

To begin with, let us note that the "pointer coordinate" G is of course one of the "collective macroscopic variables" of A Let $\{G, U, V, W, \ \}$ be a complete set of compatible observables on A, so that the collective macroscopic variables of A other than G are functions of the $G, U, V, W, \ $, let H^S and H^A be the Hilbert spaces of S and A respectively and, in the subspace E_n of H^A that corresponds to $G = g_n$, let us take as a basis the eigenvectors $|n, u_i, v_j, w_k, \ >$ common to G and $U, V, W,$

$$G|n, u_i, v_j, w_k, \quad > = g_n|n, u_i, v_j, w_k, \quad, > \qquad (15\ 1)$$

$$U|n, u_i, v_j, w_k, \quad > = u_i|n, u_i, v_j, w_k, \quad, > \qquad (15\ 2)$$

The right-hand side of Eq (10 4a) can then be expanded as

$$|\Psi_m > = \sum_{i,j,k} c_{m,i,j,k,} \ |\psi_m > \otimes |m, u_i, v_j, w_k, \quad > \qquad (15\ 3)$$

with

$$\sum_{i,j,k,} |c_{m,i,j,k,}|^2 = 1 \tag{15 4}$$

so that Eq (10 4) takes the form

$$|\Psi_f> = \sum_{m,i,j,k,} b_{m,i,j,k,} |\psi_m> \otimes |m, u_i, v_j, w_k, > \tag{15 5}$$

with

$$b_{m,i,j,k,} = a_m c_{m,i,j,k,} \tag{15 6}$$

Let then Q be one or other of the observables of S, with eigen-value equation

$$Q|q> = q|q> \tag{15 7}$$

and let us imagine that simultaneous measurements are made, of Q on S and of any one, U say, of the collective macroscopic observables of A The probability that the corresponding outcomes are q and u_i is

$$p(q, u_i) = \sum_{m,j,k,} |<q|\otimes<m, u_i, v_j, w_k, |\Psi_f>|^2 \tag{15 8}$$

$$= \sum_{m,j,k,} |<q|\psi_m>|^2 |b_{m,i,j,k,}|^2 \tag{15 9}$$

$$= \sum_m h_{m,i} |<q|\psi_m>|^2 \tag{15 10}$$

with

$$h_{m,i} = \sum_{j,k,} |b_{m,i,j,k,}|^2 \tag{15 11}$$

while the probability that a measurement of G yields g_n irrespective of what a measurement of Q yields is of course

$$p_n = |a_n|^2 \equiv \sum_i h_{n,i} \tag{15 12}$$

Note that for the special case that Q is the quantity L that is "measured" on S by A and q is the eigenvalue l_n that corresponds to $G = g_n$, Eq (15 10) yields

$$p(l_n, u_i) = h_{n,i} \qquad (15\ 12a)$$

By virtue of a general rule of probability calculus, the conditional probability p_i^n that a measurement of U yields u_i if it is known that a measurement of L yields (or has yielded) l_n then is

$$p_i^n = h_{n,i} p_n^{-1} \qquad (15\ 13)$$

With these notations Eq (15 10) reads

$$p(q, u_i) = \sum_m p_m p_i^m | <q|\psi_m> |^2 \qquad (15\ 14)$$

and it is convenient for what follows to define, for any value of the index m, an ensemble E_m described by the following (normalized) tensor product of a ket belonging to H^S with one belonging to H^A

$$|\Phi_m> = p_m^{-1/2} |\psi_m> \otimes \sum_{i,j,k,} b_{m,i,j,k,} \ |m, u_i, v_j, w_k, > \qquad (15\ 15)$$

The mean value $< Q\,U >$ of a typical product of one observable, Q, of S and one collective macroscopic observable, U, of A is

$$< Q\,U > = < \Psi_f | Q \otimes U | \Psi_f >$$
$$= \sum_{m,i,j,k,} < \psi_m | Q | \psi_m > u_i |b_{m,i,j,k,}|^2 \qquad (15\ 16)$$
$$= \sum_{m,i} < \psi_m | Q | \psi_m > h_{m,i} u_i$$

which, with the help of Eq (15 13), can also be written as

$$< Q\,U > = \sum_m p_m < Q_m > < U_m > \qquad (15\ 17)$$

where $< Q_m > = < \psi_m | Q | \psi_m >$ and $< U_m > = \sum_i p_i^m u_i$ are the mean values in E_m of Q and U respectively Equation (15 17) is identical to the one that would express the mean value $< Q\,U >$ if the overall ensemble of the $S + A$ pairs, instead of being a pure case described by $|\Psi_f >$, were a proper mixture, with weights p_m, of ensembles E_m And it should be noted that each E_m is a pure case, whose state vector $|\Phi_m >$ is a direct product of a ket of H^A with one of H^S, so that, within each E_m, Q and U are uncorrelated within any E_m the mean value $< Q\ U >_m$ of the product of Q and U is just the product $< Q_m > < U_m >$ of their mean values and correlatively the "conditional probability for outcome u_i if a measurement of Q yielded outcome q" is equal, in each E_m, to the corresponding unconditional probability This completes the proof for it shows indeed that, with respect to the kind of measurements we are here considering, the theory yields exactly the same predictions as would an *objective local theory* (see definition in Section 8 2 Remark 3), in which the role of the parameters λ specifying the "objective states" would be played by the parameter m Since objective local theories all entail the validity of the Bell inequalities (Bell theorems), it follows that if only measurements of the type in question are made on the pairs described by $|\Psi_f >$, no violation of the said inequalities can be observed (note that this was to be expected since proofs of violations of the Bell inequalities normally involve pairs on each element of which mutually incompatible observables are considered)

Scope and Limits of This Result

Let it be stressed that, of course, there is no contradiction whatsoever between the foregoing result and the fact that the violations of the Bell inequalities are experimentally established with the help of instruments of just the type considered above, that is, on which collective macroscopic variables only (positions of pointers, etc) can be observed The point is that, above, we considered $S + A$ pairs whose S and A parts play exactly the same role as the "particles" U and V in the standard developments concerning the Bell inequalities In particular, just as, in these developments, one considers measurements performed on U and on V, here we consider (at time t) measurements made *on A*, not *by (means of) A* This implies that, in order to mimick such developments, (a) we would have to split the ensemble of the $S + A$ pairs into four ensembles on which we would assume measurements are made of different pairs of observables, and (b) the experimentalist doing this would not *handle* the A's, in such a way as to change the values of some parameters (such as magnet orientations for instance) In other words, A, in all four subensembles, would always "measure," on S, the same quantity, namely the observable L (Section 10.1) associated with its "pointer coordinate" G. What would change, when going over from one subensemble to the next, is essentially the nature of the collective macroscopic quantity assumed to be measured *on A* at time t (by means of some other instruments). And, to repeat, since all these collective macroscopic quantities are mutually compatible, it is not surprising that the Bell inequalities should be obeyed.

The conclusion we reached is nevertheless philosophically quite significant for it is inherent to the empirical reality concept and to it alone. Within the "normal" standpoint—which associates "realism" with a conception of reality coinciding with the already defined "independent reality"—it is quite impossible to reconcile the experimentally verifiable predictions of quantum mechanics with those of *any* objective local theories. That, on the contrary, within the empirical reality conception this is possible therefore shows that bounds in the possibilities of knowing led mankind to build up an *allegory* of independent reality—the one in which the general ideas of the objective local theories hold good—that, efficient as it is, basically differs from the original (i.e., from whatever independent reality "genuinely" is).

• *Remark:* Recently Peres [22] and Khalfin and Tsirelson [23] proved theorems that go along the same lines as what has been shown here. Peres could show that, although, in a case in which a pair of spin j particles is created in a spin zero state, the Bell inequalities remain violated even for arbitrarily large j values, still they are *obeyed* by the outcomes of any measurement in which neighboring values of a J component are lumped together because of limited instrumental resolution. As for Khalfin and Tsirelson, they did not limit their investigations to a particular type of (macro-)systems: they reached the same result (Bell inequalities obeyed) by considering—along with Caldeira and Leggett [24]—that macroscopic systems are inherently dissipative, and taking dissipation quantitatively into account (with methods of calculation differing somewhat but not essentially from those the principles of which are reported in Sections 10.6 and 12.3). In both approaches the "specific position within nature" (to use Khalfin and Tsirelson's expression) of human beings plays a crucial role, so that their descriptions refer to empirical, not to independent, reality.

15.7 Outlook

Veiled reality is a notion that may sound hybrid, and therefore hardly attractive. Prima facie, some will consider it an unpalatable attempt at putting together the two mutually exclusive approaches of idealism and realism. The closer look we took at it in this chapter may, however, have modified such a judgment for the better. Many philosophers—not to mention most scientists—nowadays seem to consider that, undisprovable as it is, idealism still lacks something, not only from the point of view of intuitiveness but also when measured with the more precise yardstick of a strict rationality. Clearly, Bonsack's approach was, at least in part, motivated by such a feeling. As we saw when studying the Heisenberg picture in Chapter 10, the great conceptual questioning induced by quantum mechanics did not render Poincaré's structural realism obsolete; on the contrary it backed it with new arguments. On the other hand, it seems more and more certain, not only that naive realism is dead, but also that versions of realism that could hardly be labeled "naive," such as the realism of the later Einstein, meet with difficulties, even within the domain that was long seen as its stronghold, namely hard, matter-of-fact physics.

A consequence of all this is the necessity I have been stressing all along of realizing that a distinction must now be made between the two concepts to which we gave the names of "independent" and "empirical" reality.[4] But, in fact, this necessity also emerges from just a simple look at present-day texts in physics.

---

4. Anyhow, such a distinction is, of course, logically necessary (though the appropriateness of the *names* may be made the subject of endless discussions; but it is not the choice of the names that counts, only the fact the concepts must be recognized as different). Even the persons who are convinced that science is on the way of revealing to us how the world *really is* in some absolute sense of the words, must grant that their phenomenologist opponents, when they speak of reality as the set of the phenomena, refer to a concept that, as a concept, does not coincide with *their own* concepts of reality. And conversely, even the radical phenomenalists must grant that their realist opponents, when they use the word "reality," refer to something else than what *they themselves* have in mind. Hence, even though, in discussion with members of their own respective groups, both the realists and the phenomenologists can dispense with the epithets and just speak of "reality," they should realize that when they venture arguing with members of the opposing group the epithets in question are necessary (as is well known, the ancient Egyptians, who only knew the Nile as a river, could have but one word for the two concepts "north" and "downstream"; but when they discovered the Euphrates they had to enrich their vocabulary).

Upon such a perusal, the notion of empirical reality—as defined and made (reasonably) precise in Sections 15.2–15.4—is seen to remove an ambiguity present in both the scientific and popular literature (and of which few readers are truly unaware, even though some brush it away as irrelevant or nonscientific). This ambiguity is proteiform. One aspect of it is the fact that articles and textbooks that deal with such entities as wave functions of composite systems, density matrices of systems in thermal equilibrium and so on, freely use expressions such as "the system wave function *has* such and such a structure," "the system density matrix *is* this or that," but at the same time most carefully avoid explicit pronouncements on the question whether these wave functions, density matrices, and so on are "real" or not. In fact, the curious reader who investigates this question further, soon discovers arguments (complete entanglement of the wave function of a system of identical particles, "improper mixture" character of the mixtures density matrices describe in most cases etc.) that strongly favor the latter alternative. But, in the literature in question, he then finds no clue for answering the question: If so, *what* is real? or what *can* consistently be considered real? And still, in the same articles and textbooks quantum transition probabilities are interpreted without qualms as referring, not to what any of us would observe if he were there, but to what "really" takes place there.

It can be said, either that, in Bohr's time, this ambiguity was manifest to the physicists and that Bohr proposed a "philosophical way" out of it, or, equivalently, that within such a philosophy as Bohr's, the ambiguity in question just did not arise. But Bohr's writings are not always clear and, while most physicists still pay lip service to them when they write textbooks, his ideas—which anyhow are quite far from the *doxa*[5] of the present-day scientific community—do not any more, on these matters, constitute genuinely accepted references. The ambiguity is therefore again (or still) a real one. It has, for a long time, been felt as such by a number of physicists, who kept wondering about the conceptual foundations of quantum mechanics and of whom it can therefore truly be said that they were (and are) *In Search of Reality* [25]. These physicists consider that when, in our writings (popular or scientific) we obliterate the ambiguity (in the manner described

---

5. Common opinion.

above), we are indulging in some double talk. They therefore see it as a very positive fact that such misgivings were also, forcefully, convincingly and wittily, expressed by John Bell in one of his last papers [26].

Once these facts have been realized, it seems we do not have much choice. One possibility is to follow, despite all difficulties, the line of thought that was that of John Bell himself (among others), namely, to turn to ontologically interpretable theories (pilot-wave model, GRW theory, CSL theory, etc.. see Chapter 13 for a review). Is this the *only* possibility? I claim it is not. Another way exists, open to the physicists who view with great skepticism any one of these models in particular. This way is the one explained in this chapter: the way that consists in *conventionally* defining empirical reality in such a manner that the appearances are all saved and that, at the same time, the conceptual difficulties that arise when the object of physics is identified with independent reality are all removed. The analyzes in this chapter aimed at showing that this way is indeed open, and I am not far from considering that, if we choose to be as much of an "orthodox" physicist as possible, we cannot consistently escape making use of, in some way or other, the notion of empirical reality (of course, still another conceivable way is radical idealism; I have explained at length in Sections 14.3 and 14.8 why I reject this solution).

Admittedly, the notion of empirical reality is nothing like a "final answer" to all the questions and enigmas that beset the field of the conceptual foundations of quantum physics. Moreover, it demands from the part of physicists a renunciation of traditional goals of their science that some of them may find difficult. On the other hand, it seems to go in the direction indicated by the very developments of physics and it does remove difficulties. As shown above, it removes those connected with measurement theory—the Schrödinger cat paradox and the "and/or" difficulty included—through the possibility it opens of considering that within its realm the completeness assumption is meaningless. It also reconciles measurement and relativity (see end of Section 15.2). In addition, it yields a coherent qualitative picture of a phenomenal world in which, as experience shows, nonseparability does not "show up." Admittedly, this is at the price of bridling our natural tendency at uncritically enlarging, in the direction of an ontology that may prove to be too naive, the domain of validity we attribute to the epistemologically defined notions of 'having' and 'being.' Everyday experience has so much ingrained this tendency in our minds that to keep it under control is quite a difficult intellectual exercise. But the example of some prominent philosophers of the past proves it is not an impossible one. And what has—it is hoped—been shown in this chapter is that it can be performed without giving up the standards of scientific exactness.

• *Remark 1.* To prevent any misunderstanding that too quick a reading of Section 15.2 to the present one might induce, let it be forcefully stressed here that, as specified at the appropriate places above, the whole content of these sections only applies to *empirical* reality.[6] Consequently, any reference to it that would leave out this epithet as redundant would misrepresent my conceptions. This, for example, would be the case if it were asserted that I agree with the idea that *reality* is "local," or "separable," in any sense of these words. In fact, for reasons explained in detail in Chapter 14 I am quite convinced that we must consider the notion of independent reality as *meaningful* and *logically prior* to that of empirical reality, which is just a man-made allegory of the former. And I must therefore take seriously into account the information—stemming from the Bell theorems—that it is impossible to hold on to any description of independent reality (the "reality" of all the conventional "realists") that would represent it as "separable" or "local" (in the sense of not being nonseparable). Indeed (with apologies for this hair splitting) I go as far as to consider that, while, admittedly, nonseparability is a negative sort of concept, still in view of the above, any scientist who rejects pure idealism may meaningfully state, in a positive way of speaking, that independent reality *is* nonseparable (or nonlocal): for because of her very rejection of radical idealism she must consider that the "independent reality" notion is meaningful, and because she is a scientist she must consider that everything that

---

6. The argumentation of Sections 15.4 and 15.5 cannot be transferred to independent reality: not only because it is based on a man-centered definition of macroscopic systems but also because, while—obviously—independent reality is *not* exclusively composed of macroscopic systems, still, any tentative description of it can, by definition of what this reality is, only be made with the help of *true* statements.

science proves, including that feature of independent reality that nonseparability is, is meaningful information.

• *Remark 2:* If *parameter dependence* held true—more generally if the world of daily life were nonseparable—and/or if quantum superpositions of macroscopically distinct states could be directly observed, this would presumably have prevented human beings (and animals) from building up the very notion of (localized) objects (perhaps even from existing). Seen from this angle, the concepts by means of which we describe empirical reality appear especially clearly as not being much more than just tools for organizing our collective sense-data. At the same time, it becomes intuitively somewhat less surprising that these concepts should work so well, in spite of them not being faithful pictures of what "really is."

References

1. Bas van Fraassen, *Quantum Mechanics, An Empiricist View*, Clarendon Press, Oxford, 1991, p. 4.
2. H. Putnam, *Reason, Truth and History*, Cambridge U. Press, 1981.
3. F. Bonsack, "Prolegomena to a Realist Epistemology," in *Dialectica* **43**, 157, 1989.
4. H. Poincaré, *La valeur de la science*, Flammarion, Paris, 1905.
5. B. d'Espagnat, *Found. Phys.* **17**, 507 (1987).
6. H. Primas, *Chemistry, Quantum Mechanics and Reductionism*, Springer-Verlag Berlin, 1981.
7 B. d'Espagnat, *Conceptual Foundations of Quantum Mechanics*, 2nd ed., Addison-Wesley, Reading, Mass., 1976.
8. E.P. Wigner, "Remarks on the Mind-Body Question" in *The Scientist Speculates*, I.J. Good (Ed.), W. Heinemann, London, 1961.
9. J.S. Bell, *Speakable and Unspeakable in Quantum Mechanics*, Cambridge U. Press, 1987, ch. 15.
10. B. d'Espagnat, *Reality and the Physicist*, Cambridge U. Press, (1989).
11. R. Omnès, *Rev. Mod. Phys.* **64**, 339 (1992).
12. B. d'Espagnat, *J. Stat. Phys.* **56**, 747 (1989).
13. C. Bennett, G. Brassard, C. Crepeau, R. Jozsa, A. Peres, and W. Wooters, *Phys. Rev. Lett.* **70**, 1895 (1993).
14. A. Ekert, *Phys. Rev. Lett.* **67**, 661 (1991).
15. H. P. Stapp, *Nuovo Cim.* **40B**, 191 (1977); *Found. Phys.* **10**, 767 (1980).
16. P. Eberhard, *Nuovo Cim.* **38B**, 75 (1977); **46**, 392 (1978).
17 B. d'Espagnat, *Phys. Rep.* **110**, 203 (1984).
18. J.F. Clauser and A. Shimony, *Rep. Progr. Phys.* **41**, 1881 (1978).
19. R. Clifton, *Found. Phys. Lett.* **2**, 347 (1989)
20. R. Clifton, J. Butterfield, and M. Redhead, *Brit. J. Phil. Sci.* **41**, 5, (1990)
21. B. d'Espagnat, *Phys. Lett.* **A**, 171 (1992).

22. A. Peres, *Found. Phys.* **22**, 819 (1992)
23. L.A. Khalfin and B.S. Tsirelson, *Found. Phys.* **22**, 879 (1992).
24. A. Caldeira and A. Leggett, *Ann. Phys. (N.Y.)* **149**, 1059 (1983).
25. B. d'Espagnat, *In Search of Reality*, Springer-Verlag, New York, 1983.
26. J.S. Bell, "Against 'Measurement'" in *Sixty-two Years of Uncertainty*, A.I. Miller (Ed.), Plenum (NATO Series), New York, 1990.

Lessons and Hints
from Quantum
Physics

16 The two general conclusions that emerge from the queries reported in this book were drawn in Chapters 14 and 15 They are that independent reality is veiled and that physics is free from paradoxes when considered as a description of but empirical reality It remains to be seen what information these views provide in some fields we did not explore, or touched on only cursorily In this, the right thing to do is to distinguish between information proper and mere hints

16.1 Information

An Indication to Physicalist Scientists

This is an elementary point and it should not concern *all* physicalist scientists, only those who consider that the ultimate purpose of science is to discover and faithfully describe whatever "really exists," and that this process of lifting the veil of appearances is, on the whole, proceeding steadily and successfully There still are quite a number of scientists who think that way, particularly in fields other than physics, and these persons are, as a rule, quite unimpressed by the pure philosophers' objections to their opinion Needless to stress here that *contemporary physics itself*—especially the aspects of it described at length in this book!—shows such a view is most naive A special version of the view in question is even flatly contradicted by this discipline and, ironically enough, this is just the conception most persuasively suggested by the whole vocabulary of high-energy physics, with such expressions as "elementary particles," "particle states," and so on These terms strongly suggest *philosophical* atomism, that is, the idea that Ultimate Being is dispersed in myriad simple,

16. Lessons and Hints from Quantum Physics

tiny, localized elements; whereas such a conception is, as we saw, strictly incompatible with present-day knowledge. Indeed, the information we now have (see Section 14.4) makes such a picture of Being less "scientific" than its opposite: Plotinism!

A Point to Be Noted by "Nonscientists"

We all know of a lazy way of getting rid of the questions raised by the concept of reality. Not infrequently, people with a smattering of philosophy express it by such statements as: "Of course the *ultimate* elements of Reality are unknowable, but then what? Would it not be preposterous to think we could ever know all the most basic secrets of Nature? We already know a lot of things about Nature, we shall know more in the future, what more could we reasonably ask for?" This standpoint implicitly supposes that common sense, expanded and at places corrected by science, provides us with some sort of a genuinely ontological, albeit incomplete, knowledge of Reality: a knowledge that raises no basic problem, is, except on a few esoteric issues, in accordance with the general ideas commonly used in daily life, and with which we should be content. However, one of the essential lessons of contemporary physics—corroborating, on this point, the teaching of renowned thinkers of the past—is that this conservative, reassuring view cannot be kept. In Chapter 1 we pondered on the doubts expressed by great philosophers concerning the ontological significance of some of our basic notions, such as that of objects existing "out there" independently of possible knowledge about them, the notion of absolute time or spacetime, that of causality, and so on. We wondered whether these doubts were not but arbitrary suspicions, grown from Byzantine lucubrations. Having analyzed the content of contemporary quantum physics we must now grant, even if we are realists, that no matter which theory we choose to elect we must throw overboard, as these philosophers prescribed, our old commonsense ontological certainties. Indeed, even the most "realist" of the theories that are acceptable—those that facts do not falsify— are only so if "realist" is understood in the sense of what I called [1] "far realism" (see also Section 13.8): that is, as corresponding to descriptions of "what exists" that are both totally different from our daily and scientific experience and—as it seems—useless for *predicting* future experience.

An Item That Might Interest Materialist Philosophers

Nowadays, some materialist philosophers readily acknowledge
that the old, naive versions of materialism, identifying it, more
or less, with mechanicism or atomism, are not any more tenable
But, they say, materialism is still alive because it is a theory,
not of matter but of *mind* It consists in the assertion that what
thinks, hopes, apprehends, and so on emerges from what does
not think, does *not* hope, does *not* apprehend, and so on (see A
Comte-Sponville [2])

When faced with such a definition, we must ask whether or
not "what does not think, does not hope, does not apprehend" is
assumed to be knowable If the answer is *no*, then "materialism"
as just defined is a totally arbitrary postulate, extrapolating from
commonsense experience (stones, presumably, do not think) to
a Reality that, by assumption, lies beyond experience If this Reality is truly unknowable, what should make us consider that it
resembles stones more than minds? More precisely, what is the evidence that Being has "parts," or "elements," and that the most basic
of these do not think? If the answer to our query is *yes*, Reality *is* knowable, we must ask the further question: "should Reality—Being—be
described entirely in strongly objective terms?" If we are answered
that weak objectivity is enough, we must come to the conclusion
that the definition of materialism under study is incoherent, since
within a weakly objective description there are—by definition—
basic statements to which it is impossible to impart a meaning
otherwise than by referring to what is observed   and "observation" refers to mind To preserve self-consistency the materialist
philosopher must therefore demand from whatever—complete
or provisionally incomplete—description of Reality is put forth
that it should be interpretable in strongly objective terms (as classical physics was, with perhaps the exception of statistical thermodynamics) Admittedly, this demand can be met, since most
of the ontologically interpretable theories described in Chapter
13 are consistent But, with respect to them, what has already
been stated at several places in this book should once more be
stressed These theories have to cope with such considerable
inconveniences that it seems nearly impossible to scientifically
give credence to any one of them Consequently, materialism as
defined above is a conception in which what contemporary science brings in as reliable information is basically uninteresting

(just a set of working recipes) and what interesting information it yields is not reliable. This leads us far away from the notion of scientific materialism.

A Point Concerning Theorists

On the other hand, a conclusion that the theorists should derive from the matters reviewed in this book is, as it seems, that they should not let themselves be inordinately impressed by the *formalism* of quantum physics. It is quite true that, for reasons having to do with both efficiency and mathematical beauty, the Hilbert space formalism can positively be called "magnificent." And it is therefore understandable that a great number of theoretical physicists should consider this formalism not only as embodying, at least in principle, the whole of possible *operational* knowledge but even as constituting, independently of any epistemological restriction to pragmatism or operationalism, the "ultimate truth," so to speak. If this hypothesis, which partakes much of the hypothesis of "completeness," as stated in Section 4.2, is made, then of course many conceptually meaningful consequences can plausibly be inferred from it, such perhaps as a need, for would-be "realists," to give up ordinary two-valued logic, or to admit of the existence of "delayed choices" in some physically realizable experiments, or etc. But we must remember that while the completeness hypothesis seems extremely attractive, still it is only a hypothesis. The very existence of such models as the pilot-wave and the GRW theories suffices to establish this point. When we speak of positive *consequences* that quantum physics has concerning our views of the world, we therefore must be stricter.

A Point Concerning Epistemologists

This point has to do with the notion of Reality, or "Being." It is subtler than the foregoing ones and will therefore be analyzed at greater length. As a consequence of what has just been noted, we physicists must now bow to the fact that, notwithstanding superficial appearances, the notion of reality has quite unavoidable philosophical aspects; and we must accept the philosophers objections to the, most uncritical, use made of it in common practice.

In my view we should not carry their rebuttal to the extremes: In the last two chapters, arguments I consider as convincing were developed, aimed at showing that the notion of a reality whose existence and general structures do not depend on our collective mental structures is unavoidable and therefore quite necessary. But I must grant that these arguments are themselves mainly philosophical. And a (sad) consequence of this is that some brilliant philosophical minds are to be found of whom it is almost sure that they will not bow to the arguments in question and say, "Yes, I am now convinced: Reality-per-se *has* a meaning." Under these conditions, the question whether, in the field of epistemology and/or ontology, any unquestionable lesson from quantum mechanics can at all be drawn—any that would *not* depend on ontological assumptions of any sort—is a question the answer to which is not quite obvious.

To investigate it without getting trapped in a maze of side problems, one suitable method, the one to be taken up in the remaining part of this section, is to inquire as to how a leading physicist of the past having an interest in epistemology and ontological issues would have had his ideas *changed* by the more recent developments in quantum physics. Without aiming at originality, let us choose Einstein's thought as our subject of inquiry. It has often been said that this physicist began as a skeptical empiricist and ended as a realist. Admittedly there is much truth in this but—as convincingly shown, in particular, by Michel Paty [3]—in fact Einstein, who never was a radical empiricist, never was a radical, or, so-called "naive," realist either. This is what makes his views especially interesting for our purpose here.

Accordingly, what we are primarily interested in at this place is the standpoint that was Einstein's in his mature years concerning the status of the notion of reality. Far from being intuitive and dogmatic, this standpoint was subtle and elaborate. At an early stage he rejected the traditional empiricist views and insisted that no logical path exists leading from sense-data to the axioms of a general, successful theory, so that building the sciences requires the possibility of freely choosing concepts that are *arbitrary* with respect to their logical relationship with experience. But although he thus clearly parted from positivism, as we said, still, from his early reading of positivist authors such as Mach he had derived an awareness of important things that he could not simply "forget," such as the fact that realism is neither "immediately obvious" nor, in any way, provable. So that, as Paty stresses, Einstein's realism was of a more elaborate and refined nature than the realism of some of his contemporaries, notably Planck. It is true that he agreed with the latter in rejecting radical idealism and in asserting the real existence of an external world. In this respect it is not incorrect to consider him a "metaphysical realist." But still, while for Planck the sense-data, and the worldview the scientists strive to discover with their help directly refer to the real universe lying behind them, Einstein's insistence on the *free invention* of concepts made his realism distinctly more critical. According to him, while, admittedly, concepts acquire meaning only through some overall relationship with experience, the fact remains that only a *system* of concepts can be attributed a truth value. And, to repeat, he stressed that such systems are free creations of man [4], only subject to the condition that they must be governed by the project of coordinating in the most general, consistent, and economical way the totality of our sense-data. Basically this is how he understood the, to him essential, notion of *verification*: what must be verified is an unseparable set of mutually related ideas and experiments. And, he claimed, "this also holds true concerning the notions of 'physical reality,' 'reality of the external world,' and 'real state of a system.' A priori, he wrote, it is not any more legitimate to assume they are necessary than to reject them. Only verification (in the foregoing sense) can settle the issue. Behind the symbols that these words make up, there is a program" [5].[1]

Einstein also forcefully claimed, however, that there is *one* proper way of building up the valid system of concepts and thereby a valid picture of reality. Essentially it consists in relying firmly on mathematical simplicity. Since mathematics are, so to speak, the quintessence of intelligibility it follows that, for Einstein, without intelligibility the very notion of reality would be meaningless, or, at least, (in his proper words) *scientifically* meaningless. It would be a notion of a purely metaphysical nature.

Nothing in all this would make Einstein's approach very different from Poincaré's conventionalism. But it turns out that, on the subject of the notion of reality, and especially of *reality as*

---

1. This quotation as well as those from Ref. [4] are borrowed from Paty [3].

a program, Einstein, in fact, did not rely *only* on the mathematical intelligibility criterion On this matter, two other conditions also played quite a basic role in his thinking One (a), which he stated explicitly in connection with nonquantum physics, is the view that everything should be referred to conceptual objects belonging to the realm of spatiotemporal ideas [5] Another, (b), related to the latter but not identical with it, is that at least some forms of counterfactual argumentation—essentially those that make it possible to elaborate the notion of elements of reality, as discussed in Section 9 2—are valid ones It is true that Einstein does not seem to have expressed this second condition in the form in which it is stated here But he did write such phrases as "for example, nobody doubts that at any given time the center of mass of the Moon has a definite position, even in the absence of any real or potential observer" [6] Logically such a certain knowledge (absence of doubt) must be related to observations we know we *could* perform, at least in principle At other places he stressed that "physics is an effort at conceptually apprehending what exists as something independent of its observed becoming" [4 p 80], and that the scientist strives to describe a world independent from the perceiving procedures [4 p 683], which clearly go very much along the same lines

This incomplete review of Einstein's cautious approach to the problem suffices to establish an important point It shows that even if, following Einstein, we are undogmatic about the notion of reality to the extent of defining it merely programmatically, still this is a notion that we may not hastily brush away as being "empty" or "meaningless " Hence, even the philosophers inclined to consider the notion of a reality-per-se as an unwarranted metaphysical postulate should grant there is a sense in which reality is a meaningful notion in physics It then appears conceivable that the developments of physics should be liable to really teach a few things, concerning this notion, even to those who are extremely anxious to discard any metaphysical postulate whatsoever Admittedly, discarding any metaphysical postulate bars considering obvious the existence of a reality logically prior to knowledge In other words, it implies going as far as questioning the priority of existence on knowledge that, as previously noted, is, in our view, quite essential If, for the sake of the argument, this standpoint is temporarily taken up, it becomes impossible to consider nonseparability as a true lesson from physics, the reason being that when the priority in question is reversed, the borderline between "reality" and "empirical" reality becomes fuzzy; and, as we saw in Chapter 15, there exists a consistent way of conceiving of the notion of empirical reality that makes it free from nonseparability effects (since nonseparability cannot even be defined within this realm). Then does something remain, and, if yes, what?

One point that positively remains is, of course, that, even considered as mere conjectures the statements: "Physics is a description of an intelligible (macroscopic *and* microscopic) reality whose existence is independent of *our* existence" *and* "This reality is separable (or 'local')" cannot both be kept. In view of present *positive* knowledge, they cannot be true together.

But another point undoubtedly also remains. It is that even if we do not make the ontological assumption expressed by the first of these two statements; even if we adhere to Einstein's more cautious and elaborate conception of "reality as a program," still something can be asserted, and this "something" would not have pleased Einstein. It is the fact that, as we saw in this book, the conditions labeled (a) and (b) above (ultimate recourse to spatiotemporal ideas and appropriateness of counterfactual considerations) cannot be kept in complete generality and that consequently we have no rational justification for going over from the—in itself without much bearing—Reality Criterion 1 (Section 9.2) to the more far-reaching Reality Criterion 2. The rejection of the conditions in question is not an element of traditional philosophical teaching, presumably because it could hardly be grounded on purely philosophical reasons. Hence, we must grant that it is a philosophically meaningful new piece of information.

So, on the whole we here have to do with a real lesson that contemporary physics teaches us, reasonably independently from any preconceived option philosophers could brand as "metaphysical." In short, this lesson is that there is nothing like "the real, factual, noncontextual situation" of a *localized* microscopic system, and that consequently, if the scientist proposes—be it as a mere program—to describe in detail a world strictly independent of human abilities concerning observing procedures, this "world" must be nonseparable (or "nonlocal"). Thus it must radically differ from the idea the term "world" normally conveys.

16.2 On the Concept of Influence at a Distance

The question of the validity of this concept is worth some scrutinizing. When we hear about the violation of local causality our first idea may well be that it testifies of some influences at a distance that do not decrease when the distance increases—as normal forces do—and violate the relativistic finite velocity law. But when we become informed of the fact that neither energy nor information can be transmitted by such means (parameter independence, see Section 8.2), we get to realize that the issue is too delicate to be settled in such simple terms. Some physicists indeed went as far as to consider, in view of this, that, after all, nonseparability is nothing like a novel feature of physics. They stressed that even classical physics admits of some kind of "superluminal traveling" of effects carrying neither energy nor information—such as the lighthouse effect described in Section 9.3: Remark 1—and claimed that this proved their point. We saw there that this argument is flawed and that effects of the just mentioned type actually have nothing to do with nonlocality, which therefore we must go on considering, contrary to such claims, as a newly discovered and essential feature of any tentative description of independent reality. But this does not settle the question whether or not influences at a distance associated with nonlocality "really exist." In fact, this is a question, not of physics but of epistemology and semantics. Nonlocality is something new, not only in physics but also in the realm of concepts. The ideas it conveys resemble in some respects the notions of *cause* and, more generally, *influence* we are accustomed to make use of, both in daily life and in science. In other respects they differ from them. If we believe in the validity of some—presently existing or still to come—ontologically interpretable theory, describing independent reality as something immersed in space and time (or in space-time), such as the theories reviewed in Chapter 13, we are thereby led to stress the resemblances, and it is then difficult to escape the view that the influences in question really exist. If, on the contrary, we consider, either, following Kant, that the "efficient causality" concept is meaningful only within the realm of empirical reality or, as I suggested in Section 14.6, that independent reality is not embedded in space and time—or both—then we obviously cannot keep the notion of such influences at a distance and go on attaching it to independent reality

(if independent reality is not embedded in space—or in space-time—it makes no sense to speak of influences *at a distance* taking place in it)

On the other hand, the question remains whether or not there is a sense in which influences at a distance can be said to exist within *empirical* reality Of course, their existence could not be derived from our knowledge that local causality is violated—or, more generally, from nonseparability—since such knowledge concerns independent reality only But the existence in question may conceivably be derived by some other means In fact, two suggestions have been made that go more or less along such lines

Suggestion 1 (Stapp [7]; Eberhard [8])

Above (Section 15 5) the existence was noted of interesting proposals to the effect of deriving Bell's inequalities with no reference to the notion of (independent) reality These derivations are essentially based on the following premise Let R_U and R_V be two spacelike separated space-time regions in which yes-no measurements are made on some observables Let the possibility be considered that either a measurement M_U of an observable A_U or a measurement M'_U of an observable A'_U is made in R_U, and same concerning R_V with the symbol V replacing everywhere the symbol U The premise in question then reads as follows If the measurement M_U were to be performed in region R_U and the measurement M_V were to be performed in region R_V and the result "yes" were to appear in region R_U, then this same result "yes" would appear in region R_U, also in case the measurement in region R_V were to be M'_V Deriving the Bell inequalities from this assumption together with the symmetrical one (exchange of the U and V symbols) is a simple matter [7, 8] so that the observed violation of these inequalities invalidates the considered premise, and it is argued that, somehow, this implies the existence of some superluminal influence-at-a-distance between regions R_U and R_V But the very *significance* of the highly contrafactual premise just stated is not as clear as it, at first sight, seems to be[2] (see again Ref [17]–[20] of Chapter 15) In particular, a

---

2 To witt: within the realm of his own theory (see Section 11 4) Griffiths could recently (*Phys Rev Lett* to be published) build up a definition of "quantum counterfactuality" *not* based on the necessity concept But, as he showed, this

claim has been made that it is, in fact, cryptodeterministic, which would imply that independent—not just empirical—reality is involved in it. It is not possible here to go into the intricate debate that developed on this point.

Suggestion 2 (d'Espagnat [9])

It was noted in Section 15.2 that within the empirical reality approach there can be no question of an exhaustive enumeration of the facts (values of parameters, hidden or not hidden, etc.) that can play the role of causes. As pointed out in Section 15.5, this is what makes local causality irrelevant to empirical reality. More generally it implies that the existence of a direct causal link between two repeatable events never can be derived from just the observed evidence of a correlation between the two, for this correlation could always be ascribed to the existence of some unknown causes, common to both. Consequently, within the empirical reality approach a real causal link between two events A and B—enabling us to state that the former is the cause (or a cause) of the latter—cannot be defined otherwise (as several philosophers claimed, see, e.g., von Wright [10]) than by referring in some way to human possibilities of action (which, of course, is quite consonant with the whole spirit of the empirical reality approach). More precisely, this mode of definition—sometimes called the "entailment theory of causation"—goes as follows. If A and B are two repeatable events, if A is anterior to B and if A is of such a type that it can be made to happen at will, then A *causes* B if and only if it is the case that whenever A is made to happen, B happens also and whenever A is not made to happen B does not happen either.

Noteworthy in this definition is the restriction that A is of such a type that it can be made to happen at will (a generalization of this idea that does not change its basic nature is that, at least, we can *imagine* we make A happen or not happen). Such a reference to human possibilities is necessary for the just stated reason. But, on the other hand, the reference in question has the consequence that, within the standard Bohm two-spin example, it is a nontrivial question whether or not the

---

definition does *not* make it possible to go over from reality criterion 1 (in our language) to reality criterion 2, so that it entails neither the Bell inequalities nor nonseparability.

measurement of a spin component on one particle can be said to actually "cause" the corresponding component of the other particle to take up a definite value The point is that, while we are at liberty to set into position some instrument of observation operating on the first particle, we are *not* at liberty to choose the *outcome* of the measurement performed with it Hence, if we identified event *A* with the the fact that the said outcome has some given value, the condition that *A* can be made to happen at will would not be met The only event we can make happen at will is the fact that, on the first particle—*U*, say—the spin component along some given direction **n** is measured (outcome +1 or −1, in appropriate units, then automatically appears) This therefore is an event we can identify event *A* with Admittedly, because of the strict correlation implied in the Bohm standard example, the spin component S_n^V along **n** of the second particle—*V* say—then must have a definite value either −1 or +1 But is this "event *B*" (S_n^V having one of these two values) really an "event?" Since these two values are anyhow the only ones that can be found, is not the proposition that "event *B*" is true just simply a tautology?

The answer is that, quite strictly speaking, it cannot be claimed it is one In fact the structure of the proposition in question is identical to that of proposition *Q* in Eq (14 3) (see Section 14 10), where the roles of *P* and ~ *P* are played by propositions $(S_n^V = +1)$ and $(S_n^V = -1)$ respectively and that of *D* by the proposition "a Stern–Gerlach instrument oriented along direction **n** is made to interact with particle *V* " As pointed out in Section 14 10, such a modal logic proposition is not a tautology

At least, it certainly is not one when the propositions *P* and ~ *P* are either true or false Here, however, propositions $(S_n^V = +1)$ and $(S_n^V = -1)$ are merely trustworthy, and in the absence of a formal "modal logic of trustworthy propositions," the nontautological character of the proposition under study cannot truly be asserted Hence, when all is said and done, Suggestion 2 cannot be considered as conclusive either

The existence of Suggestions 1 and 2 is an indication that a sense of the expression "influences at a distance" may conceivably be defined that would render empirical reality not devoid of such influences But it is a weak one, as we saw And moreover such theoretical developments are likely to be of limited scientific and epistemological significance Indeed, parameter independence must anyhow prevent them from having a bearing extending far beyond the realm of mere semantics.

16.3 Hints from Quantum Physics

In Section 16.1 we tried to pinpoint the epistemological information that a scientist (or an epistemologist) anxious to keep away from "metaphysical" commitments of any sort—including the implicit ones that are, of course, the most insidious—may nevertheless derive from quantum physics. But, valuable as it is (mainly as a guide for discriminating the genuinely scientific questions from those of a more "mixed" sort) this standpoint is not to be raised to the status of a universal way of thinking that would be the only admissible one. As explained at length in Chapter 14, there are sound philosophical reasons, including some indirectly linked with science, to consider that the notion of a reality the existence of which is independent from our own is meaningful in itself, that is, irrespective of the nature and reliability of the kind of knowledge we can conceivably acquire concerning it. If this is granted, then a number of other points can be made, deriving from quantum physics and concerning the very relationship of knowledge and the reality in question. They constitute the subjects of the next sections. We call these points "hints" in order to make quite clear their difference in nature from the foregoing "information."

16.4 Hint 1: Extended Causality

Some of the arguments put forward in Chapter 14 against the radical idealists' rejection of the "independent reality" notion refer to physics, so that physics *does* indirectly back up the view that independent reality is indeed a valid concept. Here let us merely emphasize one particular such argument (or rather one that follows from putting several of these arguments together). It proceeds from contrasting the standpoint of the radical idealists with the one of the "physical ontologists" (Bohm, Bell, et al.) on the question of explaining the intersubjective agreement relative to contingent facts. As we saw in Chapter 1, most of the former either simply overlook the problem or else just brush it away as an unwarranted questioning *why*. There are some grounds for not considering such a procedure as fully satisfactory. In a way, the quantum mechanical formalism is more illuminating, since it does predict intersubjective agreement, as we saw (by referring either to collapse or to the correlations the total wave-function exhibits), even though this still is merely *explanation through reference to general laws* (Section 15.2). And at the other end of the chain, the ontological theories reviewed in Chapter 13 do, in a sense, better since they yield "explanations" in the common sense of the word. Without committing ourselves to one well specified theory we may view all this as constituting, on the whole, an indication that at least the *notion* of an independent reality is not totally meaningless.

On the other hand, if this indication from physics is added to all the aforementioned philosophical arguments in favor of the said notion and if the whole thus constituted set is considered convincing (which is the standpoint that will be taken up from now on), the Kant-like qualification of causality as a mere human way of organizing the phenomena our senses apprehend is thereby made too restrictive. Over and above *this* notion of causality, which underlies the notion of empirical reality (Sections 15.2–15.6), and the importance of which there is therefore no question of underestimating—we must consider as meaningful a causality of another type, operating from independent reality to phenomena. Since, because of nonseparability, independent reality cannot be conceived of as made up of localized elements embedded in space-time it is clear that this "extended causality" widely differs both from Kantian causality—which is nothing else than determinism—and from Einsteinian causality. There does not seem to be a way in which it could accommodate the notion of (in Aristotle's terms) efficient causes. But it has a place for that of structural causes, and the latter are—in this approach—more than just lawlike regularities between observed phenomena. Indeed these structural "extended causes" are nothing else than the very structures of independent reality and they constitute the ultimate explanation of the very fact that the laws—that is, physics—exist.

As is well known, according to Leibniz all contingent facts are governed by the Principle of Sufficient Reason, which he defined as follows. It is the principle "in virtue of which we judge that no fact can be found true or existent, no judgment veritable, unless there is a sufficient reason why it should be so and not otherwise, although these reasons cannot, more than often, be known to us. " (*Monadology*). And again: "Nothing happens without a sufficient reason; that is, nothing happens without its being possible for one who should know things sufficiently to give a reason showing why things are so and not otherwise" (*Principles of Nature and Grace*). It seems likely that, among the physicists, the only ones who today would endorse this principle without reservations—or at least without basic ones—are the small bunch of the advocates of what we called in Chapter 13 the ontological, deterministic theories (such as the pilot-wave theory). But still, from all that has been reviewed in this book it seems to follow that some at least of the spirit of Leibniz principle should be saved. Not determinism, certainly. Not even the kind of—at least theoretical—intelligibility he postulated when he spoke of "one who should know" (thus hinting that this "one" could in principle be, not God but just simply a human being). What should and *can* be saved of Leibniz principle is its bare "hard core" the very notion of some ultimate *reason* for—at least—the *laws* that govern the world. To make this more precise it is appropriate to state the following "basic" postulate:

Basic Postulate. Any observed regularity (statistical or otherwise) must have a cause (or a set of causes: the notion of "oneness" is not stressed here), which (i) may be or not be located in time and (ii) may be or not be discoverable by men.

Point (i) implies in particular that the general causes of an event are not necessarily located in its past. According to it, the assertion that efficient causality is a scientific notion whereas final causality is not, is meaningful only within *empirical* reality. Point (ii) in fact extends the range of the notion of cause far beyond the one of the mere notion of explanation, at least if the latter is meant in either of the senses described in Section 14.8 and labeled there (a) and (b): reduction to the known and grouping under some general law. Here it is considered that it is not meaningless to speak of causes even if these are of such a nature that they conceivably can never be discovered, and the "extended causes" we have in mind are considered as being prior to laws.[3] After all, we should keep in mind the truism that

---

3. Obviously this option amounts to discarding the rule with operationalistic flavor stating that the meaning we attach to a concept should not be more extensive than the totality of its references (direct or indirect, via some theory)

the number of brain connections is (huge but) finite and that therefore it is by no means a logical truth that mind can discover the whole of what exists and has observable effects.

• *Remark 1.* In most philosophical circles a principle inherited from Kant is held as valid, according to which cause and effect must be of the same nature; so that, having agreed that the effects are—in the philosophical sense—phenomena, we should be forced to apply the notion of causes only to phenomena. Within such a conception, the *basic postulate* stated above and the notion of extended causality are quite obviously meaningless. They are attempts at finding the cause of causality, which clearly is a vicious circle. In favor of Kant's view it has often, and quite reasonably, been argued that it contributed in a decisive way to separate science from metaphysics, thus freeing the former from a host of false problems and thereby helping in its development. This latter point is indisputable. It should therefore be clearly understood that the ideas proposed here are in no way a suggestion to the effect of reintroducing metaphysical elements *into* the development of science. It is, in fact, just the other way round. It is science itself—more precisely, *some* findings of science—that now knock at the door of metaphysics: not to borrow from it some elements that would be of help in scientific research proper but, on the contrary, to bring to it some fresh elements of information for which metaphysics can find some use. This being clarified, it must be observed that the Kantian principle of quite strictly limiting causes to the realm of phenomena is, after all, just a "principle." It is only within the realm of—radical or Kantian—idealism that it can be considered a logical truth. In other words, if we do not adhere to either of these views we must think of the Kantian restriction of causality to phenomena as merely a most efficient "rule of the game", a rule that, as already noted, we must quite strictly hold to within our scientific activities proper, but that we would have no valid reason to extend to the whole activity of human reason. To put this all in a nutshell: to, for example, agree with Kant that space and time are *not* arenas *in* which independent reality lies, does not imply we should also agree with his views on causality.

---

to observed data. I explained elsewhere (Ref. [10] of Chapter 15) in which cases I do consider that the rule in question is too restrictive.

• *Remark 2:* It is true, however, that such a notion of extended causes, which theologians might identify to that of "primary causes," should not be handled without care. A well-known objection to it is that it apparently leads to an infinite regression. "Why is it that the Earth does not indefinitely fall down in empty space?" a disciple asked his guru in a famous tale. "Because it rests on the back of an elephant," the latter replied. "And why does not the elephant itself indefinitely fall down in empty space?" the insatiable disciple asked. "Because," the tale goes on, "it rests on the shell of a giant tortoise," and so on. The fable appropriately reminds us that the radical idealists and the Kantians are justified in questioning naive forms of the quest for causes. It can, however, not be considered as carrying any decisive objection. After all, Galileo was right when he asserted that preservation by an object of the rectilinear uniform motion it has requires no cause. Similarly, it is quite natural to counter the objection embodied in the fable by considering that the general structures of independent reality, which, we claim, are the causes of the observed objective regularities, do not themselves require causes.

• *Remark 3.* It may be observed that the "extended causality" concept is implicit in the conceptions of many thinkers, including even some who do not believe in traditional realism. For example, when Henri Poincaré stated that (a) we cannot know the objects themselves and (b) we do know the relationships between them, there seems to be no way of understanding such a claim short of considering that something *causes* our knowledge and that this "something" is not a mere Kant-like phenomenon (this of course does not imply overall allegiance to Poincaré's views: the analogies and differences between the ideas this book tries to convey and those of Henri Poincaré were described in Sections 14.8 and 14.12 respectively).

• *Remark 4.* The notion of extended causality is obviously linked to the question of determining to what extent science, given that it does not yield a definite *knowledge* of independent reality, yet brings us nearer to such a knowledge. To try and analyze this point it is appropriate to bow to a common practice of philosophers and distinguish two levels in what we, up to now, called "empirical reality." Indeed, it is not necessary to be a quantum physicist to realize that the commonsense level and the scientific level are different. The limits of efficiency of our sense organs, and their dependence on intermediate agents such as sound and light waves, have been known for quite a long time and scientists are now getting better and better aware of the extent to which this has the effect of partly making our sense-data *creations from us*. If we add that the interface between the perception signals and the mind is poorly known and even more poorly understood, we realize there is no reason at all that our commonsense representations of things should be in any way *like* independent reality. Science is an attempt at correcting for all these human idiosyncrasies and this author cannot but agree with Shimony's view [11] that science—far from being just a tool—should be considered a sophisticated, self-critical, and *partly successful* enterprise, aiming at understanding in as much detail as possible the connection between the world presented to us in commonsense and ultimate reality. In my eyes, within the foregoing sentence the word "successful" refers to the fact that independent reality, especially its general structure, is not totally unknowable while the word "partly" sketchily expresses the fact that what we actually reach concerning all contingent facts is but *empirical* reality. For all this, of course, extended causality is essential. On the other hand, accepting it does not imply that we can go back from effects to causes. The reasons, drawn from physics, that make most unlikely the conjecture that we can were explained in detail in Chapter 14.

16.5 Hint 2: Space-O and Time-O

Within the veiled reality conception Einsteinian separability, as previously noted, applies to empirical, not to independent, reality; which means that special relativity theory does *not* concern independent reality. In other words, as already stated in Section 14.6, in the veiled reality conception independent reality is not in space-time, and, in particular, Einsteinian causality does not apply to it. It is true that, as pointed out in Section 13.2, the difficulties met with in reconciling the "ontological" theories with special relativity induced some authors to turn back to the old, Newtonian concepts of an absolute space and a universal time. But most physicists consider it unacceptable to justify such a regression on the mere basis of the philosophical requirement of ontological interpretability; and the veiled reality conception stands in agreement with this view. In it, therefore, independent reality is, as already said, considered as being embedded neither in space nor in time. We may then refer to Bonsack's concept of a world-O (Section 15.1) and introduce the notions of a space-O and a time-O Applying to these notions the essentials of Bonsack's analysis leads to the view that space and time, although they do not proceed from the existence-O of the subject-O, are empirical nevertheless: a conception of which, of course, philosophers like Kant had, long ago, an intuition.

In his book "Mind and Matter" [12], Schrödinger expressed the idea that, on this particular point and concerning time, Einstein's special relativity theory not only did not disprove Kant's views but, on the contrary *confirmed* them, by showing that "the whole time table is not quite as serious as it appears at first sight." This is certainly true but still, in *classical* special relativity Einsteinian causality could be viewed as a feature of independent reality defining, for each event, an absolute future and an absolute past. Now, however, with the discovery of nonseparability and the confirmation—through the Bell theorems—that no ontological escape from it is possible, it has become clear that Einsteinian causality cannot be a feature of independent reality and consequently the notion of an independently existing "timetable"— to use Schrödinger's expression again—has become even more shaky. To quote Schrödinger once more, it seems that, intellectually at least, we truly enjoy a liberation from "the tyranny of old Chronos." Since, as he writes, "what we, in our mind, construct ourselves cannot have dictatorial power over our mind" the just reached conclusion is a far-reaching one indeed.

• *Remark 1.* It is true that the astrophysicist's notion of a *cosmic time* may at first sight seem to cast some doubt on all this. General relativity and special relativity being in many respects different theories, it might prima facie be conjectured that this cosmic time is more than a "time-O." Indeed, it might be conjectured that it is an element of independent reality (or rather of the arena in which independent reality *is*) that is, a kind of Newtonian time brought up to date. This would imply that scientists did succeed in lifting up the veil on at least one—important— element of independent reality. This conjecture may be considered, but it must be observed that there are objections to it. One of them is that the very notion of cosmic time can only be defined within the *approximation* of uniformity and isotropy of the universe-as-a-whole, known as the cosmologic principle (although it is a mere observational statement and has therefore not much in common with genuine principles). The point is that it seems impossible to consider as basic a notion than can only be defined within some approximation of factual data. Admittedly the universe seems to be isotropic to such a degree that this objection cannot, at present, be considered a crucial one. Nevertheless, in this connection it should be remembered that the general relativity equations admit of other solutions, known as "exotic" cosmological models. The Gödel cosmological model is an interesting example of these. It was objected to on the basis of the fact that, in it, causality just disappears. This, it was claimed, is unacceptable since causality lies at the very basis of physics. With the advent of nonseparability and the recognition of the fact that the up-to-now accepted definitions of causality—including Einsteinian separability—do not apply to independent reality, such a criticism is inoperative. On the whole, therefore, the idea that standard model and cosmic time should apply to independent reality, remains very speculative. On the standard model issue it is perhaps significant that in one, at least, of the ontologically interpretable models, that of Bohm, the Big Bang appears as being but a "small ripple" [13].

• *Remark 2:* Concerning idealization of time, proposals have been made that reach beyond the above described, essentially qualitative considerations. This is, in particular, the case of a suggestion made by Grib [14] in connection with the "quantum logic" approach to quantum mechanics. The approach in question has not been described in this book because (a) I have already discussed it in Ref. [14] and (b), as there shown, when considered as a means of restoring the notion of a strongly objective reality immersed in space-time and obeying Einsteinian separability, this approach fails. However, Grib's conception is different. His view is that independent reality is indeed composed of truly existing quantum objects whose properties correspond to propositions obeying non-Boolean logic, but that the propositions in question do *not* possess truth values by themselves. Truth values are imparted to them by human observers and human observers are definitely Boolean, so that they impart these truth values according to Boolean logic: in other words, they must translate non-human quantum logic on their human language. Arguing along these lines, Grib is led to agree with the "veiled reality" conception. In performing the translation, the Boolean observer unavoidably meets with "paradoxes" linked with the nondistributivity of quantum logic contrasted with the distributivity of his own logic. According to Grib, it is at this point that the notion of time has to enter the scene. His idea is that, in order to remove these "paradoxes," the observer "invents" time. For example, it may be the case that, in the nondistributive quantum lattice concerning a particular problem, a, b and c are mutually exclusive atomic propositions in the sense that neither $a \wedge b$ nor $a \wedge c$ nor $b \wedge c$ can be true—so that $P_1 \equiv (a \wedge b) \vee (a \wedge c)$ cannot be true either—and however $b \vee c = I$ (here $\wedge$ means "and," $\vee$ means "or," and I means "always true"; a, b, and c could, for instance, respectively be the propositions $S_x = +$, $S_z = +$ and $S_z = -$ relative to a spin $1/2$.) For a Boolean-minded person this is a contradiction since it implies that, if a is true, $P_2 \equiv a \wedge (b \vee c)$ is true while he or she knows from Boolean logic that $P_2 \equiv P_1$ (distributivity). He or she resolves the paradox by saying: "at some time t_1 I hold that a is 'true' at another time t_2—if I measure b, say—I may see b "becoming" 'true'." And then he or she will have to consider that, at time t_2, a is not true any more. In other words, both the notion of time and the reduction of the wave packet appear here as consequences of the aforementioned "translation."

16.6 Hint 3: Independent Reality, Empirical Reality, Potentialities, and Consciousness

In his mature years Heisenberg developed a conception of the world in which quite a central role is played by the notion of *potentia*. He associated wave functions to potentia. In a sense, Heisenberg's approach is ontological, but his ontology is of a subtle kind, for along with actuality (of things, events, etc.), it leaves room for potentialities. He seems to have considered that if you wish to talk about things-in-themselves the natural language for doing so is the quantum state, viewed as some sort of a synthesis of actualities and potentialities.

Stated in such general terms this conception strikes us as liable to contain much truth. On the other hand, it is hardly more than a program for Heisenberg did not precisely define what in his views was the difference between "actual" and "potential." As long as such a definition has not been given it is clear that any explanation based on the concept of potentiality must remain vague. And consequently the transition from the potential to the actual must remain somewhat of a mystery.

Within the conceptions developed in this book, the notion of potentiality also emerges, as we saw, and takes a somewhat more precise form. In Section 15.3 it was noted that, as soon as the notion of independent reality is accepted and the view that it *might* conceivably be describable by physics is not a priori discarded, *potential* properties of objects—distinct from the "actual," or "ontological" properties with the same name—must be considered and held to be real. Such potentialities correspond to what, in allegorical language, could be called "acquired aptitudes" or "acquired fitnesses" of the system at yielding such and such a measurement result in case the appropriate measurement is or were made. When, in the standard Bohm example, S_z^U is measured—at time t_1 say—and found equal to $+1/2$, the formal operation of wave-packet reduction, imparting to V a definite spin ket (the eigenket pertaining to eigenvalue $S_z^V = -1/2$), corresponds to V acquiring a potentiality· that of inducing, with unit probability, a z-oriented Stern–Gerlach instrument to register the value $-1/2$ if and when, at a time $t > t_1$, it is made to interact with V

Another approach to the notion of potentiality is afforded by Grib's proposal, as summarized in Remark 2 of the foregoing section. In this approach also, the difference between the potential and the actual is well defined: it is identified with the one between independent and empirical reality. Quantum objects and their properties exist as *potentialities* (the latter obeying non-Boolean logic) and in this theory they are the constituents of independent reality (the plural number should not worry us: *potential* plurality can harmonize with nonseparability, as indicated by the considerations on extended pairs at the end of the last paragraph). The observed facts are elements of empirical reality and as such have "actuality." In this view, the transition from the potential to the actual has nothing in common with a physical phenomenon in the usual (i.e. tacitly ontological) sense of this word, since it just emerges from our human "ways of looking at things." But the price to be paid for such a clarification is, obviously, that the role of consciousness has to be recognized as essential.

This, of course, leads to the following question: What about the fact—illustrated by the well-known "Wigner's friend" paradox—that there are not one but many human consciousnesses (not even to mention the puzzling case of animals)? This is the problem called "arithmetical paradox" by Schrödinger, who "removed" it by asserting that the multiplicity of minds is only an appearance, hiding from us the essential oneness of eternal Mind. It is also present in what can be called the curved rope-ladder approach. This approach—this is an adequate place to summarize it—may be described by starting from a schematical description of the traditional physicalist standpoint and modifying it appropriately. The physicalist standpoint can be schematized by considering a rope ladder. Its lowest rung represents the elementary particles. The next rung represents more complex systems, such as atoms, molecules, and so on, that in a way emerge from the elementary particles through combinations due to forces. The following rungs represent even more complex systems, mostly macroscopic, such as stones, plants, nerves, brain, and so on that, in their turn, emerge from the former through combination and complexification, but nevertheless exhibit qualitatively new properties, not possessed by their elements. And then, the highest rung represents mind and thought, which, again, emerge from the foregoing ones and also exhibit qualitatively new properties. The curved rope-ladder approach, a view contemporary physics suggests, does not differ from this one, except just on one, quite decisive, point. To get a picture of it we must take one extremity of the ladder in one hand, the other one in the other hand, bend the ladder, and join the two ends. The figure thus obtained still pictures the atoms and molecules as emerging from the elementary particles, the macroscopic bodies, including brain, as emerging from the atoms and molecules, mind and thought as emerging from the brain composite structure but, going on, it shows the "elementary" particles as emerging from thought, and so on. Essentially, this "closing the circle" seems to be the guiding idea of Wheeler's "it from bit" theory [16]. It makes neither mind prior to matter nor matter prior to mind. They "generate each other" so to speak. But, of course, it does not remove the "arithmetical paradox."

The curved rope-ladder picture appears a natural and almost inescapable consequence of the veiled reality conception, since there, to repeat, it may allegorically be said that empirical reality and consciousness "generate" one another—although timelessly—*within* independent reality. But the reverse is not true. In Wheeler's theory, for example, no mention is ever made of independent reality and indeed such a notion would presumably be considered as superfluous and meaningless by this author, as it also was by Schrödinger. However, Schrödinger's objection to it—more precisely to Kant's "things-in-themselves" notion—was based as we saw on the idea that "independent reality" is but a kind of redundant and unnecessary duplication of "empirical reality"; and nowadays, since nonseparability introduces, in the "things-in-themselves" notion, an element of holism not present in empirical reality, this objection has lost its weight. More generally, we already mentioned in Chapter 14 a set of arguments that seem to show that the independent reality notion is indispensable. The question then becomes: Within the veiled reality conception, that is, when the notion of an independent reality is seriously taken into account, does the arithmetical paradox remain?

The answer to this is yes, although it is but a qualified yes. More precisely, within the veiled reality approach the arithmetical problem remains, even though it is not quite so acute as in Schrödinger's conception. It is not, because in it, if only we could accept *time* as an arena in which events-in-themselves would take place (thus setting independent reality *in* time) the multiplicity of minds would not raise any insurmountable problem. As noted in the first paragraph of Section 16.4, all the ontological theories *explain* intersubjective agreement between a multiplicity of minds. In the Bell-like version of the pilot-wave theory this agreement can be obtained simply by enlarging the configuration space: by adding to it as many dimensions as there are individual consciousnesses. The representative point then specifies at any given time, not only the "material" state of the world but also that of the individual consciousnesses of all conscious beings existing in it. A peculiarity of this model is, however, that, because of contextuality (see Chapter 13) the "content" of one consciousness, generated by the observation or measurement of some quantity, may well depend on whether or not an observation or measurement of some other quantity—even a quantity compatible with the considered one—is performed by somebody else at some other place, or on details about how the observation or measurement in question is made (i.e., whether or not it is associated with that of some other quantity, if so in what order, and so on).

This point is worth a quantitative illustration, so let us consider once more Bell's detailed description of how the pilot-wave theory accounts for the strict correlations in the Bohm standard, two spin example (Sections 13.3 and 15.3). There, the states of consciousness of the two observers must, according to the above described scheme, be associated with two new variables x_3 and x_4, so that the configuration space of the whole system is now four-dimensional. Both x_3 and x_4 can take up three possible values: 0 (nothing registered), 1 (consciousness of having seen $S_z = +1$) and -1 (consciousness of having seen $S_z = -1$). Let t_1 and t_2 be the two measurement times considered in Section 15.3. The states of consciousness of the two observers are then such that

$$x_3 = 0 \text{ if } -\varepsilon < x_1 < \varepsilon$$
$$x_3 = 1 \text{ if } x_1 > \varepsilon$$
$$x_3 = -1 \text{ if } x_1 < -\varepsilon$$

where ε is a quantity much smaller than 1, and similar inequalities hold true also concerning x_4 in relationship with x_2. Let us then consider, in the two alternative cases that $t_1 < t_2$ and $t_2 < t_1$, the interesting situation in which, at $t = 0$, x_1 and x_2 are both positive: $0 < x_i < \varepsilon$ with $i = 1, 2$.

Case (i). $t_1 < t_2$. Then:
between t_1 and t_2

$$x_1 > \varepsilon, \text{ hence } x_3 = 1$$
$$0 < x_2 < \varepsilon, \text{ hence } x_4 = 0$$

after t_2

$$x_1 > \varepsilon, \text{ hence } x_3 = 1$$
$$x_2 < -\varepsilon, \text{ hence } x_4 = -1$$

this being a consequence of Eq. (13.20) and the corresponding contextuality (nonlocality).

Case (ii). $t_2 < t_1$ Then
between t_2 and t_1

$$x_2 > \varepsilon, \quad \text{hence } x_4 = 1$$
$$0 < x_1 < \varepsilon, \quad \text{hence } x_3 = 0$$

after t_1

$$x_2 > \varepsilon, \quad \text{hence } x_4 = 1$$
$$x_1 < -\varepsilon, \quad \text{hence } x_3 = -1.$$

In both cases the two observers become aware of opposite S_z values for the spins. However, it is not the same one who gets in awareness state +1 in the two cases. And this holds true in spite of the fact that, if the two measurement events are spacelike separated, it may well be the case that cases (i) and (ii) constitute just one and the same physical process, considered in two different reference frame. In this respect, the situation is quite comparable to the one analyzed in Section 10.12 except of course for the fact that we have here to do with a deterministic theory with definite ontological claims. Note, moreover, that the foregoing formulas also have consequences not involving relativity. As they show, after time t_2 the observer, A_4, whose consciousness state is x_4, gets, in the case under study ($x_1(0)$ and $x_2(0)$ both small and positive at the start) in a state of awareness that is +1 if nothing happened between times 0 and t_2 but is −1 if, in between these two times, somebody else has—perhaps without A_4 even being informed—measured $S_z^{(1)}$ at some other, possibly quite distant, place. Consequently, while intersubjective agreement is guaranteed, the partial knowledge we acquire through some measurement or observation cannot, in general, be counted as an element to be simply added to other similarly acquired elements so as to constitute a genuine knowledge of the "outside world as it really is," its contingent features included.

A similar, only even more radical, conclusion follows from the interpretation I proposed (reported on in Section 12.2) of the relative state theory. The difference between the two descriptions is that in the first one the just mentioned partial knowledge *is* real genuine knowledge about an element of physical reality (the sign of x_4 is the sign of x_2), whereas in the second one it is, in this respect, a delusion. But in neither theory does the multiplicity of consciousnesses and the correlation between the impressions they get raise problems of consistency Thus it may be stated that, because these theories are "realist" models, which take seriously into account the independent reality notion and even put forward definite mathematical descriptions of the *general* structure of this reality, Schrödinger's "arithmetical paradox" does not arise in them

We observe that these two theories have at least two points in common The first is that, although they are conceptually quite different, they cannot be distinguished from each other (and from other nonlocal theories) by experiment The second is that, as already noted, they both require that independent reality be *immersed in time* As noticed at the end of Section 14 10, the first point strongly suggests that they should be viewed as metaphysical rather than as physical theories And the second may then perhaps be accounted for by the fact that the notion of time is so deeply ingrained in our minds that none of the elaborate, "quantitative" metaphysics human beings are capable of inventing can quite strictly do without it

On the whole, such considerations reinforce the view that we had better be as unpretentious as we can when we try and speak of independent reality In this they back up to some extent the previously (Chapter 14) introduced and justified notion of a *veiled* reality which we *apprehend* in space-time but which is not *immersed* in either space or time, or space-time We then see that when all is said and done, the vagueness of the aforementioned conception of an (atemporal) coemergence of mind and empirical reality within independent reality is not to be counted as an objection to it On the contrary, considering the basic obstacles standing in the way of more precision on the issue, it should be viewed as an argument in its favor even though, with independent reality not immersed in time, it may, admittedly, seem difficult to escape Schrödinger's somewhat disconcerting "oneness of Mind" hypothesis [4]

As a final remark, let it be noted that one of Schrödinger's general observations remains valid also within the veiled reality

---

4 Metaphorically, the relationship between the nonpersonal Mind and the individual minds could then be put in parallel with those between a hologram and its parts, the nonpersonal Mind reappearing in its entirety within each individual mind

conception Mutual generation of mind and empirical reality within a nontemporal independent reality implies that time is to be counted as a mere element of *empirical* reality, and therefore that mind creates it, just as much as it creates mind The might of old Chronos may well be, to take up again Schrödinger's image and claim, a mere intersubjective appearance (For a more elaborate, and therefore more shaded and balanced, expression of this view, see Section 16 5 again)

16.7 Spinoza Revisited

Independent reality is structured—in a way we cannot actually know—and, as explained above, via *extended causality* these structures give rise to the ordinary, observed cause-and-effect relationships As was pointed out in Section 16 4 there is no a priori reason why these same structures should not give rise to some kind of teleology In other words, Independent Reality plays, in a way, the role of the God—or "Substance"—of Spinoza, although there are some differences To stress, first of all, the similarities, they are that both Independent Reality (let it now be written with capitals) and Spinoza's God are very much distinct from the set of the material objects, the forces, and indeed all the things that we apprehend through our senses and our instruments of observation In this respect the often read assertion that by the word "God" Spinoza merely meant "nature" is, to a great extent, misleading, just because by "nature" we nowadays mean material objects, forces, and so on (and, because of this, the label "pantheism," often attached to Spinoza's doctrine, is a misleading characterization) On the contrary, as basic attributes of God Spinoza ranks both *extension* and *thought*, on equal footing And this also is a point where the two notions of Independent Reality and the Spinozan God have something in common since what has been allegorically described above as a "coemergence of mind and empirical reality within and from Independent Reality" obviously parallels the coexistence of extension and thought as God's attributes which is so essential in Spinoza's philosophy But, as previously stated, there are also some differences The main one is that, at least as interpreted by Einstein, Spinoza's God—or Substance—is *intelligible* (see Section 16 1, where it was stressed that without intelligibility the very notion of Reality would, for Einstein, have been meaningless). On the contrary, Independent Reality as conceived of here is (just as Bonsack's "World-Ω") not—or, at any rate, not totally—intelligible. It is true however that the difference is more between Independent Reality and Einstein's interpretation of Spinoza's God than between the former and Spinoza's *actual* conception of God since, along with extension and thought, Spinoza attributed to God an infinity of other attributes that are *unknowable* to man.

References

1. B. d'Espagnat, *In Search of Reality*, Springer-Verlag, New York, 1983.
2. A. Comte-Sponville, *Une èducation philosophique*, Presses Universitaires de France, Paris, 1991.
3. M. Paty, *Einstein philosophe*, Presses Universitaires de France, Paris, 1993.
4. A. Einstein, *Autobiographical Notes* in *Albert Einstein, Philosopher and Scientist*, P.A. Schilpp (Ed.), The Library of Living Philosophers, La Salle, Ill., 1949.
5. A. Einstein in: *Scientific Papers Presented to Max Born on His Retirement from the Tait Chair of Natural Philosophy in the University of Edinburgh*, Haffner, New York, 1953, p. 33.
6. A. Einstein, "Einleitende Bemerkungen über Grundbegriffe" in *Louis de Broglie, physicien et penseur*, A. George (Ed.), Albin Michel, Paris, 1953.
7 H.P. Stapp, *Nuovo Cim.* **40B**, 191 (1977); *Found. Phys.* **10**, 767 (1980).
8. P. Eberhard, *Nuovo Cim.* **38B**, 75 (1977); **46**, 392 (1978).
9. B. d'Espagnat, *Phys. Rep.* **110**, 203 (1984).
10. G.H. von Wright, *Causality and Determinism*, Columbia U. Press, New York, 1974.
11. A. Shimony in *Symposia on the Foundations of Modern Physics, Helsinki, 1992 (panel discussion)*, World Scientific, Singapore, 1993.
12. E. Schrödinger, *Mind and Matter*, Cambridge U. Press, 1958.
13. D. Bohm, *Wholeness and the Implicate Order*, Routledge and Kegan Paul, London, 1980.
14. A.A. Grib, *Quantum Logical Interpretation of Quantum Mechanics, the Role of Time*, A.A. Friedmann Laboratory, preprint, St. Petersburg, Russia, 1992.
15. B. d'Espagnat, *Conceptual Foundations of Quantum Mechanics*, 2nd ed., Addison-Wesley, Reading, Mass., 1976.
16. J.A. Wheeler, "Bits, Quanta, Meaning," in *Problems in Theoretical Physics*, A. Giovanni, F. Mancini, and M. Marinaro (Eds.), U. of Salerno Press, 1984.

Postface

By now many physicists are well aware of both the shortcomings of radical idealism and the maladjustment of physicalism to the physical knowledge of today. For want of a "third way," these scientists are on the brink of accepting the *veiled reality* concept. But some take it up only reluctantly. They ponder and mutter: "Somehow I hardly like it." Of course, being rationally minded people they do not see this as an objection. But still

My main purpose in this postface is to try and explain to such fellow-scientists why, in my views, far from being "depressing," the concept in question is not only correct but also likable. This explanation I am rather anxious to deliver. However, it requires recalling first some very general ideas. Let us go quickly over this.

One of these ideas is that scientific thought is not the only valid way of thinking. This age-old truism must today be supplemented by a simple observation. Physical science being what it now is, we can no more conceive of it as being confined within a well-defined factual domain out of which it would have nothing to say and would lead to no meaningful question. For any one of us, this rules out—or *should* rule out—drawing a strict, comfortable separation between a domain of investigation in which our scientific thought commands and some other domain in which our other valid ways of thinking would legitimately dominate. All this implies that, strong as the need may be to sort out questions and procedures and not to lightheartedly "mix everything up," a still stronger necessity exists, for a physicist interested in basic questions, not to ignore these other—nonscientific but nevertheless respectable—ways of thinking.

In this postface, some "other ways of thinking" are therefore first considered (Section 1). They are (as will appear) neither deductive nor even inductive, in the scientific sense of this word: in fact they are deeply rooted in implicit or explicit philosophical *options*. Consequently, the very fact that they are to be taken seriously implies that I must state my *own* option, which is done in Section 2. Being just an option, it is personal. Nobody is asked here to adhere to it and the readers who find it unpalatable should just simply ignore these pages. The conclusions of this book were stated in the foregoing chapters. They are therefore *independent* of the option to be explained in Section 2.

1 Other Valid Ways of Thinking

There are many of them but, for conciseness, let us mainly focus on the philosophical approaches. One prejudicial question is: Can the great philosophical systems—to most of which a philosopher's name is attached—be considered eternal? In philosophical circles the answer *yes* is often given. The claim is that, while, in science, cumulative effects make past theories either obsolete or fully integrated in present-day knowledge, in philosophy the great systems of the past are still "alive," with their own peculiar flavor and characteristics, that still make them irreplaceable sources of inspiration and valid developments.

In a sense, this view is certainly true (and corroborates the legitimacy of "taking options"). Its truth is partly due to the fact that there have hardly been any examples of a philosophical system being actually *disproved* by some other philosophical system, even though many of these systems are incompatible with one another. Admittedly, *some* elements of most systems were falsified by later scientific discoveries, but it is usually argued (by philosophers) that such refutations bear only on side issues and that the backbone and "spirit" of the system remains acceptable.

The very fact that, in investigations such as those reported on in this book, it was necessary to refer to Kant, Hume, and the like testifies, in a way, in favor of this opinion. On the other hand, nowadays such a view can definitely not be pushed to the extreme of claiming that all philosophical systems are equally tenable. One reason is that when too many "side issues" have been falsified the system loses its plausibility. Another one is that there are systems some elements of which have been falsified that are *not* merely "side issues." At the start, for example, when we know nothing of physics, we may well consider that, on the one side, the Aristotelian view that the "sensible forms" are "real"—to wit: that such familiar concepts as matter, motion, space, trajectories, and so on correspond to basic elements of independent reality—and, on the other side, the Platonic-Kantian view that they refer to mere phenomena, are equally plausible opinions. Today, as shown by the investigations reported on in this book, the second of these views remains tenable (and, to some extent, is even backed by physics) while the first, the Aristotelian one, is, if not falsified, at least relativized. It can hardly be claimed that this is but a side issue. The truth is that one of the main pillars of Aristotelianism (or, at least, of its most usual interpretation) is thereby proved to be shaky. Incidentally, this has implications concerning important philosophical systems that do significantly refer to Aristotelianism. Materialism as well as, in some of its aspects, Thomism are examples of such systems (but this is in no way a basic blow to Christianism: even less so since neither St. John the Evangelist nor St. Augustine nor many other prominent figures of the Church drew their basic inspiration from Aristotle).

Still it is true that—be it at the price of a few dramatic changes in their basic views—most of the major philosophical schemes of the past somehow "manage to survive." For what follows, let us have a quick look at the main ones.

(a) Conventional Realism

Conventional (sometimes wrongly called "naive") realism is the theory that, as a matter of principle, independent reality (the "external reality" of philosophers) is knowable in all its details, the microscopic ones included. It suffered quite serious blows from contemporary physics, as we saw all along this book. Still, it is not completely dead, as testified by the existence of the ontologically interpretable models reviewed in Chapter 13. Indeed, it is still believed in by a number of clearheaded physicists, who lucidly realize how demanding this option is, are not lured by the sham solutions mixing up two meanings of the word "phenomenon" (phenomena *per se* and phenomena *for us*), and build up models that are *truly* of the *ontologically interpretable* kind But along with all the technical difficulties we reviewed (unfruitfulness and plurality of these models, problems concerning compatibility with relativity theory) these constructions also suffer from an even more pernicious defect Originally, the rationale for conventional realism basically was the "commonsense" view that it appeared as being the most immediately plausible way of accounting for what we observe, including correlations, trajectories, etc But, as we saw, for basic reasons these models can in fact account for such things only in roundabout ways that look extremely contrived and far-fetched For these reasons, conventional realism nowdays seems to be in what can be called a state of artificial survival I shall not consider it any more in what follows

(b) The Schemes Setting Language Topmost

Between the systems of such philosophers as Carnap, Bohr (in his philosophical essays), and Wittgenstein (in his *Tractatus*), there are momentous differences, of course But one point they—and many others—have in common is the importance they attribute to language It is well illustrated by such expressions and phrases as Carnap's "linguistic framework," Bohr's "we are suspended in language," and Wittgenstein's famous "Whereof we cannot speak, thereof we must keep silent " Within this conception, Föllesdal's definition of "meaning" (quoted by J A Wheeler [1]) "Meaning is the joint product of all the evidence available to people who communicate" is adequate *and exhaustive* Communication between human beings is thereby raised to the level, if not of the absolute, at least of what is logically prior to anything whatsoever that we are allowed to think of Idealism, neo-Kantianism, phenomenalism, as well as proposals of several leading contemporary philosophers who cannot be classified in any of these "bandwagons," go more or less along these lines And this critical basis they have in common in spite of their differences is unquestionably less easily falsifiable (at least by physics) than the foundations of conventional realism

(c) The Approaches That Deny the Logical Priority of Language Bergson

A second family of conceptions opposed to conventional realism is altogether different from the one just reviewed. Instead of setting language—and, especially and prominently *discursive* language—topmost, these conceptions consider it as being a tool that, indispensable as it is for action, brings about deceptive categorizations when applied to what counts most, which, in view of their proponents, is definitely *not* action.

Most advocates of such conceptions ground them on an analysis of the nature of intelligence. Such an analysis is difficult since it is only with the tool called intelligence that intelligence can be scrutinized. The efforts of the philosophers who engaged in this thorny path should therefore be judged with benevolence, even by those among the scientists who consider some of their arguments as not being sufficiently backed up by scientific evidence to fully carry conviction.

One of these philosophers was Henri Bergson and although, on several subjects, Bergson's views were distinctly too speculative, still, those he held on this particular issue are interesting and worth considering [2]. For future reference and because they may not be familiar to most readers, let them be briefly summarized here. They were consequences of conceptions of his own concerning Evolution, which he grounded on a thoughtful analysis of the scientific knowledge and theories of his time.[1] In short, he accepted neo-Darwinism and mutation theory, but with the restriction that he could not believe these mutations are purely accidental. (One of his arguments was as follows: Even the orthodox mutation theorists seem to consider that after long periods of stagnation a whole species is seized with a tendency to change, which means that, at least, the *tendency* to change is not accidental and stochastic; and under such conditions it is logically conceivable that the *direction* of the change should also not be *totally* accidental.) Since even today, after so many years have passed, strict neo-Darwinism is viewed, in some scientific circles, as a debatable theory, this premise of Bergson's approach cannot be brushed aside as the mere fantasy of a philosopher "not in

---

1. But rather similar conceptions can be found in the works of more recent authors.

the know." It led him to the view that, admittedly, the past evolution of the organic world was far from being predetermined; that, on the contrary, chance played in it a most important role; but that nevertheless chance was not the only agent. Some kind of a vague general tendency must have been at work, not directed toward the realization of some definite project—Bergson was not a finalist—but rather consisting of a general impetus toward increasing sophistication.

This impetus led to progressive diversification. A few great directions appeared along which—with many accidents along the road—life could manage to develop. One led to the vegetable kingdom, another one to animals. With the latter, what was "looked for"—the "tool" that was elected to get certain things from matter and succeed thereby in the struggle for life—was mobility. But this "quest" for mobility took different forms, and in this respect the differentiation that took place between arthropods and vertebrates is of quite special interest. In arthropods, motricity is distributed among a large number of appendixes each of which has its own specificity. In vertebrates, activity is concentrated in only two pairs of members, the function of which is much less dependent upon their form. The independence is maximal in the case of the hand of human beings, which can perform activities of any kind.

In Bergson's general argumentation, the next step was the observation that, in a way, any activity requires knowledge, though not necessarily *conscious* knowledge. The most general, and in a way most primitive, form of knowledge Bergson called "intuition." Life, to repeat, is an endeavor at getting certain things from matter, and intuition, in its different forms, is essentially a means of using a tool to that end. But because of the just noted difference, arthropods and vertebrates developed it differently. In the case of the arthropods, the tool is an inherent and highly specialized part of the living being itself. In virtue of this, the elements of the branch that developed from arthropods to insects—and in particular to the most sophisticated of them all, the hymenopters—therefore had no real need for *conscious* deliberate choice, and their intuition evolved into instinct. In the case of the vertebrates, it is the possibility of using *external* tools that was steadily matured and, finally, mastered (by mankind). But these tools had to be invented and shapened up. This implies being able to consider different possibilities, to reject some and to keep others. In other words, it requires conscious knowledge. Bergson stressed the point that, according to him, it was a longstanding error to consider vegetative, instinctive, and reasonable life as three *successive* steps in one and the same evolutive tendency, whereas, in truth, they constitute three divergent directions of an activity that split this way while increasing.

Viewed in this light, intelligence clearly appears as not a device *primarily* aimed at knowing and understanding. Knowing and understanding are *means*, with the help of which instruments may be conceived and built. And, quite naturally, due, presumably, just to natural selection, these means developed more fully in the domains in which they were of a more immediate use. This accounts quite nicely for the fact that we have at our disposal a set of highly efficient, nonambiguous concepts for designating stones, trees, fruits, animals, and quite generally macroscopic—especially solid—bodies. But it *also* accounts for the fact that, on the contrary, we meet with severe difficulties (some of them reviewed in this book!) when we try to speak of particles, quantum states, and so on without ultimately referring to macroscopic solid things (instrument pointers for example). Bergson of course never claimed that we can meaningfully speak *only* of macroscopic objects. Through its very nature intelligence is adaptable so that it comes as no surprise that it can enlarge its domain to quite a marvellous extent (especially when we can *ultimately* refer to pointers, but not only then). However, if we follow Bergson's line we should not be taken aback by the idea that our intelligence does not enable us to describe independent reality as it really is. This is not its initial and primeval role. Indeed just because of its very nature as a means for getting tools from inorganic matter outside us, intelligence, progressively evolving from intuition, came to be a knowledge not of *things* but (as Poincaré also maintained) only of *relationships between things*. On the contrary, Bergson claimed, instinct, through its nature, remains a genuine—although virtual and unconscious—knowledge of *things*.

Bergson completed these argumented views with a belief. He speculated that although intelligence and instinct developed along diverging lines, the splitting was, in fact, never complete. According to him, a fringe of intelligence remains at the outskirts of instinct and, conversely, a fringe of the *intuition of things*, remains at the outskirts of intelligence. Since discursive language is the channel by means of which the knowledge possessed by *intelligence* is conveyed, there can be no question that it should be able to express adequately the elements of this deep intuition of things. Indeed, it even, presumably, hides them from us. Our only hope of ever approaching them is through the nondiscursive modes of thought and expression—contemplation, music, art (and also, later thinkers would add, the unconscious and archetypes)—because it is only of these modes that it can be conjectured they have some relationship with the primeval intuition of things.

A book on quantum physics is of course not a proper place to engage, on the basis of present-day knowledge concerning ontogenesis, in a critical analysis of Bergson's views on the subject. But even if taken with due reservations, these views may be useful for suggesting a few ideas. The next—purposedly short— section will make it clear in what respect I find them inspiring.

2 Own Options

There is no accounting for taste. As for me, I like Debussy more than Schönberg and Fra Angelico more than Andy Warhol. This is to say that, on matters on which no clues for decision are available, I favor freshness of approach. It is, I guess, by virtue of this turn of mind that, of the two schemes labeled (b) and (c) in Section 1 of this Postface, I spontaneously incline toward the latter. On walking at night along a seashore we all have had intuitions of some deep things lying beyond language. I must say I appreciate having grounds for considering that a few of them may be vistas to something nonillusory.

Naive, dangerous, such an idea? Not in the least.

Naive it is not. Neither its philosophical nor its scientific alleged "disproofs" really stand up: this we saw all along in the book. As we found out, such an idea is indeed no more naive than the one we all cherish, that physics has *some* relation to Being. Science and commonsense both have phenomena as their sole unquestioned domain but, we noted, physics still hopefully yields some glimpses on the *real structure* of what *is*. Why should not other approaches yield some as well? All this is quite consonant with the idea encountered when developing scheme (c), according to which the true realm of intelligence does not take in everything that is meaningful, but, on the contrary, is limited to what has some relationship with possible human action: particularly if the Bergsonian postulate is taken up as well, that a fringe of the *intuition of things* remains at the outskirts of intelligence.

Admittedly, between the veiled reality conception and the Bergsonian standpoint the similarity is only partial. What the veiled reality conception may borrow without qualms from the Bergsonian approach is merely what I just noted. It is the notion that intelligence is essentially a conscious knowledge of *relationships*, that is, *one* facet only of a more general aptitude of living beings called *intuition* by Bergson, together with the assumption that although human beings developed intuition in the form of intelligence, still they preserved a kind of glimmering halo of the original intuition, and the corresponding power of reaching—though indistinctly—at something deeper. The point on which the veiled reality conception must differ from Bergsonism has to do with the nature of this "something." For Bergson it is the "élan vital," an original impetus of life that forces its way through matter, and is curbed and dispersed by the resistance of the latter. In the veiled reality conception this something is Being. Conceptually the difference is enormous. If we cared to express both views in a language making use of the word "God" we would have to say that, for Bergson, intuition dimly reaches a "small," human, almost "Kantian" God, while according to the veiled reality conception it catches a glimpse of a great, hypercosmic God, prior to any detailed and discursively experienced reality. But this difference between the two conceptions turns out "in practice" to be less important than what they *both* bring forth in the form of a conception of human existence and possible goals. Indeed both exalt a notion of *quest* of a much wider scope than our familiar *scientific* quest, since it appeals also to affectivity. Now this enlarged "spirit of quest for some unreachable *Ultimate Real*" is what gave rise, in all times, to the most splendid works of art (the poems to Ammon, the psalms, the Greek temples, the cathedrals, Johann Sebastian Bach's music, the reader may fill up the list according to his own wonder). My claim that this notion of *Real* is valuable should therefore not be viewed as unacceptably fanciful.

Next, is veiled reality a *dangerous* notion? It must be granted that some related views *are* dangerous. When this notion is imparted the figure of a personal God, which is, in itself, a tenable move, the wishes and commandments attributed to such a God by some believers may, as experience shows, lead to slaughters and fierce wars. Likewise, credence in things lying beyond human reason may well induce belief in clairvoyance, astrology, witchery, and so forth. But all such deviations proceed from one and the same cause: identifying phenomena and Being: Just the contrary of the here propounded "theory." To the question the answer, therefore, is *no*.

To sum up, the veiled reality concept is likable. It is not naive and, when properly understood it is not dangerous either. That, to scientific ears, it sounds a little bit unfamiliar should not in principle, render it unattractive to the owners of the latter.

References

1. J.A. Wheeler, *Bits, Quanta, Meaning*, in *Problems in Theoretical Physics*, A. Giovanni, F. Mancini and M. Martinaro (Eds.), University of Salerno Press, 1954.
2. H. Bergson, *L'évolution créatrice*, Librairie Felix Alcan, Paris, 1914.

Appendix 1
Elementary Notions
on State Spaces and
Operators

Introductory note: A philosopher who, during his or her regular studies, never was acquainted with the mathematical formalism of elementary quantum physics cannot read this book in its entirety without having first acquired some knowledge concerning the subject. On the other hand, it must be granted that the available textbooks, full as they are of detailed pieces of information intended for the use of professional physicists, are not easy reading for such a philosopher In this appendix that person will find at least a rough outline of the relevant material: an outline, yet, whose only merit is shortness. It lacks everything else, and foremost rigor and completeness. It does not even quite meet the limited goal of making reading of all the sections of this book totally clear and straightforward: the more advanced formalism necessary for describing the questions some sections deal with has the effect that they are fully understandable only to people already in possession of such a background, and I must grant they were written for physicists. As it is, however, this appendix still should, I think, be of use for open-minded philosophers, unwilling to become physicists, accepting therefore to miss some computational details, but anxious all the same to grasp the general thread of the analysis and discussion.

A1.1 | Hilbert Spaces and Kets

As stressed in Chapter 2, a linear combination of two or more wave functions is also a possible wave function. Moreover, since Eq. (2.19) has a meaning only if the denominator in its right-hand side has a meaning, the wave functions that can be used in equations such as (2.19) must be square integrable. The set of all functions with this general property constitutes an abstract, infinite-dimensional vector space, which has the structure of a

Hilbert space and the same, therefore, is true concerning the set of the possible wave functions (more precisely, because of additional conditions of regularity and so on, this set is a subspace of a Hilbert space). The basic idea underlying the Dirac formalism is to abstract these quite general properties of the set of the possible wave functions from all the other features these functions may have (such as to be *functions* of such and such *variables* and so on), and to describe particles—or systems of particles—directly by means of abstract vectors—named "kets"—in an abstract vector space that has the structure of a Hilbert space (again, strictly speaking it is a subspace of a Hilbert space: but for our purposes in this book we can ignore such mathematically subtle points, which are important only in fields other than ours, and just call this space a Hilbert space). Kets are conventionally designated by the symbol $|>$, which must be conceived of as being just *one* symbol. When it is necessary to distinguish several kets from one another by means of an index it is customary to write this index inside the symbol. Examples: $|n>$, $|\psi>$ and so on.

Vector Spaces

Hilbert spaces are special cases of vector spaces. Vector spaces are abstract sets, that is, sets defined by the combination properties we decide to impart to their elements. In the case of vector spaces these properties essentially are, (i) that any linear combination

$$a_1|1> +a_2|2> + \quad + a_p|p> + \qquad\qquad (A1.1)$$

of any number of elements (called vectors) of vector space E is itself an element of E (a vector), where $a_1, a_2, \quad, a_p$ and so on are complex numbers, (ii) that there exists a special vector $|0>$ of E, called the null vector, such that $0|\psi>= |0>$ and $|\psi> +|0>= |\psi>$, for any vector $|\psi>$ of E, and (iii) (distributivity) that if $|\psi>= |1> +|2> + \quad + |k>$ and a is a complex number then $a|\psi>= a|1> +a|2> + \quad + a|k>$ The vectors $|1>, |2>, \quad, |q>$ are said to be (mutually) independent if no set of complex numbers $b_1, b_2, \quad, b_q$ exists (short of $0, 0, \quad, 0$) such that

$$b_1|1> +b_2|2> + \quad + b_q|q >= |0 > \qquad\qquad (A1.2)$$

Otherwise, they are dependent. A consequence of this is that if the vectors $|1>, |2>, \quad , |q>$ are known to be independent and if Eq.(A1.2) holds true, then necessarily $b_1 = b_2 = \quad = b_q = 0$. If a set of independent vectors $|1>, |2>, \quad , |n>$ is such that *any* vector of E can be expressed as a linear combination of $|1>, |2>, \quad , |n>$, this set is called a *basis* of E. It can be proved that all the bases of a vector space E have the same number of elements, which is called the *dimension* of E. This dimension may be finite or infinite. The dimension of the space of the square integrable functions of one or several variables is infinite.

Scalar Products

A *Hilbert space, H,* in the sense in which this expression is used in the text, essentially is a vector space on which the notion of a *scalar product* of two vectors has been defined. The scalar product of vector $|\psi>$ by vector $|\chi>$ is a complex number noted $<\chi|\psi>$, with the properties that

$$<\psi|\chi> = <\chi|\psi>^*$$
$$<\chi|(a_1|1> + a_2|2> + \quad + a_q|q>)$$
$$= a_1 <\chi|1> + a_2 <\chi|2> + \quad + a_q <\chi|q>$$
$$(<1|a_1 + <2|a_2 + \quad + <q|a_q)|\chi>$$
$$= a_1^* <1|\chi> + a_2^* <2|\chi> + \quad + a_q^* <q|\chi>$$

where the a_i are complex numbers and a^* ("star") means complex conjugation (note that wherever the bar $|$ would occur twice in succession it is written once only). Two vectors $|\psi>$ and $|\chi>$ are said to be *orthogonal* if $<\chi|\psi> = 0$. The scalar product $<\psi|\psi>$ of a ket $|\psi>$ by itself is a real, nonnegative number, denoted $\| \psi \|^2$ $\| \psi \|$ is called the *norm* of $|\psi>$ A basis $\{|e_1>, |e_2>, \quad , |e_N>\}$ of H is said to be an *orthonormal basis* if $<e_i|e_k> = \delta_{i,k}$, where $\delta_{i,k}$ is the Kronecker symbol, by definition equal to 1 if $i = k$ and zero if $i \neq k$.

Tensor Products

An important notion is that of the *tensor product* of two Hilbert spaces H_1 and H_2. By definition, the Hilbert space H is called *tensor product* of H_1 and H_2 and noted $H = H_1 \otimes H_2$, if, to any pair of vectors $|\psi(1)> \in H_1, |\chi(2)> \in H_2$, a vector of H, denoted

$|\psi(1) > \otimes|\chi(2) >$ and called the tensor product of $|\psi(1) >$ and $|\chi(2) >$, is associated, this correspondence being subjected to the following conditions: it is linear with respect to multiplication by complex numbers, distributive with respect to vector addition: $(|\psi(1) > +|\phi(1) >) \otimes |\chi(2) >= |\psi(1) > \otimes|\chi(2) > +|\phi(1) > \otimes|\chi(2) >$, and such that once a basis has been chosen in each one of the two Hilbert spaces, the set of all the tensor products of one element of the basis of H_1 by one element of the basis of H_2 is a basis of H, so that if the dimensions N_1 and N_2 of the two spaces are finite the dimension of their tensor product space is the product N_1N_2. Note that, since H is a vector space, there exist elements of H that are not themselves tensor products but just linear combinations of tensor products. For example, it can be shown that if H_1 (H_2) is the set of the square integrable functions $f^{(1)}(x_1)[f^{(2)}(x_2)]$ of variable $x_1[x_2]$, the set of the square integrable functions of the two variables x_1 and x_2 is just the tensor product $H_1 \otimes H_2$. Clearly, along with products of the type $f^{(1)}(x_1)f^{(2)}(x_2)$ this set must also contain all possible linear combinations of such products.

A1.2 Operators

Operators are symbols of operations changing a ket into another ket: thus, $|\psi >= A|\phi >$ means that operator A changes $|\phi >$ into $|\psi >$ All the operators considered in quantum mechanics are *linear*, which means that for any kets $|\psi >$ and $|\phi >$, $A(|\psi > +|\phi >) = A|\psi > +A|\phi >$ A symbol such as $< \psi|A|\phi >$ means the scalar product of ket $A|\phi >$ by ket $|\psi >$ A is said to be *Hermitean* if $< \psi|A|\phi >=< \phi|A|\psi >^*$ for any $|\psi >$ and $|\phi >$ If A is not Hermitean, $A^{\dagger}$ is said to be the adjoint of A if, for any $|\psi >$ and $|\phi >$, $< \phi|A|\psi >^*=< \psi|A^{\dagger}|\phi >$ An operator U such that the product of U and its adjoint $U^{\dagger}$ is unity is said to be *unitary*. The product of two operators A and B is defined by $AB|\psi >= A(B|\psi >)$ for any $|\psi >$ (B acts first, changing the ket into another ket, and then A acts on this latter ket to change it into the final ket). Obviously, there is no reason that $AB = BA$ and in general this is not the case. When it *is* the case, A and B are said to commute. By definition the *commutator* of two operators A and B, noted $[A, B]$ is the operator $AB - BA$. It is 0 if A and B commute.

In general the ket $A|\psi >$ is not equal to the product of ket $|\psi >$ by a number. But, given an operator A, it is in general the case that some numbers a_i and some kets $|\psi_{i,r} >$ exist such that

$$A|\psi_{i,r} >= a_i|\psi_{i,r} >$$

Such an equation is called an *eigenvalue equation*. a_i is an *eigenvalue* of A and $|\psi_{i,r} >$ is an *eigenket* of A. In general, to an eigenvalue a_i there correspond several eigenkets $|\psi_{i,r} > a_i$ is then said to be *degenerate*. Concerning Hermitean operators it can be proved:

(i) that their eigenvalues are real numbers,
(ii) that two eigenkets corresponding to different eigenvalues are necessarily orthogonal and
(iii) that, if several such operators, $A, B, C,$ commute there exists at least one set of mutually orthogonal kets that are at the same time eigenkets of A, eigenkets of B, eigenkets of C, etc.

Remark. Such sets of orthogonal kets may be used for constructing an orthonormal basis of the Hilbert space.

Definition. If the thus constructed basis is unique (to within arbitrary phase factors) the set of all the operators $A, B, C,$ is said to be a *complete set of commuting operators* (CSCO).

Subspaces, Projectors, etc.

Let H be a Hilbert space and let $|\psi >$ and $|\phi >$ be two kets belonging to H. The set H'' of all the lineal combinations $a|\psi > +b|\phi >$, where a and b are complex numbers is also a Hilbert space and is called a *subspace* of H. It is in fact a two-dimensional space with $\{|\psi >, |\phi >\}$ as one of its bases. The generalization to subspaces of higher dimensionality is trivial.

In H we can define an operator $P = |\psi >< \psi|$, where $|\psi >$ is normalized to unity ($\| \psi \| = 1$), through the convention that, for any ket $|\chi >$ of H.

$$P|\chi >= a|\psi >$$

with $a =< \psi|\chi >$ or

$$P|\chi >= |\psi > (< \psi|\chi >)$$

which implies, for any $|\chi>$, $(|\psi><\psi|)|\chi> = |\psi>(<\psi|\chi>)$, so that the bracket may be suppressed and this ket may just be written $|\psi><\psi|\chi>$ Such a possibility of introducing or suppressing brackets makes the formalism most flexible (note that the symbol $<$ | gets thereby some "autonomy" it is called a *bra* and it can be shown that if $A|\phi> = |\xi>$ then $<\xi| = <\phi|A^\dagger$) P is called a *projector* Obviously it obeys the relationship

$$P^2 = P$$

where P^2 just means PP Obviously also, when P is applied to any ket of H it changes it into a multiple of $|\psi>$, that is, into an element of the one-dimensional subspace of H made up of the set of all these multiples P is therefore a projector *onto* this subspace Projectors on subspaces with more than one dimension may also be defined For example, if $|e_1>$ and $|e_2>$ are two mutually orthogonal kets of H, both normalized to unity, the operator $P' = |e_1><e_1| + |e_2><e_2|$ is easily seen to be a projector $(P'^2 = P')$ and to project on the subspace of H defined by $|e_1>$ and $|e_2>$ If the dimension of H is N and if the set $\{|e_1>, e_2>, \quad, |e_N>\}$ is an orthonormal basis of H, the projector $\mathbb{1} = |e_1><e_1| + |e_2><e_2| + \quad + |e_N><e_N|$ projects H onto itself, that is, changes any ket of H into itself $\mathbb{1}$ is called a "decomposition of unity" The foregoing relationship is then sometimes called the "completeness relation" (it goes without saying that, in spite of an analogy in wording, this purely mathematical relationship has nothing to do with the "completeness hypothesis" defined in Chapter 4) Similarly if the Hermitean operator A has eigenvalues $\{a_i\}$ and if an (orthonormal) set $\{|\psi_i, r>\}$ of its eigenkets is a basis of H, it is easily seen that the operator

$$A' = \sum_{i,r} |\psi_{i,r}> a_i < \psi_{i,r}|$$

has the same effect as A on any ket of H and is therefore identical to A A' is called a "spectral decomposition" of A

A1.3 Matrices

Let, again, H be an N-dimensional Hilbert space, $|\psi>$ a ket of H and $\{|e_1>, e_2>, \quad, |e_N>\}$ an orthonormal basis, B, of H

The complex numbers $c_i = < e_i|\psi >$ $(i = 1, \quad ,N)$ are called the *components* of $|\psi >$ on basis B. They may be arranged in a one-column matrix:

$$\psi = \begin{pmatrix} c_1 \\ c_2 \\ \\ c_N \end{pmatrix}$$

ψ is called the *column vector* representing $|\psi >$ in basis B. It is convenient to similarly arrange the complex conjugates of the c_i in a row·

$$\psi^\dagger = (c_1^*, c_2^*, \quad , c_N^*)$$

Further, to any operator A of H let us associate the set of the complex numbers $A_{i,k} = < e_i|A|e_k >$ and let us arrange them in the form of a matrix:

$$\mathbf{A} = \begin{pmatrix} A_{11}, A_{12}, & A_{1N} \\ A_{21}, A_{22}, & A_{2N} \\ \\ A_{N1}, A_{N2}, & A_{NN} \end{pmatrix}$$

A is called the matrix representing A in basis B. It is easily checked that the main formulas of the ket formalism have their counterparts in the matrix formulation. For example, in this formulation the eigenvalue equation

$$A|\psi >= a|\psi >$$

reads

$$\mathbf{A}\psi = a\psi$$

where, in accordance with the general rules of matrix multiplication, the one-column matrix $\mathbf{A}\psi$ has the numbers $d_i = \sum_k A_{i,k} c_k$ as its elements. Consequently, the eigenvalue equation has, in fact, the form of a system of N linear and homogeneous equations with N unknown (the c_i), namely

$$\begin{cases} \dots\dots\dots\dots \\ \sum_k A_{i,k} c_k = ac_i \\ \dots\dots\dots\dots \end{cases}$$

It is well known that such a system has a nontrivial solution only if the determinant of the coefficients of the c_i is zero, which leads to a Nth degree equation for a. Its N solutions are the eigenvalues of operator A. In the special case that matrix $\mathbf{A}$ is *diagonal*—which means that $A_{i,k} = a_i \delta_{i,k}$—it is clear that the diagonal elements a_i are themselves the eigenvalues. In elementary matrix theory there exists an operation, called *diagonalization*, that transforms a nondiagonal matrix $\mathbf{A}$ into a diagonal one. It can be shown that the diagonal elements of the thus obtained diagonal matrix are just the eigenvalues a's, that is, the solutions of the eigenvalue equation for operator A.

The Trace

The *trace* of a matrix is, by definition, the sum of its diagonal elements. It can easily be proved that the matrices that represent a given operator in several different orthonormal bases all have the same trace, which is called the trace of the operator. The trace of a product can easily be seen to be invariant under circular permutation of the factors: $\text{Tr}(ABC) = \text{Tr}(CAB) = \text{Tr}(BCA)$.

A1.4 Spectrum and Spin

The set of the eigenvalues of an operator is called its *spectrum*. The Hermitean operators quantum mechanics makes use of are of three different kinds with respect to this notion. Some have a *purely continuous* spectrum (any real number within a given range is an eigenvalue); some have a *totally discrete* spectrum (the eigenvalues constitute a discrete sequence); and some have a spectrum that is partly discrete and partly continuous. As can easily be expected, the physical quantities (the so-called observables) that most clearly exhibit specific quantum properties are those that correspond to operators having a purely discrete spectrum. This is what makes spin a particularly useful tool in the analysis of the conceptual foundations of quantum mechanics. Contrary to position and momentum, which have purely continuous spectra, angular momentum and spin components have purely discontinuous ones. In this book, only the *spin* notion is made use of.

Classically a spin is but an intrinsic angular momentum: the "spin" of the Earth is a vector aligned on the pole axis. Quantum mechanically, the spin **S** of a particle is also a vector, the three components of which, S_x, S_y, S_z, along given orthogonal axes are described by Hermitean operators. But quantum spin exhibits most unfamiliar properties, one of them being that the magnitude squared, $\mathbf{S}^2$ of the spin of a particle of a given type (an electron, a photon, etc.) is a fixed quantity, whose value only depends on the nature of the particle. Conventionally, one writes:

$$\mathbf{S}^2 = s(s+1)\hbar$$

and the values of s then can only be integer or half-integer. The most often encountered particles, electrons, protons, neutrons, and so forth all have $s = 1/2$: they are said to be "spin 1/2 particles." The outcome of the measurement of a spin component S_n—along some given direction **n**—of such a particle can only be either $+1$ or -1, in $\hbar/2$ units. In other words, the eigenvalue spectrum of S_n is composed, in these units, of but the two numbers $+1$ and -1. Correlatively, the Hilbert space H_S in which the operators S_n operate is the simplest of all nontrivial conceivable ones: it is but two-dimensional. Note moreover that the spin component operators corresponding to two different directions do not commute. In particular, this is the case concerning the operators S_x, S_y and S_z.

To describe the quantum state of a particle endowed with a spin—for example, a spin 1/2—it is necessary to specify both its spatial state (for example, by means of a wave function) and its spin state. Concerning the latter, because of the just mentioned noncommutativity, no ket in H_S exists that is an eigenket of both S_n and $S_{n'}$, with $\mathbf{n} \neq \mathbf{n}'$ Physically, this implies that, at any time, only *one* component of the spin, S_z say, can have a well-defined value, which moreover is necessarily +1 or −1, as already mentioned. Indeed the eigenvalue equation for S_z reads:

$$S_z|a> = a|a>$$

with a = +1 or −1, or, more explicitly

$$S_z|+> = +|+>$$
$$S_z|-> = -|->$$

Since H_S is two-dimensional and the kets $|+>$ and $|->$ are mutually independent, these two kets necessarily constitute a basis of H_S. This implies that any ket in H_S—that is, any ket describing a possible spin state of the particle—is necessarily a linear combination of $|+>$ and $|->$ For example, the eigenket $|w_+>$ of S_x corresponding to eigenvalue +1 has the form:

$$|w_+> = (|+> + |->)/\sqrt{2}$$

Once a basis of H_S—the basis $\{|+>, |->\}$, say—has been chosen, the kets and operators in H_S can be described by, respectively, column vectors and square matrices, as we know. For example, the state S_x = +1, which, in ket notation, is described by $|w_+>$, is, in this so-called matrix representation, described by the column vector

$$\mathbf{w}_+ = \frac{1}{\sqrt{2}}\begin{pmatrix} +1 \\ +1 \end{pmatrix}$$

In this same representation the spin component S_x is described by the matrix

$$\mathbf{S}_x = \begin{pmatrix} 0 & 1 \\ 1 & 0 \end{pmatrix}$$

(by applying the known rules of matrix multiplication readers will easily verify for themselves that the eigenvalue equation

$$\mathbf{S}_x\mathbf{w}_x = +\mathbf{w}_x$$

is satisfied).

Combination of Spins

In the main text, systems Σ of two spin 1/2 particles U and V are considered According to a general rule (see Chapter 3), the spin Hilbert space H of such a system is the tensor product of the spin Hilbert spaces H_U and H_V of the two composing particles It is therefore four-dimensional Let $\{|u_+ >, |u_- >\}$ and $\{|v_+ >, |v_- >\}$ be, respectively, bases of H_U and H_V, and let them obey the eigenvalue equations

$$S_z^U |u_\pm > = \pm |u_\pm >$$

and

$$S_z^V |v_\pm > = \pm |v_\pm >$$

A basis of H then is $\{|u_+ > \otimes |v_+ >, |u_+ > \otimes |v_- >, |u_- > \otimes |v_+ >, |u_- > \otimes |v_- >,\}$ This basis is orthonormal but, of course, it is not the only orthonormal basis of H Another one is the one made up of the four kets

$$|\chi_+ > = |u_+ > \otimes |v_+ >$$
$$|\chi_0 > = (|u_+ > \otimes |v_- > + |u_- > \otimes |v_+ >)/\sqrt{2}$$
$$|\chi_- > = |u_- > \otimes |v_- >$$
$$|\psi > = (|u_+ > \otimes |v_- > - |u_- > \otimes |v_+ >)/\sqrt{2}$$

which are simple linear combinations of the former ones These four kets turn out to be eigenkets of the operator $\mathbf{S}^2 = S_x^2 + S_y^2 + S_z^2$, which represents the square of the magnitude of the total spin In particular, $|\psi >$ is the eigenket of $\mathbf{S}^2$ that corresponds to eigenvalue 0 It is called the singlet state Bohm's standard example concerning the Einstein, Podolsky, and Rosen problem is based on a consideration of this ket (see Chapters 7 and 8) The coefficients in the foregoing expressions are called the *Clebsch–Gordan coefficients* relative to the combination of two 1/2 spins Tables of *Clebsch–Gordan* coefficients relative to combinations of spins or angular momenta other than 1/2 are to be found in most textbooks on quantum mechanics

Appendix 2
Sensitive Observables
and Time Evolution

The purpose of this appendix is to show that, at least within some simple models, the proposal put forward in Section 15 2 for removing difficulty 4a does not lead to inconsistencies concerning the time evolution of the system As mentioned in this section, the question is that if we really do consider the state operator ρ' [Eq (10 5a)] as describing—at some time t after the measurement interaction has taken place—the empirical situation of the overall system, epistemological consistency requires that, for describing this same situation at a time $t_1 > t$, we should use the state operator, call it $\rho'(t_1)$, obtained by applying to the corresponding mixture the time-evolution law of quantum mechanics However, we have no guarantee that the differences between $\rho'(t_1)$ and the state operator ρ [Eq 10 5] worked out for time t_1 only have effects on predictions concerning sensitive observables, and if this is not the case some ambiguity in the predictive rules of quantum mechanics is seen to arise

Here let us show that within simple models no differences of the considered type occur This is the case concerning models such as the one of Zurek (Section 10 6), if we moreover make the, admittedly oversimplifying (see Section 15 2, Remark 3), assumption that the q first elements that compose the environment have fully measurable attributes (i e , any Hermitean operator operating in the tensor product of their Hilbert spaces corresponds to a genuine observable), whereas, on the contrary, the so-called observables corresponding to Hermitean operators involving some environment spins other than these are all sensitive and therefore "unphysical " Let, then, M' be any nonsensitive, that is genuinely physical, observable. The mean value of M' at a time t has the form:

$$< M' > = < \Psi(t)|M'|\Psi(t) > \qquad (A2.1)$$

where, in Zurek's model, $|\Psi(t) >$ is given by Eq. (10.33); and, as easily verified, with assumption stated above, Eq. (A2.1) can be written as

$$< M' > = \mathrm{Tr}[\rho_q(t)M'] \qquad (A2.2)$$

with

$$\rho_q(t) = \mathrm{Tr}^{k>q}[|\Psi(t) >< \Psi(t)|] \qquad (A2.3)$$

where $\mathrm{Tr}^{k>q}$ means partial tracing over the Hilbert spaces of the environment spins with indexes $k > q$. Let us write $\Psi(t) >$ [Eq. (10.33)] in the form:

$$|\Psi(t) >= a|\phi_+(t) > \otimes |\psi_+(t) > +b|\phi_-(t) > \otimes |\psi_-(t) >$$

where

$$|\phi_\pm(t) > = |s_\pm > \Pi^{k\leq q} \otimes |\chi_k^\pm(t) >$$
$$|\psi_\pm(t) > = \Pi^{k>q} \otimes |\chi_k^\pm(t) >$$

$|\chi_k^+(t) >$ and $|\chi_k^-(t) >$ being the state vectors within square brackets in the first and second term of the right-hand side (r.h.s.) of Eq. (10.33), respectively. A straightforward calculation then shows that $\rho_q(t)$ is equal to

$$\rho_q(t) = |a|^2|\phi_+(t) > < \phi_+(t)|$$
$$+|b|^2|\phi_-(t) >< \phi_-(t)| + Z_q(t) \qquad (A2.4)$$

with

$$Z_q(t) = z_q(t)ab^*|\phi_+(t) >< \phi_-(t)| + h.c. \qquad (A2.5)$$

where z_q is given by the r.h.s. of Eq. (10.36) with the product Π extending only from $k = q + 1$ to $k = N$ In all cases in which N is much larger than q the considerations concerning z developed in Section 10.6 just after Eq. (10.36), apply also to $z_q(t)$, and show that z_q soon become quite small and remains so At a time t at which $z_q(t)$ *is* small, the experimentally observable predictions concerning physical quantities such as M' thus cannot, as expected, appreciably differ from those yielded by a state operator $\rho_q'(t)$ identical to $\rho_q(t)$ except that, in it, $z_q(t)$ is set equal to zero

$$\rho_q'(t) = \rho_q(t) - Z_q(t)$$

This $\rho_q'(t)$ stands in correspondence (via partial tracing) with a proper mixture of the overall systems composed, in proportions $|a|^2$ and $|b|^2$ respectively, of systems in state $|\phi_+(t) >$ $\otimes|\psi_+(t) >$ and systems in state $|\phi_-(t) > \otimes|\psi_-(t) >$ More precisely, we have $\rho_q'(t) = \text{Tr}^{k>q}\rho'(t)$ with

$$\rho'(t) = |a|^2|\phi_+(t) >< \phi_+(t)| \otimes |\psi_+(t) >< \psi_+(t)|$$
$$+ |b|^2|\phi_-(t) >< \phi_-(t)| \otimes |\psi_-(t) >< \psi_-(t)|$$

and it obviously is this $\rho'(t)$ that, in the model, plays, at time t, the role of operator ρ' in the general theory The Schrödinger evolution law then entails that the state operator describing this same mixture of overall systems at time t_1 is expressible by the same formula with t replaced by t_1 Hence if we do take seriously the idea that somehow, at time t, the considered ensemble is empirically equivalent to a quantum mixture described by $\rho_q'(t_1)$, we must expect that the mean values of the genuine observables of this system must, at time t_1, be expressible by means of a formula of the type (A2 2) in which $\rho_q(t)$ is replaced by $\rho_q'(t_1) = \text{Tr}^{k>q}\rho'(t_1)$ Consequently, the looked-for consistency condition is ("robustness") that this $\rho_q'(t_1)$ should not appreciably differ from the state operator $\rho_q(t_1)$ as yielded by the r h s of (A2 3) taken at $t = t_1$ Now, we have

$$\rho_q'(t_1) = |a|^2|\phi_+(t_1) >< \phi_+(t_1)| + |b|^2\phi_-(t_1) >< \phi_-(t_1)|$$
$$= \rho_q(t_1) - Z_q(t_1)$$

which shows that the condition just alluded to is indeed fulfilled since, for any time t_1—smaller of course (our quest concerns *empirical* reality   ) than the Poincaré recurrence cycle—$z_q(t_1)$ is an extremely small number The considered objection to the removal of difficulty D_1 is thus removed for such models In

fact the foregoing argumentation should hold good for any decoherence model obeying fairly general conditions, and it can
be conjectured that its result remains valid concerning still more
general theories based on a tacit use of the empirical reality
axiom.

Appendix 3
Quantum Collapse
and Time Arrow

It is well known that the Schrödinger equation induces no privileged direction of time, this being due to the fact that if $\psi(x,t)$ is a solution of this equation, $\psi^*(x,-t)$ is also a solution If quantum mechanics reduced to the Schrödinger equation, it could therefore be stated without further ado that it is time-reversal invariant, just as classical Newtonian mechanics is However, the probability rule and the associated collapse rule also are essential parts of quantum mechanics and, at first sight at least, they seem to bring in a time arrow Do they really? It is to the discussion of this controversial question that we turn in this appendix

The question at issue is of course tightly linked with the one concerning the nature of the information a measurement actually provides on a system Is the value registered on the instrument the one the measured observable *had just before* or *has just after* the measurement? Clearly, taking the collapse theorem at face value implies deciding in favor of the second answer This is particularly obvious in the case in which the measurement is a complete one Let us, for simplicity's sake, consider a case in which the measured observable A is, by itself, a *complete set of compatible observables* (no degeneracy) Then whatever $|\psi>$, the quantum state before the measurement, may be, if the measurement yields outcome a_k, the state just *after* it took place is the corresponding eigenvector $|\phi_k>$ Knowledge of the outcome thus entails knowledge of this state, whereas it provides no information whatsoever concerning $|\psi>$

Of course, if collapse were just a postulate, the roles could rather easily be reversed We could just *assume* that, if a measurement of A is made and its outcome is a_n, this implies that, just before the measurement, the system state was $|\phi_n >$ But, as we saw in Chapter 3, collapse, particularly in the nondegeneracy case, is something more than just an arbitrary postulate It can be derived from the probability rule together with a merely qualitative and seemingly quite natural premise and, in this sense, it is a theorem The question therefore arises whether we can change anything to it It would seem that this, at any rate is not possible without making the corresponding changes in the premises Let us see whether and how this is possible

First Guess, Suggestion A

The idea here is that we should, in principle, have a free choice between the conventional description and a totally time-reversed description, which I refer to as *Suggestion* A This *Suggestion* would involve

(i) Considering as true the statement *Before a measurement of an observable A having yielded outcome a_n the system was in the corresponding eigenstate* $|\phi_n >$ (this is just the above considered assumption)
(ii) *Deriving* this statement from a new rule, call it Rule 6', which is the time-reversed of the usual probability rule (Rule 6 of Chapter 3) and reads as follows

Rule (6'): With the notations specified above ($|\phi_n > \equiv$ eigenket of *A* associated with value a_n), if, just *after* a measurement of *A*, the state vector is $|\psi >$, the (Bayesian) probability that this measurement *has* yielded outcome a_n is

$$w_n = | < \phi_n|\psi > |^2$$

The derivation in question necessitates an additional premise—call it Premise P—which parallels the one stated in Chapter 3 and consists in postulating that before the measurement the system was already describable by some ket It is carried through by considering a case in which two ideal measurements, M_1 and M_2 are performed, at times $t_0 - \epsilon$ and $t_0 + \epsilon$ respectively, of the same observable *A* Let a_n be the outcome of M_2 and let $|\chi >$ be the ket representing the system at time t_0 Since the two measurements are ideal and infinitely close in time, they must by definition (Rule 7 of Chapter 3) yield the same outcome This implies that the probability is equal to 1 that *if* the outcome of M_2 is a_n the outcome of M_1 was a_n, as well. On the other hand, according to Rule 6', this probability is $|<\phi_n|\chi>|^2$ This (just as in Chapter 3, after Rule 9) implies that

$$|\chi> = |\phi_n>$$

up to some irrelevant phase factor.

So, statement (i) does indeed follow from Rule 6' (plus Premise P). But is Rule 6' actually true? In other words, is Suggestion A correct? Let this question be asked in more general terms. In classical physics, where irreversibility is *factlike* only (that is, merely due to contingent, limiting conditions), there are natural physical phenomena—such as those of elementary, two-body celestial mechanics—that are both reversible and observable. Does this also hold true concerning quantum mechanics?

The answer must be a qualified no. In fact, Suggestion A can only be used in artificially contrived cases that, in practice are hardly ever met, not even approximately. The point is as follows. When, in ordinary language, we speak of the probability that such and such an event E should take place in the future, we implictly mean: "Under the condition that past 'is what it is' and that no restrictive condition whatsoever is added concerning what events take place after E." (Otherwise we should instead consider conditional probabilities taking such restrictions into account.) This is a natural standpoint to take since we normally consider future as being "open." And it applies, of course, to Rule 6. Now, in Suggestion A the role of the future is taken up by the past and conversely. This implies that this Suggestion A is a correct theory only if "the past is left open," that is, if no restrictive conditions whatsoever hold true concerning the past of the system. But this, generally, is not the case, and consequently quite elementary examples can be given of situations in which Rule 6' is falsified.

For instance, let A and B be two observables, with eigenvalue equations

$$A|\phi_n> = a_n|\phi_n>$$

and

$$B|\chi_k> = b_k|\chi_k>$$

that are measured on the same system at times t_1 and t_2 respectively. Assume for simplicity's sake that A and B are, respectively, the x component S_x and the z component S_z of a spin 1/2 particle and consider the values $a_n = b_k = +1/2$. Let us consider the systems on which the B measurement yields outcome b_k and let us ask for the probability that, on them, the A measurement yielded outcome a_n. According to Suggestion A we must apply Rule 6′ From statement (i) applied to the measurement of B we infer that the spin state vector of the particle between times t_1 and t_2 is $|\chi_k>$ we therefore have to consider the quantity $| < \varphi_n | \chi_k > |^2$, whose calculated value is 1/2. According to Rule 6′, this quantity expresses the probability that, under the conditions just stated, the outcome of the measurement of A at t_1 was a_n, that is, it is the limit—for an ensemble composed of arbitrarily many systems—of the fraction whose numerator is the number of particles in the ensemble that yielded both outcomes a_n at t_1 and b_k at t_2 and whose denominator is the total number of particles that yielded outcome b_k at t_2. If we have to do with an initially unpolarized beam the numerator is $N/4$ and the denominator is $N/2$ (N = total number of particles), so that this fraction is 1/2: Rule 6′ is then obeyed. But if we have to do with a beam initially polarized in some direction this is not the case in general. For example, if the incoming beam is polarized along Ox ($S_x = +1/2$) the outcome of the first measurement is the same for all the particles so that the fraction in question is equal to 1. In this case, therefore, Suggestion A (i.e. in last resort Rule 6′) is just simply false. What makes it false is the polarization of the initial beam, which can be looked at as a preselection exerted on an initially unpolarized beam. And it is worth noticing that, artificially, also Rule 6 of Chapter 3 could be rendered false, by imposing a similar *post*selection on all the elements of the ensemble, that is by requiring, for example, that the only particles to be taken into consideration are those that, at a later time t_3, yield, upon some further measurement of S_z, outcome $+1/2$.

Admittedly, such a manner of restoring a symmetry between past and future is acceptable in principle. But it remains that it implies giving up Rule 6, the probability rule, which is one of the pillars of contemporary physics. This has no counterpart in classical mechanics, where the symmetry in question pervades all the actually used basic laws. Quantum mechanics can therefore not be said to be "reversible" in the sense in which Newtonian physics is reversible. In fact, we see here that the kind of time-arrow quantum mechanics exhibits is rather tightly linked with—and therefore dependent on—our (human) abilities to do this or that (in effect, to select subensembles of given ensembles, in the future or in the past) More precisely, it is linked with the *limitations* of these abilities This is illustrated by the existence, established in general terms by Aharonov, Bergmann, and Lebowitz [1], of a fully time-symmetric "quantum mechanics" that does not actually apply to our actual "world of experience" but from which both conventional quantum mechanics *and* Suggestion A may be derived, by applying formal, so called "coherence destroying" procedures either in the future or in the past

Second Guess, Suggestion B

The shortcomings of Suggestion A may incite us to consider a milder one—let it be called Suggestion B—which has some features in common with Suggestion A but actually does *not* really make quantum mechanics lawlike reversible This Suggestion B—which may well be the one hinted at by Zeh [2 sec 4 2]—essentially consists in keeping the postulate expressed by our Statement (i) in our "First Guess," but in stressing the fact that, since this postulate is stated in ontological terms and bears on quantum states to which we unquestionably have no direct access (at which we cannot directly "look"), it cannot, in fact, be independently tested by means of any measurement In other words, it is of the nature more of a convention than of a physical assumption

Questions may then be raised as to the nature and bearing of the probability rule A probability rule of some sort is of course necessary for setting the formalism in relationship with observation But on the other hand, for the aforementioned reasons, Rule 6' cannot be taken up Hence it seems no other choice is available than just the usual probability rule, that is, Rule 6

It is true that, in a case of two successive measurements, the use of Rule 6 seems to raise a conceptual problem with our present postulate (Statement (i) of "First Guess"), it must be the case that—with the same setup and notations as above—the system must be in state $|\varphi_n >$ before t_1 and in state $|\chi_k >$ between t_1 and t_2 And it might then be wondered how this is to be reconciled with the fact, proved in Chapter 3 (Proposition A), that Rule 6 *implies*, in this case, that between times t_1 and t_2 the system must be in state $|\varphi_n >$ The answer, however, is quite simple It consists in remembering the *conventional* nature of postulate (i) within Suggestion B For indeed, this conventional nature implies that, in fact, a measurement does not reveal to us in what state the system "really was " Rather, it (its nature and outcome) determines what quantum state we attribute by convention to the system just before the measurement Under these conditions the just mentioned difficulty vanishes It is true that when the observable B that is measured at t_1 is the same as the observable A measured at t_1, which is the situation considered in the proof of Proposition A, the system must, within the time interval (t_1, t_2), be in the state $|\chi_k >$ (then identical to $|\varphi_n >$) But this fact cannot be abstracted from the considered overall experimental setup In the case in which the observable A measured at t_1 is *not* the same as observable B there is no reason that it should hold good

Clearly, Suggestion B has much in common with Bohr's approach to quantum mechanics since, in Bohr's views also, the complete experimental setup must be taken into account before we can speak of a quantum phenomenon and, correlatively, assign definite values to the corresponding physical quantities But it is difficult to see in what sense the suggestion in question could actually restore a symmetry between past and future that would make quantum mechanics time-reversible in a lawlike sense A general view that seems more in accordance with what we know is the one, advocated in this book, that makes a distinction between independent and empirical reality Independent reality is *not* embedded in space and time, and therefore the question of whether or not there is a time arrow in it is meaningless But the contingent values of physical dynamical quantities are not elements of it Empirical reality—the reality that physics, and in particular quantum physics, is mainly concerned with—is essentially the set of the phenomena, as seen and described by means of human concepts The time arrow ranks among the most primeval of these concepts not surprising then that quantum physics *as it is used* (that is, with collapse and so on) should exhibit a time arrow

References

1 Y Aharonov, P G Bergmann, and J L Lebowitz, *Phys Rev* **134B**, 1410 (1964)
2 H D Zeh, *The Physical Basis of the Direction of Time*, Springer-Verlag, Berlin, Heidelberg, 1989

Appendix 4
Illustration of Bell 2, a
Simple Example

With the object of making more explicit the substance of the premises of Bell 2, an elementary, artificial but explicit, example is given here of a deterministic, strongly objective model reproducing, in the Bohm, two-spin, standard case, the quantum mechanical predictions For the mean value $M(AB)$ of the product AB of the outcomes of measurements of S_a^U and S_b^V (notation of Section 8 1), (measured in $\hbar/2$ units) the quantum prediction is, as is well known,

$$M(AB) = -\cos(\mathbf{a} \quad \mathbf{b}) \qquad (A4\ 1)$$

The model consists in describing the spin pair by means of two opposite unit vectors λ and $\mu = -\lambda$, attached to particle U and V respectively, in assuming that the outcomes A and B of the above considered measurements are given by the formulas

$$A(\lambda, \mathbf{a}) = \mathrm{sign}(\mathbf{a} \quad \lambda) \qquad (A4\ 2a)$$
$$B(\mu, \mathbf{b}) = \mathrm{sign}(\mathbf{b} \quad \mu) \qquad (A4\ 2b)$$

and in assuming moreover that the statistical distribution $\rho(\lambda)$ depends on the azimuthal angle ϕ of λ and the bisecting line of $\mathbf{a}$ and $\mathbf{b}$, $(-\pi < \phi \le \pi)$ More precisely, it is assumed that this dependence is

$$\rho(\phi) = |\sin 2\phi|/4 \qquad (A4\ 3)$$

Let then θ be the angle between $\mathbf{a}$ and $\mathbf{b}$ and define

$$\beta = \pi/2 - \theta/2$$

The angles $\phi = -\pi + \beta, -\beta, \beta, \pi - \beta$ divide the **a, b** plane into four sectors according to the values, positive or negative, of A and B, and it is easily checked that the product AB is

$$AB = +1 \quad \text{in sectors } (\beta, \pi - \beta) \text{ and } (-\pi + \beta, -\beta)$$
$$ -1 \quad \text{elsewhere}$$

The mean value

$$M(AB) = \int AB\rho(\phi)d\phi \tag{A4.4}$$

is then readily calculated by splitting the integration domain $(-\pi, \pi)$ into these four sectors. We have

$$\int_{\beta}^{\pi/2} AB\rho(\phi)d\phi = \tfrac{1}{4}\cos^2\beta$$

hence, because of symmetry·

$$\int_{\pi/2}^{\pi-\beta} AB\rho(\phi)d\phi = \tfrac{1}{4}\cos^2\beta$$

so that the total contribution of the $(\beta, \pi - \beta)$ sector to the integral in (A4.4) is $\tfrac{1}{2}\cos^2\beta$. Because of symmetry again, that of the $(-\pi + \beta, -\beta)$ is the same. Since $\int_{-\pi}^{\pi}\rho(\phi)\,d\phi = 1$, the total contribution of the two remaining sectors is therefore $1 - \cos^2\beta$, up to the sign. Finally, therefore:

$$M(AB) = \cos^2\beta - (1 - \cos^2\beta) = -\cos\theta$$

which shows that the model does indeed reproduce the quantum predictions. The reason why the model is interesting is that, in it, A depends only on λ and **a** and B depends only on μ and **b**, so that, strictly speaking, it does not violate Local Causality as defined in Section 8.4. But ρ depends on **a** and **b** since it depends on their bisecting line. And this is why the model is not a counter-example to Bell 2. The Bell 2 premise that is violated here is not Local Causality as defined in the said section but the supplementary hypothesis mentioned in Section 9.3: that the experimentalist is free to choose at any time the directions **a** and **b**. For, if he chose them *after* the particles have left the source, their bisecting line would, in general, be in no relationship whatsoever with ρ (see, e.g., Ref. [1] for more details concerning this supplementary hypothesis) Alternatively, it may be said that the model violates parameter independence defined in a broad sense, that is, as in Section 8 2 (see footnote in Section 8 5)

Reference

1 B d'Espagnat,"Nonseparability and the Tentative Descriptions of Reality," *Phys Rep* **110**, 203 (1984)

Name Index

Subject Index

Bold-faced numbers indicate main references.

a priori forms, 334
a priori modes, 313
action at a distance, 115, 409–413
actuality, 422
algebra of observables, 89
allegory, 394, 398
alpha particle track, 276
analysis, 111, 311
"and–or" difficulty. *See* difficulty 4b
animals, 340, 399, 423, 435
antirealism, 313
appearance, 343, 366, 379
    of a classical world, 266, 320, 329
archetypes, 437
Aristotle, 414, 432
arithmetical problem ("paradox"), 18, 423
arrow (of time). *See* time arrow
art, 438
Aspect's experiments, 143
assumption Q, 66
assumption Q', 90
atomism, 316, 317
    philosophical, 401
atoms, 17
automata, 248
axiom, 328, 405
    of empirical reality, 372
axiomatic formulations of quantum
        mechanics, 89

basis
    of vector space, 442
    orthonormal, 442
    pointer, 162, 370
beables, 25, 281, 296

Being, 313, 379, 401, 403, 438
Bell theorem, 290, 298
Bell's inequalities, 141, 144, 145,153–155
Bell's theorems, 141–146, 320, 388, 461
Berkeley, 5, 16, 313
Big Bang, 420
blackbody radiation, 33
Bohm standard example, 133, 137, 226,
        411
Bohm's theory, 275–278
Bohr corollary, 224
Bohr postulate, 223
bootstrap theory, 357
Born rule, 36, 38
bra, 445
brain connections, 416
branches (of Universe), 248

C.S.L., 294
Cauchy–Schwarz inequality, 50, 82
causality, 7, 9, 19, 313, 342, 420
    Einsteinian, 418
    extended, 413–418, 428
    final, 415
    inductive, 141
    local, *xx*, 143, 126, **126**, 145, 146, 256,
        294, 312, 352, 387, 462
cause, 19, 125, 409, 415
    efficient, 414
    extended, 415
    primary, 417
    structural, 414
cause–effect relationship, 10, 19
chaos theories, 301

dynamical property, 61, 195, 221, 223, 229, 318, 368, 382
  actual versus potential, 422
dynamical quantity, 45fn, 189, 194, 217, 221, 235

E-truth. *See* truth
E.P.R., 132, 226, 388
E.P.R. criterion of reality, 62, 132, 228, 287
Effects, 205
eigenket, 444
eigenvalue, 46, 444
  degenerate, 444
eigenvalue equation, 444
élan vital, 438
elements of reality, 132, 134, 407
embedding in space-time, 322
empiricism, 3
  constructive, 360
empiricist, 307, 405
energy, 176
ensemble, 43, 57, 265
  nonquantum, 59
  Gibbsian, 123, 276
ensemble theories, 297–301
entailment theory of causation, 411
entanglement, *xx*, 51, **113**, 119, 257, 312
  in Heisenberg picture, 191
environment, 177
  theory, 163, **256**, 320, 330, 342, 355
epistemological breakings, 20
epistemological criteria (standards, sense), 62, 380–382, 386
epistemological status, 316fn
epistemology, 362, 367, 383, 409
"epsilonology," 185
error, 335, 364
esse est percipi, 5
Euclidean space, 7, 279, 296
events, 233, 273, 327
Everett's theory, 247–253
evolution, 434
existence, 14, 17, 310, 313, 315, 318, 335, 355
existence-O, 364
experience, 2, 10, 313
explanation, 17, 18, 331–332, 378, 414
  definition a of, 331

definition b of, 332
extended wholes, 115
extension, 428

facts, 239, **241**, 379fn, 384
  contingent, 16, 195
falsification, 314, 335
Feynman graphs, 316
Feynman path integrals, 318
filtration, 57
fluctuations, 122
Fock's space, 317
formalism of operations and effects, 204, 355
free choice, 28
  of experimentalist, 143, 184, 462
function (logical), 17
functionals, 211
future, 28
  (open), 184, 457

G.R.W., 292, 352
Galileo, 417
Gell–Mann and Hartle theory, 267–272, 288, 302, 355, 374
geometry, 8
GHZ, 147
God (Spinoza's), 428
Gödel cosmological model, 420
gravitation, 294
Griffiths's theory, 30, 232–239, 355
group velocity, 35

Halley's comet, 331
Hamiltonian, 36
Heisenberg field operators, 337
Heisenberg indeterminacy relationships, *xxi*, 96, 132, 294
Heisenberg operators, 193, 371
Heisenberg picture, 73, 95, 189–196, 336
hidden variable, 60
hidden variable theory, 223, 290, 332
Hidden Variables Assumption, 298
Hilbert space, 44, 442
history, 268, 288
  consistent, 233
holism, 424
holistic aspects, 110, 424